AMERICAN MATHEMATICAL SOCIETY
COLLOQUIUM PUBLICATIONS
VOLUME XXVIII

ANALYTIC TOPOLOGY

BY
GORDON THOMAS WHYBURN
PROFESSOR OF MATHEMATICS
THE UNIVERSITY OF VIRGINIA

PUBLISHED BY THE
AMERICAN MATHEMATICAL SOCIETY
Providence, Rhode Island
1942

First Printing 1942

Second Printing 1955

Third Printing 1967

Fourth Printing with corrections 1971

Library of Congress Catalog Card Number 63-21794
International Standard Book Number 0-8218-1028-6

PREFACE

The material here presented represents an elaboration on my Colloquium Lectures delivered before the American Mathematical Society at its September, 1940 meeting at Dartmouth College. The results of some of my own efforts together with a selection of those of other mathematicians relative to the subject chosen are presented in what is intended to be a coherent and approximately self-contained exposition, framed in the familiar topology of separable metric spaces. No attempt is made at comprehensive coverage either of the known work embraced by the title of the book or of the work of others, or even myself, which may be closely related to that included.

I wish to express my appreciation to the American Mathematical Society for the opportunity of delivering the lectures and publishing in its Colloquium Series. Thanks are due also to the Waverly Press for its careful and sympathetic handling of the manuscript.

On the personal side, it has been my privilege and good fortune to stand in the middle ground between distinguished teachers on the one hand and a group of distinguished students and associates on the other and receive stimulus and inspiration from both. Of the former, the influence of R. L. Moore will be apparent and his invaluable contribution in this way is gratefully though inadequately acknowledged. Of the other group, too numerous to mention, Hubert A. Arnold, M. Garcia and Paul A. White have helped directly by reading and correcting parts of the manuscript and proof. Finally, I wish to acknowledge the generous assistance rendered by my wife, Lucille Whyburn, who contributed materially to the content and organization of the lectures and manuscript and assisted greatly in the preparation of both and whose unfailing encouragement transcends all attempts at evaluation.

G. T. Whyburn

Charlottesville, Va.
November, 1941

TABLE OF CONTENTS

INTRODUCTION

As used here the term "Analytic Topology" is meant to cover those phases of topology which are being developed advantageously by methods in which continuous transformations play the essential role. In the process of evolving, coming of age and assuming more stable form, topology, through interaction with other branches of mathematics, not only is leaving its mark on them but is itself adopting more and more the language and symbolism of the older fields. Thus, for example, we have not only a topological function theory giving the results of analysis which are essentially topological in character, but also a function-theoretic topology dealing with topological situations with the aid and principal use of some of the basic tools of analysis. Without drawing the lines too sharply or giving too clear cut a definition, let us say in a general way that analytic topology deals with topological situations with the aid of analytical language and tools, and to some extent conversely, just as analytic geometry handles geometric situations by analytic methods. I hope this concept will be made clearer as the treatment progresses and actual examples are given illustrating the type of relationship which has been so vaguely defined.

The major questions to be dealt with are, first, the existence of transformations of various sorts from a space A to the same or another space B and, second, the analysis of the action of these transformations on A to produce B. Since thus we are dealing with the transition from A to itself or to something else possibly quite different topologically, our subject exhibits kinship with earlier work on dynamics in the Colloquium Series. This is especially true of the final chapter on periodicity which connects directly with many of the concepts of this subject as discussed by G. D. Birkhoff. However, the even closer kinship with other purely topological treatises, notably that of R. L. Moore in the Colloquium Series and that of K. Menger on "Kurventheorie", will be too obvious to require comment.

The book divides roughly into two parts, corresponding to the first six and last six chapters, respectively. In the first part there is developed the necessary topological machinery and framework for the latter part, which is devoted to pure analytic topology. Even in the second chapter, however, notably in §§3, 4, there emerge some of the fruits of the application of analytic or transformation methods to topological situations. For here a variety of results, some classic and others quite recent, are brought together in what seems their proper relationship and derived in a simple and novel way from one central mapping theorem.

The book is meant to be largely self-contained, at least in so far as topological developments are concerned. In the later stages some use is made of a few notions of combinatorial topology and of the theory of groups without any attempt at adequate introduction. Since these appear largely in end-results and applications, there seems little need or justification for taking the space to develop them here.

At the beginning we assume once for all a set of axioms sufficient to make all spaces considered separable and metrizable. Once the metric is introduced, however, attention is no longer focused on the axioms, but rather on the (equivalent) standpoint that we are operating always in a given separable metric space.

Cross references are given in brackets, with the roman numeral for the chapter followed by the section and number of the theorem, lemma, or corollary referred to, e.g., [IV, (3.2)] refers to result (3.2) of §3 in Chapter IV, which would be the *second* main result in this section. If only the number in parenthesis is given, as (3.2) for example, the reference is to the result of that number in the *present chapter*, i.e., the one being read at the time. To assist in locating results referred to, the chapter number and section number appear at the heading of each double page.

References to the literature in the main are held to a minimum. For convenience these are made at the ends of the chapters in the form of author's name followed by numerals in brackets referring to his books or papers by that number in the bibliography at the close of the book. In some cases one or more authors' names have been used in connection with a theorem though by no means in all where this might well, or possibly should, be done. A considerable amount of the material in the first part is of such a classical nature and so well known that specific citations to sources in the literature are not made. Later on more attempt is made to cite the original author and source. In some cases, also, closely related material not actually covered is mentioned in the references.

CHAPTER I

INTRODUCTORY TOPOLOGY

1. **Operations with sets.** We shall have occasion to use sets of points and sets or collections of point sets of various sorts. Capital letters A, B, C, $\cdots$ will be used to designate sets and, in general, small letters stand for points. $a \in A$ means "a is an element of the set A" or "a is a point of A" if A is a point set. a *non* $\in A$ means that a is not an element of A.

If A and B are sets,

$A = B$ means that every point in the set A is also a point in the set B, and conversely every point in B is also in A.

$A \subset B$—read "A is a subset of B" or "A is contained in B"—means that every point of A is a point of B.

$A \supset B$ means $B \subset A$, "A contains B." $A = B$ is equivalent to $A \subset B$ and $B \subset A$.

$A + B$ (sum of A and B) means the set of all points belonging either to A or to B. In general, if $[G]$ is a collection of sets, $\sum G$ is the set of all points x such that x belongs to at least one element (or set) of the collection $[G]$.

$A \cdot B$ (intersection or product) means the set of all points belonging to both A and B. For any collection of sets $[G]$, $\prod G$ is the set of all points x such that x belongs to every set of $[G]$.

If $B \subset A$, then $A - B$ is by definition the set of all points which belong to A but not to B.

If $[G]$ is a collection of sets, any collection of sets each of which is an element of the collection $[G]$ is called a *subcollection* of $[G]$.

Real and complex numbers and their properties will be used freely. A set or collection whose elements can be put into (1-1) correspondence with a subset of the set of all positive integers will be called *countable, or enumerable.* If such a correspondence is established and the elements arranged in order of ascending integers, e.g., a_1, a_2, a_3, $\cdots$, the resulting *arranged set* is called a sequence.

The empty or vacuous set is designated by 0. Two sets A and B are said to be *disjoint* if their intersection is empty, i.e., $A \cdot B = 0$.

2. **Topological spaces.** A class S of elements or points in which there is determined a class of subsets called neighborhoods satisfying the following four conditions or axioms is called a *topological space.*

(1) **For any $x \in S$, any neighborhood of x is a set of points containing x.**

(2) **If U is a neighborhood of a point x and y is any point of U, then U is a neighborhood also of y.**

(3) If U and V are neighborhoods* and $x \in U \cdot V$, then there exists a neighborhood W of x such that $W \subset U \cdot V$.

(4) If x and y are distinct points, there exists a neighborhood of x not containing y.

A point p in such a space is said to be a *limit point* of a point set M provided every neighborhood of p contains at least one point of M distinct from p.

If A is a set, the *closure* of A, written $\bar{A}$, is the set consisting of all points of A and all limit points of A.

We shall need also the two following additional axioms.

(5) If V is any neighborhood of a point x, there exists a neighborhood U_x of x with $\bar{U}_x \subset V$.

(6) There exists a countable (fundamental) sequence of neighborhoods $R_1, R_2, R_3, \cdots$ such that if p is any point and U_p is any neighborhood of p, then for some integer m we have $p \subset R_m \subset U_p$.

A topological space satisfying (5) is said to be *regular* and such a space satisfying (6) is said to be *perfectly separable.*

The following propositions are now readily provable in a regular, perfectly separable topological space S.

(2.1) *If p and q are distinct points, there exist neighborhoods U_p and U_q of p and q, respectively, with $U_p \cdot U_q = 0$.*

(2.2) *If p is a limit point of $A + B$, then p is a limit point either of A or of B (possibly of both).*

(2.3) *If p is a limit point of a set M, every neighborhood of p contains infinitely many distinct points of M.*

(2.4) *Given any point p which is in at least one neighborhood, there exists a monotone decreasing sequence of neighborhoods $Q_1 \supset Q_2 \supset Q_3 \supset \cdots$ closing down on p, i.e., such that p belongs to every Q_i and if U is any neighborhood of p, then for some n we have $p \subset Q_n \subset U$.*

The proofs of (2.1), (2.2), (2.3) are left as exercises. To prove (2.4), let $[R_{n_i}]$ be the collection of all neighborhoods of the fundamental sequence in Axiom (6) which contain the point p. Set $Q_1 = R_{n_1}$. By Axiom (3), since $p \in Q_1 \cdot R_{n_2}$, there exists a neighborhood Q_2 such that $p \subset Q_2 \subset Q_1 \cdot R_{n_2}$. Similarly, there exists a neighborhood Q_3 satisfying $p \subset Q_3 \subset Q_2 \cdot R_{n_3} \subset R_{n_1} \cdot R_{n_2} \cdot R_{n_3}$. Continuing this process indefinitely we obtain a sequence

$$Q_1 \supset Q_2 \supset Q_3 \supset \cdots \supset p$$

such that for each i,

$$Q_i \subset R_{n_1} \cdot R_{n_2} \cdots R_{n_i}.$$

* Note that in view of Axiom 2, it is no longer essential to associate a neighborhood with any particular one of its points.

Now if U is any neighborhood of p, there exists an integer m such that $p \subset R_m \subset U$. For some i, $n_i = m$. Whence, $p \subset Q_i \subset R_{n_i} \subset U$, as was to be shown.

DEFINITION. *A sequence of points* $p_1, p_2, p_3, \cdots$ *(not necessarily distinct) is said to converge to a point* p*, written* $\left\{\begin{matrix} p_n \to p, \text{ or} \\ \lim p_n = p \end{matrix}\right\}$*, provided every neighborhood of* p *contains almost all (i.e., all but a finite number) of the points in the sequence.*

(2.5) *If* p *is a limit point of a set* M *and* p *is in some neighborhood,* M *contains an infinite sequence of distinct points converging to* p.

To prove this, let $Q_1 \supset Q_2 \supset Q_3 \supset \cdots$ be a sequence of neighborhoods closing down on p [see (2.4)]. Since p is a limit point of M, for each n, Q_n contains a point p_n of M distinct from p. For any n, there exists a neighborhood U of p not containing p_n and an m such that $Q_m \subset U$. Hence $p_n \neq p_{m+i}$ $(i = 1, 2, \cdots)$. Thus any given point p_n occurs in the sequence $[p_n]$ only a finite number of times. Accordingly there are infinitely many distinct points p_n. Any neighborhood of p contains almost all the sets $Q_1, Q_2, \cdots$ and hence almost all the points p_n. Accordingly, $p_n \to p$; and if we eliminate duplicates we retain an infinite sequence of distinct points $p_{n_i} \to p$.

3. **Open and closed sets. Compact sets. Separability.** A set of points G is said to be *open* provided that for every point $p \in G$ there exists at least one neighborhood V_p of p which is contained wholly in G.

A set of points F is said to be *closed* provided F contains all of its limit points.

The *complement* of a set G is the aggregate of points not contained in G.

(3.1) *The closure of any set is closed, i.e.,*

$$\bar{\bar{X}} = \bar{X}.$$

(3.2) (a) *If a set is open, its complement is closed.*
(b) *If a set is closed, its complement is open.*

(3.3) (a) *The sum of any collection of open sets is open.*
(b) *The product of any collection of closed sets is closed.*

(3.4) (a) *The product of any finite number of open sets is open.*
(b) *The sum of any finite number of closed sets is closed.*

These propositions are valid in any topological space. The proofs are left as exercises. In the remaining theorems of this section, we assume a space satisfying all of the Axioms (1)–(6).

A set K is said to be *compact* provided every infinite subset of K has at least one limit point which belongs to K.

(3.5) *Every compact set is closed.*

For suppose K is compact but not closed. Then there exists a limit point p of K which does not belong to K. By (2.5), K contains an infinite sequence $[p_n]$ of distinct points converging to p. By the compactness of K, $[p_n]$ must have a limit point $q \subset K$. But if U_p and U_q are disjoint neighborhoods of p and q, respectively, almost all the points p_n must be in U_p, thus only a finite number can be in U_q contrary to (2.3).

(3.6) *If $K_1 \supset K_2 \supset K_3 \supset \cdots$ is a monotone decreasing sequence of nonvacuous compact sets, $\prod_1^\infty K_i$ is nonvacuous.*

For each n, let p_n be a point of K_n. If some p_n, say p_k, is chosen infinitely many times, we have $p_k \in \prod_1^\infty K_i$. If not, the set $[p_n]$ is an infinite subset of K_1. Hence it has a limit point, say p. Then for each n, p is a limit point of the subset $\sum_n^\infty p_i$ of K_n. Since each K_n is closed, by (3.5), this gives $p \in K_n$ for each n. Whence, $p \in \prod_1^\infty K_i$ so that $\prod_1^\infty K_i \neq 0$.

DEFINITION. *A set M is said to be separable provided there exists a countable subset P of M such that $\overline{P} \supset M$, i.e., every point of M is either a point or a limit point of P.*

(3.7) *Every set is separable.*

To prove this, let M be any set and let $[R_{n_i}]$ be the collection of all neighborhoods in the fundamental sequence (Axiom (6)) which intersect M. For each i, let $p_i \in M \cdot R_{n_i}$ and set $P = \sum p_i$. Then if x is any point of M and x non $\in P$, we must show that x is a limit point of P. To this end, let U be any neighborhood of x. There exists a neighborhood R_{n_k} of the fundamental sequence such that $x \subset R_{n_k} \subset U$. This gives $p_k \subset R_{n_k} \subset U$ and $p_k \neq x$. Thus x is a limit point of P.

DEFINITION. *A point p is said to be a condensation point of a set M provided every neighborhood of p contains uncountably many points of M.*

(3.8) *For any set M, all save possibly a countable number of points of M are condensation points of M.*

For let R be the sum of all those sets R_i of the fundamental sequence such that $M \cdot R_i$ is countable. Then $M \cdot R$ is countable and every point of $M - M \cdot R$ is a condensation point of M since any neighborhood R_n of the fundamental sequence which contains a point of $M - M \cdot R$ contains uncountably many points of M.

4. **Covering theorems.** A point set M is said to be *covered* by a collection of open sets G provided that each point of M lies in at least one set of the collection $[G]$.

An immediate consequence of the perfect separability of our space is the following proposition, known as the

(4.1) LINDELÖF THEOREM. *If $[G]$ is any collection of open sets covering a point set M, some countable subcollection $[G_i]$ of $[G]$ covers M.*

For, referring to the fundamental sequence of neighborhoods $R_1, R_2, \cdots$, it follows that for each point p of M there exists an integer δ so that we have (i) $p \subset R_\delta \subseteq G \in [G]$, where G is some set of the collection $[G]$ containing p. Let $[R_{n_i}]$ be the collection of those neighborhoods R_{n_i} of the fundamental sequence having the property that some $G \in [G]$ exists such that $R_{n_i} \subset G$. For each i, select one such set G and call it G_i. Then $[G_i]$ is a countable subcollection of $[G]$ and by virtue of (i) it follows that $[R_{n_i}]$ and thus also $[G_i]$ covers M.

If restrictions are placed on the set M, we can prove the stronger covering property embodied in the

(4.2) BOREL THEOREM. *If $[G]$ is any collection of open sets covering a compact set M, then some finite subcollection of $[G]$ covers M.*

Proof. By the Lindelöf Theorem, some countable subcollection $[G_i]$ of $[G]$ covers M. Suppose, contrary to our theorem, that no finite subcollection of $[G_i]$ covers M. Then if for each n, we set $N_n = M - M \cdot \sum_1^n G_i$, it follows that for each n, N_n is nonvacuous, closed and compact and contains N_{n+1}. Hence $N = \prod_1^\infty N_n \neq 0$. But this is impossible because $N \subset M$ and $N \cdot \sum_1^\infty G_i = 0$ whereas $[G_i]$ covers M. Thus the Borel Theorem is demonstrated.

DEFINITION. *A set M is said to be conditionally compact provided every infinite subset of M has a limit point (which may or may not belong to M). Obviously any subset of a compact set is conditionally compact.*

As an application of the Lindelöf Theorem we prove the following converse proposition.

(4.3) *If M is conditionally compact, $\bar{M}$ is compact.*

Suppose, on the contrary, that $\bar{M}$ contains an infinite subset X which has no limit point in $\bar{M}$. Since $\bar{M}$ is closed, X can have no limit point. Thus with the aid of Axiom (5) it follows that for each point p of our space, there exists a neighborhood V_p of p such that $\bar{V}_p$ contains at most one point of X. By the Lindelöf Theorem there exists a countable sequence $V_1, V_2, \cdots$ of the neighborhoods V_p covering the space. Now since for each n, $\sum_1^n \bar{V}_i$ contains at most n points of X, there exists a point p_n of M not belonging to $\sum_1^n \bar{V}_i$. Since the set $[p_n]$ is infinite and M is conditionally compact, there exists a limit point p of this set $[p_n]$; but for some k, $p \in V_k$; and this is impossible since V_k can contain at most k of the points p_n.

5. Metric spaces. Metrization Theorem.

By a *metric space* is meant a class of elements, or points, in which a *distance function* or *metric* is defined, i.e., to each pair of elements x, y of S a non-negative real number $\rho(x, y)$ is associated satisfying the conditions:

(1) $\rho(x, y) = 0$ if and only if $x = y$,

(2) $\rho(x, y) = \rho(y, x)$ (symmetry),

(3) $\rho(x, y) + \rho(y, z) \geqq \rho(x, z)$ (triangle inequality).

The following examples of metric spaces are of fundamental importance.

(i) The *real number system* R in which the distance function is defined as

$$\rho(x, y) = | x - y |, \qquad x, y \in R.$$

(ii) *Euclidean n-space* R^n with the ordinary distance function

$$\rho(x, y) = \sqrt{\sum_1^n (x_i - y_i)^2}, \qquad x = (x_1, x_2, \cdots, x_n), \qquad y = (y_1, y_2, \cdots, y_n),$$

$$x_i, y_i \in R.$$

(iii) *Hilbert space* H consisting of all sequences of real numbers $x = (x_1, x_2, \cdots)$ with $\sum_1^\infty x_i^2$ convergent as points and with the metric

$$\rho(x, y) = \sqrt{\sum_1^\infty (x_i - y_i)^2}.$$

(iv) The *Hilbert fundamental parallelotope* Q_ω consisting of all sequences $x = (x_1, x_2, \cdots)$ of real numbers satisfying $0 \leqq x_i \leqq 1/i$ as points and the same metric as in (iii). Clearly $Q_\omega \subset H$.

(v) The *space* Q'_ω consisting of all sequences of real numbers $x = (x_1, x_2, \cdots)$ with $0 \leqq x_i \leqq 1$ as points and with the metric

$$\rho(x, y) = \sum_1^\infty 2^{-i} | x_i - y_i |.$$

(vi) The *topological* (or *cartesian*) *product* $X \times Y$ of two metric spaces X and Y, consisting of all ordered pairs (x, y) where $x \in X$, $y \in Y$ and where the distance $\rho(p_1, p_2)$ between two points $p_1 = (x_1, y_1)$ and $p_2 = (x_2, y_2)$ is defined by

$$\rho(p_1, p_2) = \sqrt{\rho(x_1, x_2)^2 + \rho(y_1, y_2)^2}$$

or by

$$\rho(p_1, p_2) = \max [\rho(x_1, x_2), \rho(y_1, y_2)].$$

It results that either of these is a distance function and, no matter which is used, it follows that a sequence $p_1, p_2, p_3, \cdots$ in $X \times Y$ will converge to a point p in $X \times Y$ (i.e., $\lim \rho(p_i, p) = 0$) if and only if both $\lim \rho(x_i, x) = 0$ and $\lim \rho(y_i, y) = 0$, where $p_i = (x_i, y_i)$ and $p = (x, y)$.

A topological space is said to be *metrizable* provided it is possible to define a distance function satisfying conditions (1), (2), and (3) and such that a point p will be a limit point of a set M if and only if there exist points of M distinct from p as close to p as we please in the sense of this distance. Our principal object in this section will be to prove that *any regular, perfectly separable topological space is metrizable.*

To this end we first establish two lemmas.

(5.1) (Tychonoff) *Any regular perfectly separable topological space is normal,* i.e., any two disjoint closed sets are contained in disjoint open sets.

Proof. Let A and B be any two disjoint closed sets lying in our space. By the regularity property it follows that for each $x \in A$ there exists a neighborhood U of x such that $\overline{U} \cdot B = 0$. Applying the Lindelöf Theorem to the collection $[U]$ we obtain a countable sequence of neighborhoods $U_1, U_2, \cdots$ covering A and such that $\overline{U}_i \cdot B = 0$ $(i = 1, 2, \cdots)$. Similarly there exists a countable collection of neighborhoods $V_1, V_2, \cdots$ covering B such that $\bar{V}_i \cdot A = 0$ $(i = 1, 2, \cdots)$. Now let

$$\begin{aligned} U_1^* &= U_1, & V_1^* &= V_1 - V_1 \cdot \bar{U}_1, \\ U_2^* &= U_2 - U_2 \cdot \bar{V}_1, & V_2^* &= V_2 - V_2(\bar{U}_1 + \bar{U}_2), \\ &\cdots\cdots\cdots & &\cdots\cdots\cdots, \\ U_n^* &= U_n - U_n \cdot \sum_1^{n-1} \bar{V}_i, & V_n^* &= V_n - V_n \cdot \sum_1^{n} \bar{U}_i \\ &\cdots\cdots\cdots & &\cdots\cdots\cdots \end{aligned}$$

Then for each n, U_n^* and V_n^* are open and $U_n^* \cdot A = U_n \cdot A$ and $V_n^* \cdot B = V_n \cdot B$. Thus if $U^* = \sum_1^\infty U_n^*$ and $V^* = \sum_1^\infty V_n^*$, U^* and V^* are open sets containing A and B, respectively. Also $U^* \cdot V^* = 0$. For if not we would have an m and an n such that $U_m^* \cdot V_n^* \neq 0$. But if $m > n$, the definition of $U_m^* = U_m - U_m \cdot \sum_1^{m-1} \bar{V}_i$ gives $U_m^* \cdot V_n^* = 0$ since $V_n^* \subset V_n$; and if $m \leqq n$, the definition of $V_n^* = V_n - V_n \cdot \sum_1^n \bar{U}_i$ gives $U_m^* \cdot V_n^* = 0$ since $U_m^* \subset U_m$.

(5.2) *If A and B are disjoint closed subsets of a perfectly separable, regular, topological space S, then there exists a continuous real function $f(x)$ defined for all $x \epsilon S$ and such that*:

(i) $0 \leqq f(x) \leqq 1$ *on* S.

(ii) $f(x) = 0$ *on* A.

(iii) $f(x) = 1$ *on* B.

Proof. For each fraction of the form $r = m/2^n$ $(m = 0, 1, 2, \cdots, 2^n)$ we shall define an open set $G(r)$ such that (a) $A \subset G(0)$, $G(1) = S - B$, (b) $r_1 < r_2$ always implies $\overline{G(r_1)} \subset G(r_2)$. By (5.1) there exists an open set $G(0)$ such that $A \subset G(0) \subset \overline{G(0)} \subset G(1) = S - B$. Thus we have the sets $G(r)$ defined and conditions (a) and (b) satisfied for $n = 0$. Let us suppose this has been done for $n = k - 1$ and proceed to do it for $n = k$. Thus we have to define $G(i/2^k)$ for odd integers $i < 2^k$, since for i even, $i/2^k$ reduces to the form $(i/2)/2^{k-1}$. Now by (b) we have $\overline{G[(i - 1)/2^k]} \subset G[(i + 1)/2^k]$. Thus applying (5.1) with $A = \overline{G[(i-1)/2^k]}$, $B = S - G[(i + 1)/2^k]$, we obtain an open set $G(i/2^k)$ such that $\overline{G[(i - 1)/2^k]} \subset G(i/2^k) \subset \overline{G(i/2^k)} \subset G[(i + 1)/2^k]$. Thus we obtain the sets $G(r)$ by complete induction for all numbers r of the form $m/2^n$.

Now define $f(x) = 0$, for $x \epsilon G(0)$, and $f(x) =$ the least upper bound of the numbers r such that $x \epsilon [S - G(r)]$, for x non $\epsilon\, G(0)$. Then $f(x) = 0$ on A, $f(x) = 1$ on B, $0 \leqq f(x) \leqq 1$ on S. It remains to show that f is continuous. Let $p \epsilon S$, and $\epsilon > 0$. If $0 < f(p) < 1$, let r_1 and r_2 be fractions of the form $m/2^n$ such that $f(p) - \epsilon/2 < r_1 < f(p) < r_2 < f(p) + \epsilon/2$. Thus in the open set $G(r_2) - \overline{G(r_1)}$ we have $r_1 \leqq f(x) \leqq r_2$ which gives $|f(x) - f(p)| < \epsilon$. If $f(p) = 0$ or 1, a similar argument proves the continuity of $f(x)$ using only r_2 when $f(p) = 0$ and only r_1 when $f(p) = 1$.

(5.3) METRIZATION THEOREM. *Any perfectly separable, regular, topological space S is metrizable.*

Proof. Let $R_1, R_2, \cdots$ be a fundamental sequence of neighborhoods in S. Order all pairs R_i, R_j such that $\bar{R}_i \subset R_j$ into a sequence $P_1, P_2, \cdots$. For each such pair $P_n = R_i, R_j$, we take $\bar{R}_i = A$ and $S - R_j = B$ and, applying (5.2), obtain a continuous function $f_n(x)$ defined on S and satisfying (i), (ii), and (iii) of (5.2).

For each pair of points $x, y \in S$ we now define

$$\rho(x, y) = \sum_1^\infty 2^{-n} | f_n(x) - f_n(y) |,$$

and proceed to show that this function $\rho(x, y)$ is a distance function effecting the desired metrization.

We first show that conditions (1), (2), and (3) on a distance function are satisfied by ρ. Now if $x = y$, we have $f_n(x) = f_n(y)$ for all n, and hence $\rho(x, y) = 0$. On the other hand if $x \neq y$, then for some n and $P_n = R_i, R_j$, we have $x \in R_i$, $y \in (S - R_j)$. Whence $f_n(x) = 0$, $f_n(y) = 1$ and thus $\rho(x, y) \geqq 2^{-n}$. This proves (1). Now (2) is obvious and (3) results at once from the following

$$\rho(x, y) + \rho(y, z) = \sum_1^\infty 2^{-n}[| f_n(x) - f_n(y) | + | f_n(y) - f_n(z) |]$$

$$\geqq \sum_1^\infty 2^{-n} | f_n(x) - f_n(z) | = \rho(x, z).$$

To complete the proof we have to show that $p \in S$ is a limit point of a point set M if and only if there are points of M distinct from p but arbitrarily near p. Let p be a limit point of M and let $\epsilon > 0$ be arbitrary. Taking N so large that $2^{-N} < \epsilon/2$, we can find a neighborhood U of p throughout which the oscillation of $\sum_1^N 2^{-n} | f_n(x) - f_n(y) |$ is less than $\epsilon/2$. Then if $q \in U \cdot M$, we have $\rho(p, q) < \epsilon/2 + \sum_{N+1}^\infty 2^{-n} < \epsilon$. On the other hand suppose that a point p is not a limit point of a set M. Then there exists a pair $P_n = R_i, R_j$ such that $p \in R_i$, $M - p \cdot M \subset S - R_j$. This gives $f_n(p) = 0$, $f_n(x) = 2^{-n}$ on $M - p \cdot M$. Whence $\rho(p, x) \geqq 2^{-n}$ for all $x \in M - p \cdot M$. Thus there are no points of M different from p at a distance less than 2^{-n} from p, and the proof is complete.

Now if in any metric space S we define neighborhoods to be sets of the form $V_r(x)$, where x is a point of S, r is a positive number and $V_r(x)$ is the set of all points y of S with $\rho(x, y) < r$ clearly Axioms (1)–(5) of §2 are satisfied. If in addition we suppose S separable and take a countable set $P = \sum p_i$ such that $\bar{P} = S$, it is readily seen that the countable set of neighborhoods $[V_r(p_i)]$, r rational, satisfies Axiom (6). Accordingly:

(5.4) *Any separable metric space is a regular perfectly separable topological space, and conversely any regular perfectly separable topological space is metrizable.*

Henceforth we shall assume all our spaces separable and metric. Further, the terms "neighborhood" and "open set" will be used synonomously since they give equivalent topologies to our spaces.

Note. In the proof of the Metrization Theorem we have really assigned to each point x of S the point $f(x) = [f_1(x), f_2(x), \cdots]$ of the space Q'_ω (see example (v), p. 6) in such a way that this correspondence is (1-1) (i.e., for each x, $f(x)$ is uniquely determined and $f(x) = f(y)$ implies $x = y$). Furthermore we also showed that this correspondence and its "inverse" are continuous, i.e., $x_i \rightarrow x$ implies $f(x_i) \rightarrow f(x)$, and conversely. Such a correspondence or transformation is said to be *topological* or to be a *homeomorphism* (see Chapter II below). Thus we have shown that *any space S satisfying Axioms* (1)–(6) *is homeomorphic with a subset of* Q'_ω.

Also it is seen at once that if to each point $x = (x_1, x_2, \cdots)$ of Q'_ω we make correspond the point $y = g(x) = (x_1, x_2/2, x_3/3, \cdots)$ of Q_ω, each point $y = (y_1, y_2, \cdots)$ of Q_ω will correspond uniquely to the point $x = (y_1, 2y_2, 3y_3, \cdots)$ of Q'_ω and this correspondence is also topological. Hence Q_ω and Q'_ω are homeomorphic. Thus *any space S as above is homeomorphic with a subset of* Q_ω (and hence of H), because clearly two sets each homeomorphic with a given third set are homeomorphic with each other.

Finally, *the spaces* Q_ω *and* Q'_ω *are both compact.* For in either space clearly "$p_n \rightarrow p$" is equivalent to "for every i, the ith coordinates of p_n converge to the ith coordinate of p." Thus if $[p_n]$ is an infinite sequence of distinct points of Q'_ω (or Q_ω), we choose an infinite subsequence (p_n^1) whose first coordinates converge to a number p^1. Then from (p_n^1) choose an infinite sequence (p_n^2) whose second coordinates converge to p^2, and so on. Thus clearly the sequence (p_i^i) will converge to the point $(p^1, p^2, \cdots)$.

6. **Diameters and distances.** As an immediate consequence of the definition we have

(6.1) *Any distance function* $\rho(x, y)$ *is continuous.*

That is, for any two points a and b and any $\epsilon > 0$ there exist neighborhoods U_a and U_b of a and b, respectively, such that for $x \in U_a$, $y \in U_b$

(i) $| \rho(a, b) - \rho(x, y) | < \epsilon$.

To prove this we have only to take U_a and U_b so that for $x \in U_a$, $y \in U_b$

(ii) $\rho(a, x) < \epsilon/2$, $\rho(b, y) < \epsilon/2$.

This gives by the triangle inequality

$$\rho(x, y) \leqq \rho(x, a) + \rho(a, b) + \rho(b, y) < \rho(a, b) + \epsilon,$$

$$\rho(a, b) \leqq \rho(a, x) + \rho(x, y) + \rho(y, b) < \rho(x, y) + \epsilon,$$

or

$$\rho(a, b) - \epsilon < \rho(x, y) < \rho(a, b) + \epsilon.$$

which is equivalent to (i).

Definitions. *By the diameter* $\delta(N)$ *of any set* N *is meant the least upper bound, finite or infinite, of the aggregate* $[\rho(x, y)]$ *where* $x, y \in N$. *By the distance* $\rho(X, Y)$ *between the two sets* X *and* Y *is meant the greatest lower bound of the aggregate* $[\rho(x, y)]$ *for* $x \in X$, $y \in Y$.

Obviously $\delta(N) = \delta(\bar{N})$ and $\rho(X, Y) = \rho(\bar{X}, \bar{Y})$ for any sets N, X, Y.

(6.2) *If N is compact, there exist points* $x, y \in N$ *such that* $\rho(x, y) = \delta(N) < \infty$.

For let $(x_1, y_1), (x_2, y_2), \cdots$ be a sequence of pairs of points of N such that $\lim \rho(x_n, y_n) = \delta(N)$, finite or infinite. Since N is compact, the sequence $x_1, x_2, x_3, \cdots$ contains a subsequence which converges, in the sense of §2, to a point x. We may suppose the notation adjusted so that $x_n \to x$. Similarly the sequence $[y_n]$, after the adjustment for $[x_n]$ has been made, contains a subsequence converging to a point y. Again we can adjust the notation so that $x_n \to x, y_n \to y$. Since $\lim \rho(x_n, y_n) = \delta(N)$, it results from (6.1) that $\rho(x, y) = \delta(N) < \infty$.

(6.3) *If X and Y are disjoint compact sets, there exist points* $x \in X$ *and* $y \in Y$ *such that*

$$\rho(x, y) = \rho(X, Y) > 0.$$

To prove this we choose a sequence of pairs of points $[(x_n, y_n)]$ as in (6.2) so that $x_n \in X, y_n \in Y, x_n \to x \in X, y_n \to y \in Y$, and $\lim \rho(x_n, y_n) = \rho(X, Y)$. The continuity of ρ gives $\rho(X, Y) = \rho(x, y)$ and since $x \neq y, \rho(x, y) > 0$.

7. Superior and inferior limits. Convergence. Let G be any infinite collection of point sets, not necessarily different. The set of all points x of our space S such that every neighborhood of x contains points of infinitely many sets of G is called the *superior limit* or *limit superior* of G and is written lim sup G. The set of all points y such that every neighborhood of y contains points of all but a finite number of the sets of G is called the *inferior limit* or *limit inferior* of G and is written lim inf G. If for a given system G, lim sup G = lim inf G, then the system (collection, or sequence) G is said to be *convergent* and we write lim G = lim sup G = lim inf G. Under these conditions we say that *G converges* to the limit lim G.

For example, let G be the collection of all positive integers. Then lim sup G = lim inf G = lim G = 0. Thus G is convergent and has a vacuous limit. Again, let G be the system of sets $[L_n]_{n=1}^{\infty}$, where L_n is the straight line interval joining the points $[(-1)^n(1 - (1/n)), 0]$ and $[(-1)^n(1 - (1/n)), 1]$. Then lim sup G is the sum of the interval from $(-1, 0)$ to $(-1, 1)$ and the one from $(1, 0)$ to $(1, 1)$, lim inf G = 0, and thus lim G does not exist.

From the definitions, we have at once for any system G

(i) $$\liminf G \subset \limsup G.$$

Furthermore, lim inf G *and* lim sup G *are always closed point sets*. For if x is a limit point of lim inf G, then any neighborhood V of x contains a point y of lim inf G; and since V is a neighborhood also of y, then V contains points of all save a finite number of the sets of G and thus x belongs to lim inf G. Similarly if x is a limit point of lim sup G, any neighborhood V of x contains a point z of lim sup G and thus V, a neighborhood of z, contains points of infinitely many of the sets of G. Therefore x belongs to lim sup G.

It is a consequence also of our definitions that any system G having a vacuous lim sup is convergent and has a vacuous limit. In this case G is necessarily a countable system, and its elements may be ordered into a sequence A_1, A_2, A_3, $\cdots$.

If G^* is any infinite subcollection of a collection G, we have at once

(ii) $$\lim\inf G \subset \lim\inf G^* \subset \lim\sup G^* \subset \lim\sup G.$$

Therefore if G is a convergent system, then every infinite subcollection G^* of G is convergent and has the same limit as G. Now since every infinite system contains a countably infinite subsystem, it follows that any convergent system contains an infinite *sequence* G_1, G_2, G_3, $\cdots$ of elements having the same limit as G. *In general, any system G contains a countable sequence $[G_i]$ of elements such that*

$$\lim\inf [G_i] = \lim\inf [G] \quad \text{and} \quad \lim\sup [G_i] = \lim\sup [G].$$

To obtain the sequence $[G_i]$ we proceed as follows. Let R_1, R_2, R_3, $\cdots$ be the fundamental sequence of neighborhoods, and let R_{n_1}, R_{n_2}, $\cdots$ be those sets of this sequence which contain points of infinitely many elements of G. Let R_{m_1}, R_{m_2}, R_{m_3}, $\cdots$ be those sets of the fundamental sequence whose complements contain infinitely many elements of G. For each i, let $[F_j^i]$ be an infinite sequence of elements of G such that for every j, $F_j^i \cdot R_{n_i} \neq 0$ and let $[E_j^i]_{j=1}^{\infty}$ be an infinite sequence of elements of G each of which is contained in the complement of R_{n_i}. Then the sum $[G_n]$ of the two collections $[F_j^i]_{i,j=1}^{\infty}$ and $[E_j^i]_{i,j=1}^{\infty}$ is countable and has the desired properties. For if x is any point of lim sup G, then any neighborhood V of x contains a set R_{n_i} and thus contains points of infinitely many of the sets of $[G_n]$, since it contains points from each of the sets $[F_j^i]_{j=1}^{\infty}$. Thus, lim sup $G_i \supset$ lim sup G and by (ii) we have lim sup $[G_n]$ = lim sup G. Likewise if x is any point not belonging to lim inf G and V is a neighborhood of x whose complement contains points of infinitely many of the elements of G, then V contains a set R_{m_i} and thus the complement of V contains the sets $[E_j^i]_{j=1}^{\infty}$ all of which belong to $[G_n]$; therefore no such point x can belong to lim inf $[G_n]$ and we have proved that lim inf $[G_n] \subset$ lim inf G. Then by (ii) we get that lim inf $[G_n]$ = lim inf G, which completes the proof.

It is obvious in view of the theorem just proved that in studying convergence properties of collections of sets we may restrict ourselves to countable collections or sequences without loss of generality. We proceed then to some fundamental theorems relating to convergence properties of sequences of sets.

(7.1) Theorem. *Every infinite sequence of sets contains a convergent subsequence.*

Proof. Let $[A_i]$ be any infinite sequence of sets and let us set up the following array of sequences:

$$[A_i^1] = A_1^1, A_2^1, A_3^1, \cdots,$$
$$[A_i^2] = A_1^2, A_2^2, A_3^2, \cdots,$$
$$[A_i^3] = A_1^3, A_2^3, A_3^3, \cdots,$$
$$\cdots\cdots\cdots\cdots\cdots\cdots$$

In this array, the first sequence $[A_i^1]$ is identical with the given sequence $[A_i]$; and in general, for each n, the sequence $[A_i^{n+1}]$ is obtained from the sequence $[A_i^n]$ in the following manner. If the sequence $[A_i^n]$ contains any infinite subsequence whose elements occur in the same order as in $[A_i^n]$ and whose limit superior has no point in the set R_n, where R_n is the nth set in the fundamental sequence $R_1, R_2, R_3, \cdots$, then we pick out one such subsequence of $[A_i^n]$ and call it $[A_i^{n+1}]$. If, on the other hand, the limit superior of every infinite subsequence of $[A_i^n]$ has a point in R_n, then we take for $[A_i^{n+1}]$ exactly the sequence $[A_i^n]$.

We shall now prove that the diagonal sequence $[A_n^n]$ in the above array is convergent. Clearly it is an infinite subsequence of $[A_i] = [A_i^1]$, since for $k \neq j$, $A_k^k \neq A_j^j$. Suppose, on the contrary, that it is not convergent and thus that there exists a point x belonging to lim sup $[A_n^n]$ but not to lim inf $[A_n^n]$. Then there exists a neighborhood V of x, which we may take $\equiv R_m$ for some m, and an infinite subsequence $[A_{n_i}^{n_i}]$ of $[A_n^n]$ all elements of which are contained in the complement of R_m. Now the sequence $[A_n^n]$ for $n > m$ is an infinite subsequence of $[A_i^m]$ and therefore so also is $[A_{n_i}^{n_i}]$ for $i > m$. Thus $[A_i^m]$ does contain an infinite subsequence, namely $[A_{n_i}^{n_i}]$, $i > m$, whose limit superior has no point in R_m. Therefore by the method of choice of $[A_i^{n+1}]$, lim sup $[A_i^{m+1}] \cdot R_m = 0$. But $[A_n^n]$, for $n > m$, is a subsequence of $[A_i^{m+1}]$ and hence lim sup $[A_n^n] \cdot R_m = 0$, which is absurd since x belongs to both R_m and lim sup $[A_n^n]$. Thus the supposition that the sequence $[A_n^n]$ is not convergent leads to a contradiction.

Notation. If X is any point set and ϵ is any positive number, we shall denote by $V_\epsilon(X)$ the set of all points p such that the distance from p to some point of X is less than ϵ; and this set will be called the *spherical neighborhood* of X with radius ϵ.

(7.2) THEOREM. *If $[A_i]$ is a sequence of sets whose limit superior is L and the sum of whose elements is a conditionally compact set, then for each $\epsilon > 0$ there exists an m such that for every $n > m$, $A_n \subset V_\epsilon(L)$.*

Proof. Suppose this is not so. Then there exists an $\epsilon > 0$ and an infinite sequence of points $p_1, p_2, p_3, \cdots$ such that for each i, p_i belongs to A_{n_i} but not to $V_\epsilon(L)$ and such that $n_i \neq n_j$ for $i \neq j$. But since $\sum A_i$ is conditionally compact, there exists a point p which either is a limit point of the set $p_1 + p_2 + p_3 + \cdots$ or is identical with p_i for infinitely many i's. In either case p must belong to $L = \text{lim sup } [A_i]$. But clearly this is impossible, since $V_\epsilon(L)$ contains no one of the points p_i.

COROLLARY. *If $[A_i]$ is convergent and has limit L and if $\sum_1^\infty A_i$ is conditionally compact, then for each ϵ, there exists an m such that for every $n > m$, $\rho(A_n, L) < \epsilon$.*

8. Connected sets. Well-chained sets. A set of points M is said to be *connected* provided that however it be expressed as the sum of two disjoint nonvacuous sets M_1 and M_2, at least one of these sets will contain a limit point of the other. In other words, a set M is connected if it is not the sum of two sets A and B such that $A \cdot \bar{B} = \bar{A} \cdot B = 0$.

Two such sets A and B are said to be *mutually separated*, i.e., two sets A and B are mutually separated provided they are mutually exclusive (= disjoint) and neither of them contains a limit point of the other. Any division of a set M of the form $M = A + B$ where A and B are nonvacuous mutually separated sets is called a *separation* of M.

It follows at once from the definition that the sum of any number of connected sets whose product does not vanish is connected. Also it follows that if M is connected, so also is any set M_0 such that $M \subset M_0 \subset \bar{M}$. For if there were a separation $M_0 = A + B$, M would be wholly in A or wholly in B, since otherwise $M = M \cdot A + M \cdot B$ would be a separation of M. This is impossible, because if $M \subset A$, every point of B would be a limit point of A and similarly if $M \subset B$.

It follows from this in particular that the closure of any connected set is connected.

By a *component* of a set M is meant a maximal connected subset of M, i.e., a connected subset of M which is not contained in any other connected subset of M. Thus for any point a of M, the component of M containing a consists of a together with all points of M in connected subsets of M containing a.

If E is any set of points, any subset F of E is called a *closed subset of* E and is said to be "closed in E" or "closed relative to E" provided that no point of $E - F$ is a limit point of F. A subset O of E is called an *open subset of* E and is said to be "open in E" provided that no point of O is a limit point of $E - O$. Now it is seen at once that a connected set may be defined as a set M no proper subset of which is both open and closed in M.

For if M is not connected and $M = A + B$ is a separation, A is both open and closed in M; and on the other hand, if a proper subset X of M is both open and closed in M, $M = X + (M - X)$ is a separation of M. Also *any component of a set M is closed in M.*

Definition. *If a and b are points, then by an ϵ-chain of points joining a and b is meant a finite sequence of points*

$$a = x_1, x_2, x_3, \cdots, x_n = b$$

such that the distance between any two successive points in this sequence is less than ϵ. A set of points M is said to be well-chained provided that for every $\epsilon > 0$, any two points a and b can be joined by an ϵ-chain of points all lying in the set M.

(8.1) Lemma. *If A is any subset of any set M and ϵ is any positive number, the set M_a of all points of M which can be joined to A by an ϵ-chain of points of M is both open and closed in M.*

To prove this lemma, set $M_b = M - M_a$. Then M_a cannot contain a limit point x of M_b; for if so then some point z of M_b is at a distance less than ϵ from x and if $[a = x_1, x_2, \cdots, x_n = x]$ is an ϵ-chain in M from a to x, $a \in A$, clearly $[a = x_1, x_2, \cdots, x_n, z]$ is an ϵ-chain in M from a to z, contrary to the fact that z does not belong to M_a. Thus M_a is open in M. Likewise no point z of M_b is a limit point of M_a; for if so then some point x of M_a is at a distance less than ϵ from z; and if $[a = x_1, x_2, \cdots, x_n = x]$ is an ϵ-chain in M from a to x, $a \in A$, then $[a = x_1, x_2, \cdots, x_n, z]$ is an ϵ-chain in M from a to z, contrary to the fact that z does not belong to M_a. Thus M_a is closed in M and the lemma is proved.

(8.2) *Every connected set is well-chained.*

Suppose on the contrary that some connected set M contains two points a and b which, for some $\epsilon > 0$, cannot be joined by an ϵ-chain of points of M. But then by the lemma the set M_a of all points which can be so joined to b is both open and closed in M, which clearly contradicts the fact that M is connected.

Now it is not true, conversely, that every well-chained set is connected. For clearly the set R of all rational points on the unit interval (0, 1) is well-chained but not connected. *However, if a set K is compact and well-chained, it is connected,* a fact which will be deduced a little later from a more general proposition.

(8.3) *If N is any connected subset of a connected set M such that $M - N$ is disconnected, then for any separation $M - N = M_1 + M_2$, $M_1 + N$ and $M_2 + N$ are connected.*

For if there existed a separation $M_1 + N = A + B$, N would have to lie wholly either in A or B, say in A, since otherwise $N = N \cdot A + N \cdot B$ would be a separation of N. This would give $B \subset M_1$. Whence $M = (M_2 + A) + B$ would be a separation of M, contrary to the connectedness of M. Similarly $M_2 + N$ is connected.

9. Limit Theorem. Applications.

(9.1) THEOREM. *If $[A_i]$ is an infinite sequence of sets such that* (a) $\sum A_i$ *is conditionally compact,* (b) *for each i, any pair of points of A_i can be joined in A_i by an ϵ_i-chain and $\epsilon_i \to 0$ with $1/i$,* (c) $\liminf [A_i] \neq 0$, *then* $\limsup [A_i]$ *is connected.*

Proof. Let $L = \limsup [A_i]$, $l = \liminf [A_i]$. Since L is closed and contained in the compact set $\overline{\sum A_i}$, it follows that L is compact. Thus if L is not connected and we have a separation $L = A + B$, A and B are compact disjoint sets at least one of which, say A, intersects l. Thus by (6.2) $\rho(A, B) = 3d > 0$. This gives

$$\rho[V_d(A), V_d(B)] > d.$$

By (7.2) there exists an integer N such that for $n > N$, $A_n \subset V_d(L) = V_d(A) + V_d(B)$; and since $l \cdot A \neq 0$ we may suppose also $A_n \cdot V_d(A) \neq 0$. Thus there exists an integer $k > N$ such that $\epsilon_k < d$, $A_k \cdot V_d(A) \neq 0 \neq A_k \cdot V_d(B)$. Clearly this is impossible, since if x is the last point in $V_d(A)$ of an ϵ_k-chain in A_k from a point of $V_d(A)$ to a point of $V_d(B)$ and y is the successor of x, then since $y \subset V_d(B)$ we have $\rho(x, y) > d > \epsilon_k$.

(9.11) *If* $[A_i]$ *is a convergent sequence satisfying* (a) *and* (b) *of* (9.1), lim $[A_i]$ *is connected.*

(9.12) *If* $[A_i]$ *is a sequence of connected sets satisfying* (a) *and* (c) *of* (9.1), lim sup $[A_i]$ *is connected.*

(9.2) *If two points a and b of a compact set K can be joined in K by an* ϵ*-chain for every* $\epsilon > 0$, *they lie together in the same component of K.*

Proof. For each i, let A_i be the set of all points of K which can be joined to a by an $1/i$-chain in K. Then since $\sum A_i \subset K$, condition (a) of (9.1) is satisfied. Condition (b) is satisfied by choosing $\epsilon_i = 1/i$; and since lim inf $[A_i] \supset a + b$, (c) is satisfied. Accordingly $L =$ lim sup $[A_i]$ is connected. Since $a + b \subset L \subset K$ clearly our conclusion follows. It is of interest to note that as here defined the set L actually will be a component of K.

(9.21) *Every compact well-chained set is connected.*

(9.22) *Every interval of real numbers is connected.* (Hence the real number space R is connected.)

(9.3) *If A and B are disjoint closed subsets of a compact set K such that no component of K intersects both A and B, there exists a separation* $K = K_a + K_b$, *where* K_a *and* K_b *are disjoint compact sets containing A and B, respectively.*

Proof. We first show that there exists an $\epsilon > 0$ such that no ϵ-chain in K joins a point of A to a point of B. If this is not so, then for each i there exists an $1/i$-chain A_i in K joining a point a_i of A to a point b_i of B. Since, (7.1), the sequence $[A_i]$ contains a convergent subsequence, there is no loss of generality in supposing the whole sequence converges. Clearly if we take $\epsilon_i = 1/i$, condition (b) of (9.1) is satisfied; and since $\sum A_i \subset K$, (a) is satisfied. Accordingly, by (9.11), $L =$ lim $[A_i]$ is connected. But since $A_i \cdot A \supset a_i$, $A_i \cdot B \supset b_i$, and A and B are compact, we have $L \cdot A \neq 0 \neq L \cdot B$, contrary to the hypothesis that no component of K intersects both A and B. Thus an ϵ satisfying the above statement exists.

Now let K_a be the set of all points of K which can be joined to some point of A by an ϵ-chain in K and let $K_b = K - K_a$. Then $A \subset K_a$, $B \subset K_b$. By (8.1), K_a is both open and closed in K. Accordingly K_b is both open and closed in K. Since K is compact, K_a and K_b are compact.

(9.4) *If* $M_1 \supset M_2 \supset M_3 \supset \cdots$ *is a monotone decreasing sequence of nonvacuous, compact connected sets,* $\prod M_i$ *is a nonvacuous compact connected set.*

For we have only to note that under these conditions $\prod_1^\infty M_i =$ lim $[M_i] =$ lim sup $M_i \neq 0$, and $\sum M_i = M_1$. Accordingly our conclusion follows from (9.12).

10. **Continua.** A compact connected set will be called a *continuum*. A set L will be said to be *locally compact* provided that for any $p \in L$ there exists an open set G containing p such that $L \cdot \bar{G}$ is compact. If L is the whole space,

this obviously is the same as saying that there exists a conditionally compact open set containing p. A locally compact connected set will be called a *generalized continuum.*

If G is an open set, the set $\bar{G} - G$ will be called the *boundary* or *frontier* of G and will be denoted by $F(G)$. Since $F(G) = \bar{G}\cdot(S - G)$, where S is the whole space, *the boundary of every open set is closed.*

(10.1) *If N is a generalized continuum and G is an open set such that $N\cdot G$ is nonvacuous and different from N and $N\cdot\bar{G}$ is compact, every component of $N\cdot\bar{G}$ intersects $F(G)$.*

For suppose some component A of $N\cdot\bar{G}$ fails to intersect $F(G)$. Let $K = N\cdot\bar{G}$, $B = N\cdot F(G)$. Since K is compact, A and B are closed subsets of K [B is nonvacuous because otherwise we would have the separation $N = N\cdot G + N\cdot(S - \bar{G})$, where S is the entire space], we may apply (9.3) and obtain a separation $K = K_a + K_b$ where $K_a \supset A$, $K_b \supset B$. But then $K_a \subset G$, $K_b \supset N\cdot F(G)$. Hence $N = K_a + (K_b + N - K)$ would be a separation since K_a and K_b are compact and $\overline{N - K}\cdot\bar{G} \subset F(G)$. This contradicts the connectedness of N.

(10.2) *If N is a generalized continuum and G is an open set such that $N\cdot G$ is nonvacuous and different from N and $N\cdot\bar{G}$ is compact, $F(G)$ contains at least one limit point of each component of $N\cdot G$.*

For suppose there exists a component A of $N\cdot G$ having no limit point in $F(G)$. Then since A is closed in $N\cdot G$, it follows that A is closed. Thus since A and $S - G$, (S denotes the space) are disjoint closed sets it follows by normality, (5.1), that there exists an open set H such that

$$A \subset H \subset \bar{H} \subset G.$$

Now by (10.1) it follows that the component C of $N\cdot\bar{H}$ containing A intersects $F(H)$. Whence $C \neq A$. But since $C \subset N\cdot\bar{H} \subset N\cdot G$, this contradicts the fact that C is a component of $N\cdot G$.

(10.3) *No continuum is the sum of a countable number (greater than 1) of disjoint closed sets.*

Suppose, on the contrary that some continuum M is expressible in the form

$$M = K_1 + K_2 + K_3 + \cdots$$

where the sets K_i are closed and disjoint. Let G_1 be an open set containing K_2 and such that $\bar{G}_1\cdot K_1 = 0$ and let M_1 be a component of $M\cdot\bar{G}_1$ intersecting K_2. Then since by (10.1) $M_1\cdot F(G_1) \neq 0$, it follows that M_1 does not intersect K_1 but does intersect some K_i with $i > 2$. Let K_{n_2} be the first set after K_2 in the sequence $[K_i]$ which interesects M_1 and let G_2 be an open set containing K_{n_2} and such that $\bar{G}_2\cdot K_2 = 0$. Let M_2 be a component of $M_1\cdot\bar{G}_2$ containing a point of K_{n_2}. Since $M_2\cdot F(G_2) \neq 0$ and $M_2 \subset M_1$, M_2 intersects some K_i with $i > n_2$ but no K_i with $i < n_2$. Let K_{n_3} be the first K_i after K_{n_2} which intersects M_2 and let G_3 be an open set containing K_{n_3} such that $\bar{G}_3\cdot K_{n_2} = 0$. Then if M_3 is a component containing a point of K_{n_3}, M_3 intersects some set K_i with $i > n_3$ but no K_i with $i < n_3$.

Continuing this process indefinitely, we obtain a sequence of subcontinua of $M: M_1 \supset M_2 \supset M_3 \supset \cdots$ such that for each i, M_i intersects no set K_j for $j < n_i$ and $n_i \rightarrow \infty$ with i. Now by (3.6), $P = \prod_1^\infty M_i \neq 0$. But $P \cdot K_i = 0$ for all i. Hence $P \cdot \sum K_i = P \cdot M = 0$, contrary to $P \subset M$.

11. **Irreducible continua. Reduction Theorem.** A set of points H is said to be *irreducible* with respect to a given property P provided the set H has property P but no non-empty closed proper subset of H has property P. A set M which is irreducible with respect to the property of being a continuum containing two points a and b (or more generally a closed set K) is called an *irreducible continuum* from a to b (or about K) or a continuum irreducible *between* a and b. This means, of course, that M is a continuum containing both a and b but no proper subcontinum of M can contain both a and b. In this section we shall prove that an arbitrary continuum M contains an irreducible continuum between any two points a and b of M or, indeed, about any given closed subset K of M. This will follow easily from a general theorem known as the Brouwer Reduction Theorem, which we now proceed to establish.

A property P is said to be *inducible* (or *inductive*) provided that when each set of a monotone decreasing sequence $A_1, A_2, A_3, \cdots$ of compact sets has property P, so also does their product $A = \prod_1^\infty A_i$. For example, the property of being nonvacuous is inducible, as was shown in (3.6). Also the property of being connected is inducible, as was proved in (9.4).

(11.1) Brouwer reduction theorem. *If P is an inducible property, then any nonvacuous compact set K having property P contains a nonvacuous closed subset which is irreducible with respect to property P.*

Proof. Let $R_1, R_2, \cdots$ be a fundamental sequence of neighborhoods in the space. Let n_1 be the least integer such that K contains a closed subset A_1 having property P and not interesecting R_{n_1}. Let n_2 be the least integer greater than n_1 such that A_1 contains a closed subset A_2 having property P and not intersecting R_{n_2}. Similarly, let n_3 be the least integer greater than n_2 such that A_2 contains a closed subset A_3 having property P and not intersecting R_{n_3}, and so on. Continuing this process indefinitely we obtain an infinite monotone decreasing sequence of compact sets $A_1, A_2, A_3, \cdots$ each having property P. Now by (3.6), if A denotes $\prod A_i$, then A is nonvacuous and compact; and since P is inducible, it follows that A has property P. Furthermore, A is irreducible with respect to property P. For if some proper closed subset B of A had property P, there would exist an integer k such that $R_k \cdot A \neq 0$ but $R_k \cdot B = 0$. But since $R_{n_i} \cdot A = 0$ for all i's we have $k \neq n_i$ for all i's, whereas $R_k \cdot B = 0$ would give $k = n_i$ for some $i < k$ by the definition of the integers n_i. (Note. We have assumed throughout that K itself is not already irreducible relative to property P.)

(11.2) *If K is any closed subset of a continuum, M, then M contains an irreducible subcontinuum about K.*

To see this we have only to note that the property of being a subcontinuum of M containing K is inducible.

12. Continua of convergence. Locally connected sets. If the continuum K is a subset of a set M, then K will be called a *continuum of convergence* of M provided there exists in M a sequence of mutually exclusive continua K_1, K_2, K_3, $\cdots$ no one of which contains a point of K and which converges to K as a limit, i.e., $K \cdot \sum_1^\infty K_i = 0$ and $\lim [K_i] = K$.

For example, let $M = Q + \sum_1^\infty K_i$, where Q is the square with vertices (0, 0), (1, 0), (1, 1) and (0, 1) and, for each i, K_i is the straight line interval from $(1/i, 0)$ to $(1/i, 1)$ and let K be the interval from (0, 0) to (0, 1). Then K is a continuum of convergence of M, for $K = \lim [K_i]$. It is to be noted that this continuum M is not *locally connected*, that is, it contains points x, e.g., the point (0, 1/2) in every neighborhood of which there are points of M which cannot be joined to x by a connected subset of M of diameter less than some positive number given in advance. It is in connection with the study of the property of *local connectedness* that we find the principal applications for the notion of continuum of convergence. Indeed we shall show presently that any non-locally connected continuum always has continua of convergence of a particular type.

DEFINITION. *A point set M is said to be locally connected at a point p of M if for every $\epsilon > 0$ a $\delta > 0$ exists such that every point x of M whose distance from p is less than δ lies together with p in a connected subset of M of diameter less than ϵ, or, in other words, if for each neighborhood U of p a neighborhood V of p exists such that every point of $M \cdot V$ lies in the component of $M \cdot U$ containing p. A set M which is locally connected at every one of its points is said to be locally connected.*

(12.1) THEOREM. *If the generalized continuum M is not locally connected at one of its points p, then there exists a spherical neighborhood R with center p and an infinite sequence of distinct components N_1, N_2, N_3, $\cdots$ of $M \cdot \bar{R}$ converging to a limit continuum N which contains p and has no point in common with any of the continua N_1, N_2, N_3, $\cdots$.*

Proof. Since M is not locally connected at p, there exists some spherical neighborhood R with center p and radius ϵ such that $M \cdot \bar{R}$ is compact and for every positive δ, $V_\delta(p)$ contains points of M which do not belong to the component C of $M \cdot \bar{R}$ containing p. Let x_1 be such a point lying in $V_{\epsilon/2}(p)$ and let C_1 be the component of $M \cdot \bar{R}$ containing x_1. Since C_1 is closed and does not contain p, there exists a point x_2 in $M \cdot V_{\epsilon/4}(p)$ which does not belong to $C_1 + C$. Let C_2 be the component of $M \cdot \bar{R}$ containing x_2. Likewise since $C_1 + C_2$ is closed and does not contain p, there exists a point x_3 in $M \cdot V_{\epsilon/8}(p)$ which does not belong to $C + C_1 + C_2$. Let C_3 be the component of $M \cdot \bar{R}$ containing x_3, and so on. Continuing this process indefinitely, we obtain a sequence C_1, C_2, C_3, $\cdots$ of distinct [since for each n, $C_n \supset x_n$ and $x_n \cdot (C_1 + C_2 + \cdots + C_{n-1}) = 0$] components of $M \cdot \bar{R}$ whose inferior limit contains p.

Now, by (7.1) the sequence $[C_i]$ contains a convergent subsequence C_{n_i} with limit N which necessarily contains p. For each i, set $N_i \equiv C_{n_i}$. Now, by §9, N is connected and hence is a continuum. Furthermore $N \subset C$, because $N \supset p$. Therefore $N \cdot \sum_1^\infty N_i = 0$. This completes the proof.

(12.2) *Every set M which contains a generalized continuum which is not locally connected has a continuum of convergence.*

(12.3) *If the generalized continuum M is not locally connected at a point p, then there exists a non-degenerate subcontinuum H of M containing p and such that M is not locally connected at any point of H.*

For if ϵ is the radius of R and if H denotes the component of $N \cdot \overline{V_{\epsilon/2}(p)}$ containing p as in (12.1), then it is clear that M is not locally connected at any point of H, and H is a continuum containing more than one point because it contains p and by (10.1) it must contain at least one point of $F[V_{\epsilon/2}(p)]$.

13. **Semi-locally-connected sets. Regular sets.** A connected set M will be said to be $\left\{\begin{matrix}\textit{semi-locally-connected} \text{ (abbreviated s.l.c.)} \\ \textit{regular}\end{matrix}\right\}$ at a point x of M provided that for any $\epsilon > 0$ there exists a neighborhood V of x in M of diameter less than ϵ such that $\left\{\begin{matrix}M - V \text{ has only a finite number of components} \\ F(V) \text{ is a finite set of points}\end{matrix}\right\}$. If M is s.l.c. or regular at each of its points, it is said to be *s.l.c.* or *regular*, respectively.

(13.1) *If a connected set M is regular at a point x of M, it is both semi-locally-connected and locally connected at x.*

For any separation $M - V = M_1 + M_2$, both M_1 and M_2 must contain a point of $F(V)$. Thus if $F(V)$ is a finite set, $M - V$ can have no more components than there are points in $F(V)$. Thus M is s.l.c. at x.

Now in the same way, $V + F(V)$ can have no more components than there are points in $F(V)$. Thus x is not a limit point of $M - C$, where C is the component of $V + F(V)$ containing x. Hence M is locally connected at x.

(13.2) *If a generalized continuum M is not s.l.c. at a point $x \in M$, x lies on a continuum of convergence of M.*

Proof. For some $\epsilon > 0$, it is true that for any neighborhood V of x in M of diameter less than 2ϵ, $\bar{V}$ is compact and $M - V$ must have infinitely many components. Now there must be infinitely many components of $M - V_{\epsilon/i}(x)$ each intersecting both $F[V_{\epsilon/i}(x)]$ and $M - V_\epsilon(x)$ $(i = 2, 3, \cdots)$. For if, for some i, only a finite number of such components existed, let N denote their sum. Let $V = M - N = V_{\epsilon/i}(x) +$ all components of $M - V_{\epsilon/i}(x)$ lying entirely in $V_\epsilon(x)$. Then V is open (since N is closed) and $\delta(V) < 2\epsilon$. This is impossible because $M - V = N$ and N has only a finite number of components.

Now let C_1 be a component of $M - V_{\epsilon/2}(x)$ intersecting both $F[V_{\epsilon/2}(x)]$ and $M - V_\epsilon(x)$ and let K_1 be a component of $C_1 \cdot \overline{V_\epsilon(x)}$ containing a point of $F[V_{\epsilon/2}(x)]$. Similarly let C_2 be a component of $M - V_{\epsilon/3}(x)$ not containing C_1 but intersecting both $F[V_{\epsilon/3}(x)]$ and $M - V_\epsilon(x)$ and let K_2 be a component of $C_2 \cdot \overline{V_\epsilon(x)}$ containing a point of $F[V_{\epsilon/2}(x)]$. Similarly let C_3 be a component of $M - V_{\epsilon/4}(x)$ containing neither C_1 nor C_2 but intersecting both $F[V_{\epsilon/4}(x)]$ and $M - V_\epsilon(x)$ and so on indefinitely. Let $[K_{n_i}]$ be a convergent subsequence of $[K_n]$ and let K be its limit set. Then K is a continuum of convergence which contains x and at least one point of $F[V_\epsilon(x)]$. See **Errata,** page 280.

Since clearly M can be locally connected at no point of $K - x$ we have

(13.21) COROLLARY. *Any locally connected generalized continuum is semi-locally-connected.*

14. Locally connected sets. It will be recalled that a set M is *locally connected* provided it is locally connected at each of its points, i.e., for each $p \epsilon M$ and each neighborhood U of p there exists a neighborhood V of p such that $M \cdot V$ lies in a single component of $M \cdot U$ (see §12).

A connected open subset of a set M will be called a *region* in M.

(14.1) *A set M is locally connected if and only if each component of an arbitrary open subset of M is itself open in M.*

(14.2) *A set M is locally connected if and only if each point of M is contained in arbitrarily small regions in M.*

The proofs of these propositions result at once from the definition of local connectedness and are left as exercises. (Note that (13.21) is also readily obtained from either (14.2) or (14.3) by an application of the Borel Theorem.)

(14.3) THEOREM. *Every connected open subset (or region) of a locally connected generalized continuum is itself a locally connected generalized continuum.*

For let R be a region in a locally connected generalized continuum M and let $p \epsilon R$. There exists a neighborhood U of p so that $M \cdot \bar{U}$ is compact and such that $M \cdot \bar{U} = R \cdot \bar{U}$. Thus R is connected and locally compact. Finally, by local connectedness of M, there exists, for any $\epsilon > 0$, a region Q in M with $p \subset Q \subset R$ and $\delta(Q) < \epsilon$. Since clearly Q is also a region in R, it follows by (14.2) that R is locally connected.

15. Property S. Uniformly locally connected sets. A point set M is said to have *property S* provided that for each $\epsilon > 0$, M is the sum of a finite number of connected sets each of diameter less than ϵ.

(15.1) *If M has property S, it is locally connected.*

For let x be any point of M and ϵ any positive number. Let $M = M_1 + M_2 + \cdots + M_n$, where $\delta(M_i) < \epsilon/2$. Let K be the sum of all those sets M_i which either contain x or have x for a limit point. Then clearly K is connected and $\delta(K) < \epsilon$. Thus since x is not a limit point of $M - K$ it follows at once that M is locally connected at x.

(15.2) *If M has property S, so also does every set M_0 such that $M \subset M_0 \subset \bar{M}$.*

For let ϵ be any positive number and let $M = M_1 + M_2 + \cdots + M_n$ where M_i is connected and of diameter less than ϵ for $1 \leqq i \leqq n$. Then $M_0 = \bar{M}_1 \cdot M_0 + \bar{M}_2 \cdot M_0 + \cdots + \bar{M}_n \cdot M_0$ and clearly $\bar{M}_i \cdot M_0$ is connected (since $M_i \subset \bar{M}_i \cdot M_0 \subset \bar{M}_i$) and of diameter less than ϵ.

The two facts just established yield at once the following proposition:

(15.3) *If M has property S, then every set M_0 such that $M \subset M_0 \subset \bar{M}$ is locally connected.*

DEFINITION. *If N is any subset of a metric space D and ϵ is any positive number, we shall denote by $T_\epsilon(N)$ the set of all points x of D which can be joined to N by a chain of connected subsets $L_1, L_2, \cdots, L_n$ of D such that for each i, $\delta(L_i) < \epsilon/2^i$, $L_1 \cdot N \neq 0$, $L_n \supset x$, and any two successive sets (links) L_i and L_{i+1} have at least one common point. Such a chain will be called a chain of type T_ϵ or simply type T.*

(15.4) THEOREM. *If the metric space D has property S, then every set $T_\epsilon(N)$ has property S.*

For let δ be any positive number, and let us choose an integer k such that $\sum_k^\infty \epsilon/2^i < \delta/4$. Let E be the set of all points in $T_\epsilon(N)$ which can be joined to N (i.e., to points of N) by a chain of type T which has at most k links. Let us express D as the sum of a finite number of connected sets each of diameter less than $\epsilon/2^{k+1}$; and of these sets, let $Q_1, Q_2, \cdots, Q_n$ be the ones which contain at least one point of E. Then we have $E \subset \sum_1^n Q_i$. Now for each i, $Q_i \subset T_\epsilon(N)$; for Q_i contains a point x of E, and x can be joined to N by a chain L_1, $L_2, \cdots, L_r$ of type T having k links or less; and since $\delta(Q_i) < \epsilon/2^{k+1}$, therefore $[L_1, L_2, \cdots, L_r, Q_i]$ is a chain of type T, and hence $Q_i \subset T_\epsilon(N)$. For each i $(1 \leqq i \leqq n)$ let W_i be the set of all points of $T_\epsilon(N)$ which can be joined to some point of Q_i by a connected subset of $T_\epsilon(N)$ of diameter less than $\delta/4$. Then for each i, W_i is a connected subset of $T_\epsilon(N)$ of diameter less than δ. It remains only to show that $T_\epsilon(N) \subset \sum_1^n W_i$. To this end let x be any point of $T_\epsilon(N)$ and let $L_1, L_2, \cdots, L_m$ be a chain of type T joining x to N. Obviously we need consider only the case in which x does not belong to E, and in this case $m > k$. Then since $L_k \subset E$, it follows that for some j, $L_K \cdot Q_j \neq 0$; and since $\sum_k^\infty \epsilon/2^i < \delta/4$, it follows that $\delta(\sum_k^m L_i) \leqq \sum_k^m \delta(L_i) < \delta/4$. Hence $\sum_k^m L_i$ is a connected subset of $T_\epsilon(N)$ of diameter $K < \delta/4$ which joins x to a point of Q_j. Therefore $x \subset W_j$, and our theorem is proved.

(15.41) COROLLARY. *Any metric space D having property S is the sum of a finite number of arbitrarily small connected subsets each having property S. Furthermore these subsets may be chosen either as open sets or as closed sets.*

For let δ be any positive number, let $\epsilon = \delta/3$, and let $D = \sum_1^n D_i$, where each D_i is connected and of diameter less than ϵ. Then, for each i, $T_\epsilon(D_i)$ is connected and of diameter less than δ and clearly $D = \sum_1^n T_\epsilon(D_i)$. Now the sets $[T_\epsilon(D_i)]$ themselves are open; and since it is true that if a set E has property S, so does every set E_0 such that $E \subset E_0 \subset \bar{E}$, it follows that the sets $[\overline{T_\epsilon(D_i)}]$ have property S, and of course they are closed.

From this corollary it follows that in any metric space having property S there exists a monotone decreasing *fine* subdivision into connected sets. In other words, we can subdivide such a space into a finite number of connected sets each having property S and being of diameter less than 1; then we can subdivide each of these sets into a finite number of connected sets of diameter less than 1/2, and so on indefinitely.

(15.42) COROLLARY. *Any point p of a metric space D having property S is contained in an arbitrarily small connected open set (region) which has property S.*

To see this we have only to take $N = p$, and then the set $T_\epsilon(p)$ is the desired region. Since $\overline{T_\epsilon(p)}$ also has property S and hence is locally connected (because any set having property S is locally connected), we have shown that p is contained in an arbitrarily small region whose closure is locally connected.

(15.43) *Every locally connected generalized continuum has property S locally, i.e., each $p \in M$ is contained in an arbitrarily small region in M having property S.*

For let $p \in M$ and let $\epsilon > 0$ be chosen so that $M \cdot \overline{V_\epsilon(p)} = K$ is compact. Then $T_\epsilon(p)$ has property S. This is proved by the same argument as given for (15.4), substituting p for N and changing the first part of the third sentence to read: "By the Borel Theorem we can cover K by a finite number of regions in M each of diameter less than $\epsilon/(k+1)$."

DEFINITION. *A set M is said to be uniformly locally connected if for each $\epsilon > 0$ a $\delta_\epsilon > 0$ exists such that every two points x and y of M whose distance apart is less than δ_ϵ lie together in a connected subset of M of diameter less than ϵ.*

Obviously any uniformly locally connected set is also locally connected. In case the set is compact, the converse is also true, that is:

(15.5) *Every compact locally connected set M is uniformly locally connected.*

For if not, then for some $\epsilon > 0$ it is true that, for every positive integer n, some two points x_n and y_n of M exist with $\rho(x_n, y_n) < 1/n$ but which lie together in no connected subset of M of diameter less than ϵ. Since M is compact, the sequence $[x_n]$ contains a convergent subsequence $[x_{n_i}]$ with limit point p in M. Clearly the sequence $[y_{n_i}]$ also converges to p, since $\rho(x_{n_i}, y_{n_i}) < 1/n_i \leqq 1/i$. But since M is locally connected at p, there exists a δ such that $V_\delta(p)$ lies in a region R of diameter less than ϵ. And for n_i sufficiently large, $x_{n_i} + y_{n_i} \subset R$, contrary to the definition of x_{n_i} and y_{n_i}.

We proceed now to show that the property of being uniformly locally connected is stronger for conditionally compact sets than property S. In the first place it is seen at once that, for example, if C is a circle and p is a point of C, then the set $C - p$ has property S but is not uniformly locally connected. Thus there exist sets having property S but which are not uniformly locally connected.

(15.6) *Every conditionally compact and uniformly locally connected set M has property S.*

To prove this, let ϵ be any positive number, let δ be a number greater than 0 such that every two points x and y with $\rho(x, y) < \delta$ lie together in a connected subset of M of diameter less than $\epsilon/3$, and let $P = p_1 + p_2 + \cdots$ be a countable set of points dense in M (i.e., such that $\overline{P} \supset M$). For each n, let R_n be the set of all points of M which lie together with p_n in a connected subset of M of diameter less than $\epsilon/3$. Then for each n, R_n is connected and $\delta(R_n) < \epsilon$. We now show that for some k, $M = \sum_1^k R_n$. If this is not so, then an infinite sequence $[p_{n_i}]$ of the points of P exist such that for each i, p_{n_i} is not contained in $\sum_1^{n_i-1} R_n$. Since M is conditionally compact, $[p_{n_i}]$ has a limit point p. But then some two points, say p_{n_r} and p_{n_s} $(s > r)$, are such that $\rho(p_{n_r}, p_{n_s}) < \delta$ and hence $p_{n_s} \subset R_{n_r} \subset \sum_1^{n_r-1} R_n$, contrary to the definition of the sequence $[p_{n_i}]$. Therefore, for some k, $M = \sum_1^k R_n$, and hence M has property S.

The two propositions just established yield at once the following characterization of locally connected continua:

(15.7) *In order that a continuum M should be locally connected it is necessary and sufficient that M should have property S.*

The condition is necessary, because of (15.5) every locally connected continuum is uniformly locally connected and hence, by (15.6) has property S. It is sufficient, because by (15.1) any set having property S is locally connected.

In concluding this section we note

(15.8) *The spaces R^n $(n > 0)$, H, Q_ω and Q'_ω (§5) are connected and uniformly locally connected.*

This results at once from the facts that these spaces are "convex" and that an interval is connected (9.22). To exhibit the convexity we have only to note that if $x = (x_1, x_2, \cdots)$, $y = (y_1, y_2, \cdots)$ are two points of any one of these spaces, every point of the interval $[x_i + \lambda(y_i - x_i)]$ $(0 \leqq \lambda \leqq 1)$ belongs to the space. Since any such interval is obviously equivalent to the real number interval $0 \leqq \lambda \leqq 1$, our result follows.

References*

§§1–3. For material and further references see Hausdorff [1], Kuratowski [1], and Moore [1].

§4. For material, history, and references see Hildebrandt [1].

§§5, 6. Alexandroff [1], Fréchet [1], Tychonoff [1], Urysohn [1].

§7. (7.1) Zarankiewicz [1]. Also see Lubben [1].

§8. Hausdorff [1].

§9. (9.1) For special case, see Zoretti [1]. Also Lubben [1].

§10. Janiszewski [1], Mullikin [1].
(10.3) Sierpinski [1]. Related result, Moore [2].

§11. Janiszewski [1], Brouwer [1].

§12. Zarankiewicz [1], Moore [3, 4], Wilder [1].

§13. s.l.c.: Whyburn [1]; regularity: Menger [1], Urysohn [2].

§14. Hahn [1], Kuratowski [2].

§15. Sierpinski [2], Moore [5], Whyburn [2].

---

* See the statement concerning references in the Introduction. In general, the citations are not necessarily to the original sources nor are they in any sense complete. Instead they are meant as a guide to more exhaustive treatments of the same and related topics where references to original sources may be found.

CHAPTER II

CONTINUOUS TRANSFORMATIONS. JUNCTION PROPERTIES OF LOCALLY CONNECTED SETS

1. **Continuous transformations.** If A and B are sets, any law which assigns to each point $x \in A$ a unique point $f(x) \in B$ is called a (single-valued) *transformation* or *mapping* of A *into* B. If, in addition, for each point y of B there is at least one $x \in A$ such that $f(x) = y$, f is said to be a transformation of A *onto* B and we indicate this by writing $f(A) = B$. Further, if for each $y \in B$ there is one and only one $x \in A$ with $f(x) = y$, f is said to be *one-to-one*, written 1-1 or (1-1). In this case if for each $y \in B$ we set $f^{-1}(y) = x$ we have a single-valued "inverse" f^{-1} for f.

Let $f(x)$ be a transformation of A into B. For any subset A_1 of A, $f(A_1)$ denotes the set of all points $y \in B$ such that for some $x \in A_1$, $f(x) = y$ and is called the *image* or the *transform* of A_1 under f. For any subset B_1 of B, $f^{-1}(B_1)$ denotes the set of all $x \in A$ such that $f(x) \in B_1$ and is called the *inverse* of B_1 under f.

A transformation f of A into B is said to be *continuous* at a point $x \in A$ provided any one of the following four equivalent conditions is satisfied:

(i) For any neighborhood U of $f(x)$ there exists a neighborhood V of x such that $f(A \cdot V) \subset U$.

(ii) For any $\epsilon > 0$ a $\delta > 0$ exists such that if $p \in A$ and $\rho(p, x) < \delta$, then $\rho(f(p), f(x)) < \epsilon$.

(iii) If (x_i) is any sequence of points in A with $x_i \to x$, then $f(x_i) \to f(x)$.

(iv) If x is a limit point of a subset M of A, $f(x)$ is either a point or a limit point of $f(M)$, i.e., if $x \in \bar{M}$, $f(x) \in \overline{f(M)}$.

The proof that these conditions are equivalent is left as an exercise.

If $f(x)$ is continuous at all points of A_1, $A_1 \subset A$, it is said to be *continuous on* A_1. That $f(x)$ is continuous on the whole set A on which it is defined is indicated merely by the statement that "f is continuous."

If for the (1-1) transformation $f(A) = B$, both f and its inverse f^{-1} are continuous, f is called a *topological transformation* or a *homeomorphism* (compare §5, note). If such a transformation exists for two sets A and B, these sets are said to be homeomorphic.

(1.1) *The image under any continuous transformation of a* $\left\{\begin{matrix}\textit{compact}\\ \textit{connected}\end{matrix}\right\}$ *set is* $\left\{\begin{matrix}\textit{compact}\\ \textit{connected}\end{matrix}\right\}$.

For suppose A is compact and $f(A) = B$ is continuous. Let M be any infinite subset of B and for each $y \in M$ choose a point $x \in f^{-1}(y)$ and let N be the set of all such chosen points x. Since A is compact and N is infinite, there exists a point $p \epsilon A$ which is a limit point of N. But then $f(p)$ is a limit point of $M = f(N)$. For if (x_i) is a sequence of distinct points in N with $x_i \to p$, the points $[f(x_i)]$ are distinct and $f(x_i) \to f(p)$ by continuity.

Now if A is connected and $f(A) = B$ is continuous, any separation $B = B_1 + B_2$ of B would yield a separation $A = f^{-1}(B_1) + f^{-1}(B_2)$ of A. The latter two sets are separated, since if $f^{-1}(B_1)$, say, contained a limit point p of $f^{-1}(B_2)$, by continuity $f(p)$ would belong to B_1 and be a limit point of B_2 which is impossible. Thus B is connected.

DEFINITION. *A transformation f is said to be uniformly continuous on a set A provided that for any $\epsilon > 0$ a $\delta > 0$ exists such that if x and y are any two points of A with $\rho(x, y) < \delta, \rho(f(x), f(y)) < \epsilon$.*

(1.2) *Any transformation f which is continuous on a compact set A is uniformly continuous on A.*

Suppose on the contrary that f is continuous on A but not uniformly continuous. Then for some $\epsilon > 0$ there exist two sequences of points $[x_n]$ and $[y_n]$ in A such that, for each n, $\rho(x_n, y_n) < 1/n$, but $\rho[f(x_n), f(y_n)] \geqq \epsilon$; and since A is compact we can suppose them so chosen that $[x_n] \to x$ where $x \epsilon A$. But then also $[y_n] \to x$ and thus by continuity $[f(x_n)] \to x$ and $[f(y_n)] \to x$, so that, for n sufficiently large, $\rho[f(x_n), f(y_n)] < \epsilon$, contrary to the definition of $[x_n]$ and $[y_n]$.

(1.3) *In order that a transformation $f(A) = B$ be continuous it is necessary and sufficient that for every closed subset K of B, $f^{-1}(K)$ be closed in A.*

To prove the necessity, let K be any closed subset of B and let x be any limit point of $f^{-1}(K)$. Then since by continuity, $f(x)$ is a point or a limit point of K and K is closed, we have $f(x) \epsilon K$ so that $x \epsilon f^{-1}(K)$. Thus $f^{-1}(K)$ is closed.

To show the sufficiency, let us suppose the condition is satisfied but that f is not continuous at some point $x \epsilon A$. Then there exists a sequence $[x_i] \to x$ in A and some neighborhood V of $f(x)$ in B such that for infinitely many n_i's, $f(x_{n_i}) \epsilon (B - V)$. But $B - V$ is closed in B whereas $f^{-1}(B - V) \supset \sum x_{n_i}$ and thus $f^{-1}(B - V)$ is not closed since x is a limit point of it. This contradiction proves f continuous.

(1.31) *If A, B and C are sets and $f_1(A) = B$ and $f_2(B) = C$ are continuous transformations, the transformation $f(A) = f_2f_1(A) = f_2[f_1(A)] = C$ is continuous.*

For let K be any closed subset of C. Then $f_2^{-1}(K)$ is closed in B since f_2 is continuous, and $f_1^{-1}[f_2^{-1}(K)] = f^{-1}(K)$ is closed in A since f_1 is continuous. Hence f is continuous.

(1.4) *If A is compact, f is continuous and (1-1) and $f(A) = B$, then f^{-1} is continuous and thus f is a homeomorphism sending A into B.*

Suppose, on the contrary, that f^{-1} is not continuous at some point $y \epsilon B$, where $y = f(x)$. Then there exists in B a sequence $[y_i] \to y$ and a neighborhood U of x in A such that if $y_i = f(x_i)$, then infinitely many of the points $[x_{n_i}]$ of $[x_i]$ are in $A - U$. But since A is compact, the sequence $[x_{n_i}]$ has at least one limit point z and since $z \subset A - U$, $z \neq x$; thus $f(z) \neq y$, whereas $[f(x_{n_i})] \to y$. This contradicts the continuity of f, and thus our theorem follows.

Alternate proof. We have $f^{-1}(B) = A$. Let K be any closed set in A. Then since f is (1-1) we have

$$(f^{-1})^{-1}(K) = f(K)$$

and $f(K)$ is compact by (1.1). Thus $(f^{-1})^{-1}(K)$ is closed and hence by (1.3) f^{-1} is continuous.

(1.5) *If A has property S and $f(A) = B$ is uniformly continuous, B has property S.*

For let $\epsilon > 0$ be given. By uniform continuity of f there exists a $\delta > 0$ such that any set in A of diameter less than δ maps into a set in B of diameter less than ϵ. Thus if $A = \sum_1^k A_i$ where A_i is connected and $\delta(A_i) < \delta$, we have $B = \sum_1^k f(A_i)$ where each set $f(A_i)$ is connected by (1.1) and of diameter less than ϵ.

(1.51) *The image of a locally connected continuum under any continuous transformation is itself a locally connected continuum.*

For (1.2) gives uniform continuity in this situation. Hence this result follows from (1.5) and [I, (15.7)]. In concluding this section we note

(1.6) *The space Q'_ω is homeomorphic with the "plane" P'_ω in Q'_ω consisting of all points of the form $(0, x_2, x_3, \cdots)$.*

For if for each $x = (x_1, x_2, \cdots) \in Q'_\omega$ we define

$$f(x) = (0, x_1, x_2, \cdots),$$

then since for $x, y \in Q'_\omega$ we have

$$\rho[f(x), f(y)] = \sum_1^\infty 2^{-i-1} | x_i - y_i | = \rho(x, y)/2$$

it follows that f is (1-1) and both f and f^{-1} are continuous.

Similarly, if in the Hilbert space H we define

$$f(x_1, x_2, \cdots) = (1, x_1/2, 2x_2/3, 3x_3/4, \cdots),$$

then f maps H topologically onto the "plane" H_1 of points of the form $x_1 = 1$ in H and at the same time carries Q_ω topologically onto $Q_\omega \cdot H_1$. Furthermore, if for each $x = (1, x_2, x_3, \cdots) \in H_1$ we define

$$g(x) = (x_1/\sigma, x_2/\sigma, \cdots) \qquad \text{where } \sigma = \sqrt{\sum x_i^2}\,;\ x_1 = 1,$$

then g maps H_1 topologically onto a subset of the unit sphere S: $\sum x_i^2 = 1$ in H. Hence gf maps H topologically onto a subset of S and of course maps Q_ω topologically onto a subset of $Q_\omega \cdot S$. Thus, combining this with (1.6) and [I, §5 (note)] we have

(1.7) *Any separable metric space can be "topologically imbedded in" (i.e., mapped topologically onto a subset of) a plane in Q'_ω, in H or in Q_ω or in a sphere in H or in Q_ω.*

DEFINITION. *Let X and Y be metric spaces and suppose Y is bounded, i.e., $\rho(y_1, y_2) < M < \infty$ for $y_1, y_2 \in Y$. Let Y^X denote the set of all continuous mappings $f(X) \subset Y$ of X into Y, metrized by the formula*

$$\rho(f_1, f_2) = \underset{x \in X}{\text{l.u.b.}} \{\rho[f_1(x), f_2(x)]\}, \qquad f_1, f_2 \in Y^X.$$

As thus set up, clearly Y^X is a metric space and a sequence $f_1, f_2, f_3, \cdots$ in Y^X converges to a point $f \in Y^X$ if and only if the transformations $f_1(x), f_2(x), \cdots$ converge uniformly to the transformation $f(x)$ in the usual sense.

2. Complete spaces. Extension of transformations. A sequence of points $x_1, x_2, \cdots$ is called a *fundamental sequence* or a *Cauchy sequence* provided that for any $\epsilon > 0$ an integer N exists such that if $m, n > N$, $\rho(x_m, x_n) < \epsilon$.

A metric space or set D is said to be *complete* provided every fundamental sequence in D converges to a point in D. A space or set is *topologically complete* provided it is homeomorphic with some complete space.

We have immediately

(2.1) *Every compact set is complete. Every closed subset of a complete set is complete.*

Thus in particular the spaces Q_ω and Q'_ω are complete. Also, since the real number system R is complete, it results at once that *for every $n > 0$, the Euclidean n-space R^n is complete.* For if $p_1, p_2, \cdots$ is a fundamental sequence in R^n, $p_i = (x_1^i, x_2^i, \cdots, x_n^i)$, the sequences (x_1^i), (x_2^i), $\cdots$, (x_n^i) are fundamental sequences in R. Hence they converge to limits $x_1, x_2, \cdots, x_n$, respectively, in R; and accordingly the sequence $p_1, p_2, \cdots$ converges to the point $p = (x_1, x_2, \cdots, x_n)$ in R^n.

DEFINITION. *For any metric space or set X, by the complete enclosure $\bar{\bar{X}}$ of X will be meant the space whose points are the fundamental sequences in X, any point $x \in X$ being identified with the sequence $(x, x, x, \cdots) \in \bar{\bar{X}}$ and two fundamental sequences $(x_1, x_2, \cdots)$ and $(y_1, y_2, \cdots)$ being the same or equivalent and determine the same point provided $(x_1, y_1, x_2, y_2, \cdots)$ is a fundamental sequence, and with the distance function*

$$\rho(x, y) = \lim \rho(x_n, y_n),$$

where $x = (x_1, x_2, \cdots) \in \bar{\bar{X}}$, $y = (y_1, y_2, \cdots) \in \bar{\bar{X}}$.

It results at once that $\bar{\bar{X}}$ is a metric space and that X is imbedded in $\bar{\bar{X}}$ *isometrically*, i.e., if $x, y \in X$, then $\rho(x, y) = \rho(x^1, y^1)$ where $x^1 = (x, x, x, \cdots) \in \bar{\bar{X}}$, $y^1 = (y, y, y, \cdots) \in \bar{\bar{X}}$. Furthermore, *$X$ is dense in $\bar{\bar{X}}$*, that is, any point of $\bar{\bar{X}}$ is a point or a limit point of X or, in the space $\bar{\bar{X}}$, we have $\bar{X} = \bar{\bar{X}}$. For clearly if $x = (x_1, x_2, \cdots) \in \bar{\bar{X}}$ where $x_i \in X$, we have $x = \lim x_i^1$ where $x_i^1 = (x_i, x_i, \cdots) \in \bar{\bar{X}}$, because $\rho(x, x_i^1) = \lim_{n \to \infty} \rho(x_n, x_i)$ and for n and i sufficiently large, $\rho(x_n, x_i)$ is arbitrarily small by the Cauchy condition.

Finally, the *space $\bar{\bar{X}}$ is complete.* For let $x_1, x_2, x_3, \cdots$ be a fundamental sequence in $\tilde{X}$ where $x_n = (x_i^n)$, $x_i^n \in X$ and where $\delta(\sum_{i=1}^{\infty} x_i^n) < 1/n$. Then the sequence (x_n^n) is a fundamental sequence. For given $\epsilon > 0$ an N exists so that for $m, n > N$,

$$\rho(x_m, x_n) = \lim_{k\to\infty} \rho(x_k^m, x_k^n) < \epsilon/3.$$

Thus if N is chosen also so that $1/N < \epsilon/3$, then for $m, n > N$ we have $\rho(x_k^m, x_k^n) < \epsilon/3$ for a definite k sufficiently large. Whence

$$\rho(x_m^m, x_n^n) \leqq \rho(x_m^m, x_k^m) + \rho(x_k^m, x_k^n) + \rho(x_k^n, x_n^n) < \epsilon/3 + \epsilon/3 + \epsilon/3 = \epsilon.$$

Accordingly $p = (x_n^n)$ is a point of $\tilde{X}$. Furthermore, since $\rho(x_k, p) = \lim_{n\to\infty} \rho(x_n^k, x_n^n) < \epsilon$, we have $x_k \to p$ so that $\tilde{X}$ is complete. Thus we have proved

(2.2) THEOREM. *Any metric space X can be isometrically imbedded in a complete space $\tilde{X}$ (called the complete enclosure of X) in which X is dense.*

(2.21) *A metric space X is complete if and only if $\tilde{X} = X$.*

We next prove an extension theorem for uniformly continuous transformations.

(2.3) *If the transformation f is defined and uniformly continuous on a set D and if $f(D)$ is a subset of a complete space C, the definition of f may be extended to all limit points of D in one and only one way so that the extended transformation is uniformly continuous on $\bar{D}$ and is identical with f on D.*

Proof. For each $x \in \bar{D}$ let us select from D some sequence (x_n) with $x_n \to x$, agreeing that if $x \in D$, (x_n) is simply the sequence $x, x, x, \cdots$. Then, by uniform continuity, for any $\epsilon > 0$ an integer N exists so that for $m, n > N$, $\rho[f(x_m), f(x_n)] < \epsilon$. Accordingly $[f(x_i)]$ is a fundamental sequence, and since C is complete, $[f(x_i)]$ converges to some point of C which we define to be $f(x)$.

Now if (y_n) is any other sequence in D with $y_n \to x$, then since the sequence $x_1, y_1, x_2, y_2, \cdots$ also converges to x, it follows in the same way that the sequence $f(x_1), f(y_1), f(x_2), f(y_2), \cdots$ converges to $f(x)$ so that surely $f(y_n) \to f(x)$. Hence $f(x)$ is independent of the particular sequence (x_n).

Now to prove uniform continuity of f on $\bar{D}$, let $\epsilon > 0$ and let $\delta > 0$ be given by the uniform continuity of f on D. Let $x, y \in \bar{D}$ with $\rho(x, y) < \delta$, and let (x_i) and (y_i) be the sequences selected above for x and y, respectively. Since $x_i \to x$, $y_i \to y$, for n sufficiently large we have $\rho(x_n, y_n) < \delta$ and hence $\rho[f(x_n), f(y_n)] < \epsilon$. Accordingly f is uniquely determined and uniformly continuous on $\bar{D}$. (The uniqueness results at once from continuity, since D is dense in $\bar{D}$.)

EXERCISE. Prove that if X and Y are metric spaces and Y is bounded and complete, then Y^X is complete.

DEFINITION. *Any set which is the product of a countable number of open sets is called a G_δ-set. A set which is the sum of a countable number of closed sets is called an F_σ-set. Clearly a set is a G_δ-set if and only if its complement is an F_σ.*

We proceed to prove the important

(2.4) THEOREM. *Any G_δ-set in a complete space is topologically complete.*

Proof. Let E be a complete metric space and let A be a G_δ-set in E.

Let F be any closed set in E. For the purpose of this proof, to abbreviate formulas, we denote distances $\rho(x, y)$ simply by xy and let

$$xF = \rho(x, F) = \underset{t \in F}{\text{g.l.b.}}\ (xt), \qquad xyF = \underset{t \in F}{\text{g.l.b.}}\ (xt + yt). \tag{1}$$

Now from

$$xt + yt \geqq xF + yF, \qquad xt + yt \leqq 2 \cdot xt + xy$$

we get immediately

$$xyF \geqq xF + yF, \tag{2}$$

$$xyF \leqq 2 \cdot xF + xy. \tag{3}$$

Now if x and y do not belong to F, $xF > 0$ and by (2), $xyF > 0$. In this case let

$$\varphi(x, y) = \frac{xy}{xy + xyF} = \underset{t \in F}{\text{l.u.b.}}\ \frac{xy}{xy + xt + yt}. \tag{4}$$

Then φ is a distance function. Obviously it is symmetric and vanishes if and only if $x = y$. To prove the triangle inequality

$$\varphi(x, y) + \varphi(y, z) \geqq (x, z), \tag{5}$$

we note that since $yt \leqq yz + zt$, $yt \leqq xy + xt$, for $t \in F$ (and hence $x, y, z \neq t$) we have

$$\varphi(x, y) + \varphi(y, z) \geqq \frac{xy}{xy + xt + yt} + \frac{yz}{yz + yt + zt}$$

$$\geqq \frac{xy + yz}{xy + yz + xt + zt} \geqq \frac{xz}{xz + xt + zt}.$$

Now let

$$B = E - A = F_1 + F_2 + \cdots, \qquad \text{where each } F_n \text{ is closed.}$$

Let $\sum c_n$ be a convergent series of positive numbers and define, for $x, y \in A$,

$$\widetilde{xy} = \sum c_n \frac{xy}{xy + xyF_n} = \sum c_n \varphi_n(x, y). \tag{6}$$

By (5) it follows that this is a distance function in A. If we let $\widetilde{A}$ be the new space thus formed from A in replacing xy by $\widetilde{xy}$, it results that A and $\widetilde{A}$ are homeomorphic. For let x be a fixed point in A and let $xy \to 0$. Then since by (2)

$$\widetilde{xy} \leqq \sum c_n \frac{xy}{xy + xF_n},$$

and since each term in this sum approaches 0 as $xy \to 0$, by uniform convergence it results that $\widetilde{xy} \to 0$. Now conversely, if $\widetilde{xy} \to 0$, then since by (3)

$$\widetilde{xy} \geqq \frac{c_1}{2} \frac{xy}{xy + xF_1} \tag{7}$$

it follows that $xy \to 0$. Hence A and $\widetilde{A}$ are homeomorphic.

Finally, let

$$\widehat{xy} = \max [xy, \widetilde{xy}], \qquad x, y \in A. \tag{8}$$

Again this is a distance function and if $\widehat{A}$ denotes the space obtained from A in replacing xy by $\widehat{xy}$ it follows that $\widehat{A}$ is homeomorphic with A.

It remains to show that $\widehat{A}$ is complete. To this end let (x_n) be any fundamental sequence in $\widehat{A}$. Since $\widehat{xy} \geqq xy$, (x_n) is likewise a fundamental sequence in A and hence converges to a limit p in E. Now p cannot belong to B. For if $x_n p \to 0$, $p \in B$, say $p \in F_1$, by (7) we have for $m < n$,

$$\widetilde{x_m x_n} \geqq \frac{c_1}{2} \frac{x_m x_n}{x_m x_n + x_n F_1}$$

thus since as $n \to \infty$, $x_m x_n \to x_m p > 0$, $x_n F_1 \to 0$, we have

$$\liminf_{n \to \infty} [\widetilde{x_m x_n}] \geqq c_1/2.$$

Hence (x_n) cannot be a fundamental sequence in $\widetilde{A}$ and hence surely not in $\widehat{A}$, contrary to hypothesis. Therefore $p \in A$, so that there is a point $p \in A$ such that $x_n \to p$ in $\widehat{A}$. Accordingly, A is complete.

3. Junction properties of connected and locally connected sets. A set of points P will be said to be *dense* in a set M provided $\bar{P} \supset M$. A set is *non-degenerate* provided it contains more than one point.

(3.1) *If M is any connected set and G is any collection of open sets covering M, then any two points a and b of M can be joined by a finite chain $[G_1, G_2, \cdots, G_n]$ of sets of G such that $G_1 \supset a$, $G_n \supset b$ and any two successive sets (links) G_i and G_{i+1} have at least one common point.*

For let M_a be the set of all points of M which can be so joined to a. Then since M_a is the product of M times the sum G^* of all such finite chains in G starting from a, it follows that M_a is open in M. But M_a is also closed in M; for if $y \in M$ is a limit point of M_a any set Q of G containing y contains a point x of M_a and there exists a chain $[G_1, G_2, \cdots, G_n]$ in G joining a to x; and clearly $[G_1, G_2, \cdots, G_n, Q]$ is a chain in G joining a to y. Hence y belongs to M_a and M_a is closed. Since M is connected, we have $M_a = M$ which gives $b \subset M_a$ and proves our result.

DEFINITION. *If X is a set of points and ϵ is a number such that $\delta(X) < \epsilon$*

we shall call X an ε-set. For example, if R is a region such that $\delta(R) < \epsilon$, *we call R an ε-region.*

Now from (3.1) we have as a corollary that

If M is connected and locally connected, then any two points of M can be joined, for any $\epsilon > 0$, by a chain of ϵ-regions of the same type as in (3.1).

(3.2) *If F is any finite subset of a connected and locally connected set M and P is any subset of M which is dense in M and contains F, any two points a and b of F can be joined, for every* $\epsilon > 0$, *by an ε-chain of points of P containing F and each pair of successive points of which lie together in an ε-region of M.*

For let the points of F be $a = p_1, p_2, \cdots, p_n = b$. Now we can join p_1 and p_2 by an ϵ-chain each pair of successive points of which lie together in an ϵ-region of M. For this purpose it is only necessary to join p_1 and p_2 by a chain of $\epsilon/2$-regions as in (3.1), and take one point of P in each link in this chain. Similarly we join p_2 to p_3 by such a chain of points, and then join p_3 to p_4, and so on until we reach p_n. Then clearly the sequence of points taken in the order in which they are obtained in this complete process is an ϵ-chain of points from a to b which contains F and is such that any two successive points lie together in an ϵ-region of M.

(3.3) THEOREM. *If K is any compact subset of a nondegenerate connected and locally connected set M and P is any subset of M which is dense in M, there exists a subset E of P which is dense in K and which is the image under a* (1-1) *and uniformly continuous transformation of the set of all dyadic rational* numbers on the unit interval.*

Proof. Let $\epsilon_i = 1/2^i$ $(i = 1, 2, 3, \cdots)$. It follows by the Borel Theorem that, for each i, K is contained in the sum of a finite number of $\epsilon_i/4$-regions in M. In each of these regions let us choose a point of P and call P_i the finite set thus chosen. Then for each i, each point of K lies together with some point of P_i in an $\epsilon_i/4$-region and each point of P_i lies together with some point of K in an $\epsilon_i/4$-region. Clearly we may assume the P_i chosen so that $P_i \cdot P_j = 0$ for $i \neq j$, since each set P_i is finite. Now let a and b be any two points of M. By (3.2), there exists in P an $\epsilon_1/2$-chain C^1 of distinct points from a to b which contains P_1, and is such that any two successive points lie in an $\epsilon_1/2$-region and which, furthermore, we may suppose contains exactly $2^{v_1} + 1$ points

$$a = x_0^1, x_1^1, \cdots, x_{2^{v_1}}^1 = b.$$

For each integer i, $0 \leqq i \leqq 2^{v_1}$, let F_i^1 be the set of all those points of P_2 which lie together with x_i^1 in an $\epsilon_1/2$-region. Since C^1 contains P_1, it follows that every point of P_2 belongs to some set F_i^1. Now for each i, there exists an ϵ_1-region M_i which contains $x_i^1 + F_i^1 + x_{i+1}^1$, because each point of $F_i^1 + x_{i+1}^1$ lies together with x_i^1 in some $\epsilon_1/2$-region. Since the region M_i is itself a connected and locally connected space, it follows by (3.2) that $P \cdot M_i$ contains an

* That is, all numbers on (0, 1) which can be written as fractions having powers of 2 for denominators.

$\epsilon_2/2$-chain C_i^1 of points from x_i^1 to x_{i+1}^1 which contains $F_i^1 - \sum_0^{i-1} F_j^1$ and is such that any two successive points lie together in an $\epsilon_2/2$-region. Furthermore, we may suppose that all these chains $[C_i^1]$ contain the same number, say $2^{v_2} + 1$, of points which we denote by

$$x_i^1 = x_{i\cdot 2^{v_2}}^2,\ x_{i\cdot 2^{v_2}+1}^2,\ \cdots,\ x^2_{(i+1)\cdot 2^{v_2}} = x_{i+1}^1,$$

and that they overlap only in the end points shared by successive chains. Clearly the chains $[C_i^1]$ taken in order form an $\epsilon_2/2$-chain C^2 in P from a to b which contains all points of P_2 and is such that any two successive points lie together in an $\epsilon_2/2$-region.

Let us continue in this manner. In general, for each k, and each i, $0 \leqq i \leqq 2^{v_1+\cdots+v_{k-1}}$, let F_i^{k-1} be the set of all points of P_k which lie together with x_i^{k-1} in an $\epsilon_{k-1}/2$-region. Then there exists an ϵ_{k-1}-region M_i^{k-1} containing $x_i^{k-1} + F_i^{k-1} + x_{i+1}^{k-1}$. By (3.2), $P\cdot M_i^{k-1}$ contains an $\epsilon_k/2$-chain C_i^{k-1} from x_i^{k-1} to x_{i+1}^{k-1} which contains all points of $F_i^{k-1} - \sum_0^{i-1} F_j^{k-1}$ and is such that any two successive points lie together in an $\epsilon_k/2$-region. We may suppose that all these chains $[C_i^{k-1}]$ contain the same number, say $2^{v_k} + 1$, of points and that they overlap only in common end points of successive chains. Clearly, then, these form an $\epsilon_k/2$-chain C^k of distinct points in P from a to b which contains P_k and is such that any two successive points lie together in an $\epsilon_k/2$-region and which consists of exactly $2^{v_1+v_2+\cdots+v_k} + 1$ points:

$$a = x_0^k,\ x_1^k,\ \cdots,\ x_{2^{v_1+v_2+\cdots+v_k}}^k = b.$$

Furthermore, in this notation we have always:

(i) $$x_{i\cdot 2^{v_k}}^k = x_i^{k-1}.$$

Now let D be the set of all dyadic rational numbers on the interval (0, 1) and let t be any number of D. Then for some k we can write

$$t = i/2^{v_1+v_2+\cdots+v_k}, \qquad 0 \leqq i \leqq 2^{v_1+v_2+\cdots+v_k}.$$

To this value of t we make correspond the point x_i^k and write $f(t) = x_i^k$. Thus we define the transformation f over D. It is a consequence of (i) that f is single-valued on D; and since $x_i^k \neq x_j^k$ for $i \neq j$, f is (1-1). Also, if $E = f(D)$, then $E \supset \sum_1^\infty P_i$, so that E is dense in K.

Furthermore f is uniformly continuous on D. For let ϵ be any positive number and let us choose k so large that

(ii) $$\sum_k^\infty \epsilon_i < \epsilon/2$$

and set

(iii) $$\delta = 1/2^v, \quad \text{where } v = v_1 + v_2 + \cdots + v_k.$$

Then if t_1 and t_2 are any two values of t in D such that $|t_1 - t_2| < \delta$, they must lie between three successive values,

$$(j-1)/2^v, \qquad j/2^v, \qquad (j+1)/2^v$$

of t in D. Let us suppose t_1 lies between the last two of these values. We can write $t_1 = m/2^{v+u}$, where $u = v_{k+1} + \cdots + v_{k+w}$. Thus we have $f(t_1) = x_m^{k+w}$. Now the point x_m^{k+w} was arrived at in the following manner. We defined the chain C_j^k of points joining the two points x_j^k and x_{j+1}^k and lying wholly in the ϵ_k-neighborhood of the point x_j^k; then joining some two points of C_j^k there was set up the chain $C_{j_1}^{k+1}$ so that it lay wholly in the ϵ_{k+1}-neighborhood of each of these points and hence lay in the $(\epsilon_k + \epsilon_{k+1})$-neighborhood of x_j^k. Between some two points of $C_{j_1}^{k+1}$ there was set up the chain of points $C_{j_2}^{k+2}$ lying wholly in the ϵ_{k+2}-neighborhood of each of these points and hence in the $(\epsilon_k + \epsilon_{k+1} + \epsilon_{k+2})$-neighborhood of x_j^k, and so on. After w steps of this sort we reach the point x_m^{k+w}, which therefore lies in the $(\epsilon_k + \epsilon_{k+1} + \cdots + \epsilon_{k+w})$-neighborhood of x_j^k. By virtue of (ii), we have

$$\rho[f(t_1), x_j^k] = \rho[x_m^{k+w}, x_j^k] < \epsilon/2.$$

In exactly the same manner we prove that

$$\rho[f(t_2), x_j^k] < \epsilon/2.$$

These two relations give at once that $\rho[f(t_1), f(t_2)] < \epsilon$, which proves that f is uniformly continuous on D.

4. **Mapping theorems.** We now apply the theorem of §3 to obtain a number of important mapping and imbedding results.

(4.1) Theorem (Hahn-Mazurkiewicz). *In order that a metric space M be the image under a continuous transformation of the unit interval I it is necessary and sufficient that M be a locally connected continuum.*

Proof. Since I is a locally connected continuum, the necessity of the condition is a direct consequence of (1.51). To prove the sufficiency, let M be a locally connected continuum. Then, taking $K = P = M$ and applying (3.3), we obtain a subset E of M which is the image under a uniformly continuous transformation f of the set D of dyadic rational numbers and which is dense in M. Since D is dense in I, by (2.3) we can extend the definition of f to all points of I so that f is continuous on I. Since $f(I) \supset f(D) = E$ and E is dense in M whereas $f(I)$ is compact, we have $f(I) = M$ and our proof is complete.

(4.2) Imbedding theorem. *If K is any compact subset of a connected and locally connected set M, there exists a locally connected continuum in $\tilde{M}$ containing K.*

For, taking $P = M \subset \tilde{M}$ and applying (3.3) and extending the transformation thus obtained as in the proof of (4.1), we obtain a continuous transformation f of I into a subset L of $\tilde{M}$ containing a set E dense in K. Since L is necessarily compact we have $L \supset K$; and by (4.1), L is a locally connected continuum.

It need hardly be remarked that in case M is complete so that $\tilde{M} = M$, we have $L \subset M$.

DEFINITIONS. *A set A is said to be totally disconnected provided every connected subset of A reduces to a single point, i.e., provided the components of A are all single points. A set is said to be degenerate or nondegenerate accordingly as it does or does not consist of a single point.*

(4.3) *If N is any compact metric space, there exists a compact totally disconnected set C on the interval I and a continuous transformation $f(C) = N$.*

Proof. Let $h(N) = N_1$ be a topological transformation where $N_1 \subset Q_\omega \subset H =$ Hilbert space [see I, §5]. Let us now take $M = H$ [see I, (15.8)], $K = N_1$, $P = H - N_1$ and apply (3.3).* We thus obtain a uniformly continuous mapping $g(D) = E$ where $E \subset P$ and E is dense in N_1. Thus if we extend D continuously to I we get a continuous mapping $g(I) = L \supset E + N_1$. Since $g^{-1}(E) = D$, the set $C = g^{-1}(N_1)$ can contain no interval of I. Thus if we set

$$f(C) = h^{-1}g(C) = h^{-1}(N_1) = N$$

the conditions in our theorem are satisfied.

The methods used in this proof yield at once

(4.4) *Given any set A which can be mapped continuously onto the unit interval I and any compact set N, there exists a closed subset F of A and a continuous transformation $f(F) = N$.*

For we have only to take $h(N) = N_1 \subset Q_\omega$ as before, then map A onto I continuously by $g_1(x)$ and I onto Q_ω continuously by $g_2(x)$ so that we have $g(A) = g_2g_1(A) = g_2(I) = Q_\omega$. Then if we let $F = g^{-1}(N_1)$, F is closed in A and we have $f(F) = h^{-1}g(F) = h^{-1}(N_1) = N$.

Now it was shown in [I, (5.2)] that any space satisfying Axioms (1)–(6) can be mapped continuously *into* I. However with the aid of the metric we can readily show

(4.5) *Any connected set M containing more than one point can be mapped continuously onto the unit interval I.*

For if $a, b \in M$, $a \neq b$, we have only to define

$$f(x) = \frac{\rho(a, x)}{\rho(a, x) + \rho(x, b)} \qquad \text{for } x \in M.$$

This yields at once

(4.6) *Given any two nondegenerate locally connected continua A and B, there exists a continuous transformation of A onto B.*

* $H - N_1$ is dense in N_1 since N_1 is compact and no compact set can contain an open subset of H. For if for some $d > 0$ we had a point $p \in H$ so that $V_{2d}(p) \subset N_1$, the set of points $x_n = (x_i^n)$ where $x_i^n = d$ for $n = i$, $x_i^n = 0$ for $n \neq i$ would be contained in N_1 but clearly it can have no limit point.

We could equally well take N_1 in the plane H_1 of H, in the plane $H_1 \cdot Q_\omega$ of Q_ω or the plane P'_ω of Q'_ω and use $M = H$, Q_ω or Q'_ω, respectively, by virtue of (1.7). In these cases obviously $M - N_1$ is dense in N_1.

For we have only to map A onto I and I onto B continuously and take the "product" of these two transformations.

In connection with (4.3) and (4.4) it is of interest to note that we can easily construct "totally disconnected" compact sets on I, i.e., sets containing no interval of I, which can be mapped continuously onto I. One of the simplest and best known of such sets is the so-called *"Cantor discontinuum"* or "Cantor ternary set." This is obtained by first dividing I into three equal subintervals and removing the interior of the middle one. Then each of the two remaining intervals is divided into three equal intervals and the interior points of the middle ones removed. Then each of the remaining four intervals is treated similarly and so on indefinitely. The set C remaining is the required set.

This set C is readily mapped continuously onto I as follows: the points 0 and 1 are mapped into themselves. The end points of the open interval removed at the first stage are both mapped into 1/2. The end points of the two open intervals removed at the second stage are mapped into 1/4 and 3/4, respectively, (left-most into 1/4, right-most into 3/4). The end points of the four open intervals removed at the next stage are mapped into 1/8, 3/8, 5/8, 7/8, respectively, and so on. Continuing in this manner, it is clear that we obtain a uniformly continuous mapping of a subset E of C, dense in C, onto the set D of dyadic rational numbers. Extending this to $\bar{E} = C$ yields the required mapping of C onto I.*

Since, also, it is readily verified that C can be mapped continuously onto any closed subset of C, from (4.4) we get

(4.7) *The Cantor discontinuum C can be mapped continuously onto any compact set whatever.*

It is an interesting exercise to prove (4.7) directly without making use of (3.3) or its consequences.

* This same mapping of C onto I may be described in the following interesting way (due to J. L. Kelley). As is easily verified, the set C consists of exactly those numbers on I which can be expanded in the triadic number system using only the digits 0 and 2. If for each $x \in C$ we take the triadic expansion: $x = (.a_1a_2a_3 \cdots) = a_1/3 + a_2/3^2 + \cdots$, $a_i = 0, 2$ and define $f(x) = (1/2)(a_1/2 + a_2/2^2 + \cdots) = a_1/2^2 + a_2/2^3 + \cdots =$ the number whose dyadic expansion is

$$\left\{.\frac{a_1}{2}\frac{a_2}{2}\frac{a_3}{2}\cdots\right\},$$

we see that since $0 < x - y < 1/3^k$ implies $0 < f(x) - f(y) < 1/2^{k+1} + 1/3^k$, this mapping is continuous; and since if $z \in I$, $z = (.b_1b_2 \cdots)$ (dyadic expansion), then $f(x) = z$ where $x = [.(2b_1)(2b_2) \cdots]$ (triadic expansion) $\in C$, we have $f(C) = I$. It is readily seen that this mapping is identical with the one described above.

It is of interest to note in this connection that if E is any closed set on I of positive measure $[m(E) > 0]$ and for each x, $I_x = (0, x)$, the function $f(x) = m(E \cdot I_x)/m(E)$ maps E continuously onto I. For since $f(0) = 0, f(1) = 1$ and f is continuous, it therefore takes every value between 0 and 1 on the set I, and since f is constant in any component of $I - E$, it therefore takes every one of its values on the set E.

As a final application of (3.3) we state here without complete proof an interesting mapping theorem having to do with locally connected sets in which the non-local separating points are everywhere dense. A point x of a locally connected set M is a *local separating point* of M (see Chapter III below) provided there is some region R in M containing x such that $R - x$ is not connected.

(4.8) THEOREM. *If N is any countable set of non-local separating points of a nondegenerate locally connected continuum M which is dense in M, there exists a continuous mapping $f(I) = M$ of the interval onto M such that f^{-1} is single-valued on N and $f^{-1}(N)$ is dense in I.*

To obtain this theorem from (3.3), the proof of (3.3) is modified somewhat so as to obtain the additional conclusion that the set E can be made to contain an arbitrary countable subset of K and f can be so chosen that when extended to I, the extended mapping will be such that $f^{-1}(y)$ is a single point for every non-local-separating point $y \in E$. Then, choosing $K = M$ and $P = E = N$ and applying this strengthened form of (3.3), we get our required mapping $f(I) = \bar{E} = M$. Since $f^{-1}(E) \supset D$, it surely is dense in I.

5. **Arcwise connectedness.** A continuum T will be called a *simple continuous arc*, or merely an *arc* or a *simple arc* from a to b where $a, b \in T$, provided that if x is any point of T distinct from a and from b, then $T - x$ is the sum of two mutually separated sets T_a and T_b containing a and b, respectively; or, in other words, provided that every point of $T - a - b$ separates a and b in T. The points a, b are called the *end points* of the arc T and T is said to *join* a and b. For example, any linear interval such as $(0, 1)$ is a simple arc between its extremities, since if $0 < x < 1$, then $(0, 1) - x$ breaks up into $(0, x) - x$ and $(x, 1) - x$, which are mutually separated. Also any (proper) arc of a circle is a simple arc, as is also the graph of any single-valued continuous function of the real variable x on any interval (α, β). It will be shown later, (see Chapter III) that any simple arc is homeomorphic with the unit interval $(0, 1)$, a fact which justifies our use of the term *simple arc*. All the properties of arcs which we use, however, will be proved directly from the definition as given here.

DEFINITION. *By a regular chain of sets joining two points a and b is meant a finite sequence $X_1, X_2, \cdots, X_n$ of sets (links) such that $a \subset X_1$, $b \subset X_n$, $X_i \cdot X_{i+1}$ consists of exactly one point whereas $X_i \cdot X_j = 0$ if $|i - j| > 1$. Clearly the sum C of the links in a regular chain of continua is itself a continuum.*

A point x is said to be *interior to* a set K relative to a set $M \supset K$ provided x is not a limit point of $M - K$. We proceed to prove

(5.1) ARCWISE CONNECTEDNESS THEOREM. *Every two points a and b of a locally connected continuum M can be joined in M by a simple continuous arc.*

Proof. We first note that since by [I, (15.6)], M has property S, then by [I, (15.42)] we have

(i) For any $\epsilon > 0$, any point x of M is an interior point (rel. M) of some locally connected subcontinuum $K_\epsilon(x)$ of M of diameter less than ϵ.

We next prove

(ii) Any two points a and b of a region R in M can, for each $\epsilon > 0$, be joined in R by a regular ϵ-chain of locally connected continua.

For let S be the set of all points of R which can be so joined to a. Then S is open in R; for if x is any point of S, if C is a chain of the desired type in R from a to x, if X is the link of C containing x, and if $\sigma > 0$ is a number less than $\epsilon - \delta(X)$ and also less than $\rho[x, F(R) + C - X]$, then the chain obtained from C by replacing the link X by the set $X + K_\sigma(x)$ is an ϵ-chain of the desired type in R having x as an interior point. But S is also closed in R. For let y be any limit point of S in R. Let us choose $\sigma < \epsilon$ so that $K_\sigma(y) \subset R$. As y is an interior point of $K_\sigma(y)$ and thus $K_\sigma(y) \cdot S \neq 0$, hence there exists a regular ϵ-chain C of locally connected continua $X_1, X_2, \cdots, X_n$ in R from a to some point x of $K_\sigma(y)$ such that $X_i \cdot K_\sigma(y) = 0$ for $i < n$. Now if C contains y, then $y \subset S$. If not, let $K = K_\sigma(y)$ and let f be a continuous mapping of the unit interval $I = (0, 1)$ onto K [see I, (4.1)]. Let u be a point of I such that $f(u) = y$. Since the set $D = f^{-1}(K \cdot X_n)$ is compact but does not contain u, there exists a point $v \in D$ such that the interval uv of I has only v in common with D. [For, we have only to take $v \in D$ so that $\rho(u, D) = \rho(u, v)$, where ρ is the Euclidean distance function.] Then $f(uv) = X_{n+1}$ is a locally connected continuum contained in K and such that $X_{n+1} \cdot X_n = f(v)$. Thus $X_1, X_2, \cdots, X_n, X_{n+1}$ is a regular ϵ-chain of locally connected continua in R from a to y. Hence every limit point y of S in R belongs to S and thus S is closed in R. As S is both open and closed in R and R is connected, we have $S = R$. Thus $b \subset S$ and (ii) is proved.

We proceed now to construct an arc in M from a to b. By (ii), a and b can be joined in M by a simple regular 1-chain C_1 of compact locally connected continua. Likewise for each $n > 1$, it follows by (ii), since C_{n-1} is a locally connected continuum, that a and b can be joined in C_{n-1} by a simple regular $1/n$-chain of locally connected continua. Let $C = \prod_1^\infty C_n$. Then C is a simple arc from a to b, as will now be demonstrated. In the first place, since each C_n is a continuum containing $a + b$ and $C_n \supset C_{n+1}$, it follows that C is a continuum containing $a + b$.

Furthermore, C is locally connected. This results at once from the fact that for any n and any link L of C_n, $C \cdot L$ is connected. For we can write $C_n = A + L + B$ where A and B are continua with $a \in A$, $b \in B$, $L \cdot A = p$, $L \cdot B = q$, $A \cdot B = 0$ and p and q are single points. Since $(A + L) \cdot B = q$ it follows by [I, (8.3)] that $(A + L) \cdot C$ is connected. Again, since $A \cdot L = p$, it follows in the same way that $L \cdot C$ is connected. (Note: the cases $A = a$ and/or $B = b$ require no special treatment even though they are actually simpler.) Thus C has property S and accordingly it is locally connected. Finally, each point x of $C - (a + b)$ separates a and b in C. [That is, $C - x = C_a + C_b$, where C_a and C_b are mutually separated and contain a and b, respectively.] For let

P_n be the set of all points common to at least two links of C_n and let $P = \sum_1^\infty P_n$. Then clearly $P \subset C$ and every point of P separates a and b in C. Let x be any point of $C - P - a - b$. Then if x does not separate a and b in C, some component R of $C - x$ contains $a + b$. But by (ii), R contains a continuum K containing $a + b$, and hence $K \supset P$; but this is impossible, because x is a limit point of P but not of K. Therefore x separates a and b in C and hence C is a simple arc from a to b.

Note. A set X will be said to be *arcwise connected* provided every two points of X can be joined by a simple arc lying in X.

(5.2) *Every locally connected generalized continuum is arcwise connected.*

For by [I, (15.43)] and (3.1), it follows at once that any two points of such a generalized continuum K lie together in a locally connected continuum $M \subset K$; and by the theorem just proved, M is arcwise connected.

(5.3) *Any region in a locally connected generalized continuum is arcwise connected.*

(5.4) *Any complete, connected and locally connected set or space is arcwise connected.*

For by (4.2) any two points of such a space C may be imbedded in a locally connected continuum M in C and M is arcwise connected.

This, together with (2.4) gives at once

(5.5) *Any connected and locally connected G_δ-set in a complete space is arcwise connected.*

REMARKS. *Alternate proofs for* (5.1).

(1) Although in the above proof we showed that the set C is a simple arc from a to b by proving that it satisfies our definition of an arc, it is interesting to note that the method of construction of C lends itself readily to yield a direct proof that C is homeomorphic with the unit interval. For clearly the chain C_1 can, by choosing a sufficiently small norm and by combining the proper number of links, be so chosen that the set P_1 of all points common to at least two links of C_1 contains exactly $2^{v_1} - 1$ distinct points, where v_1 is some integer. Similarly the chain C_2 may be so chosen that each of its links lies wholly in some link of C_1 and each link of C_1 contains exactly 2^{v_2} links of C_2, where v_2 is some integer—uniform for all links of C_1, and thus so that the set P_2 of all points common to at least two links of C_2 contains exactly $2^{v_1+v_2} - 1$ points. We can continue in this way so that the set P_n of all points common to at least two links of C_n contains exactly $2^{v_1+v_2+\cdots+v_n} - 1$ points, and such that each link of C_{n-1} contains exactly 2^{v_n} links of C_n. Now if for each n we place the points of $a + P_n + b$ into correspondence with the numbers $m/2^{v_1+v_2+\cdots+v_n}$, $(m = 0, 1, 2, \cdots, 2^{v_1+v_2+\cdots+v_n})$, in the order in which they occur in C_n from a to b, we obtain a (1-1) correspondence f between the set $\sum_1^\infty P_n$ and the set D of rational numbers of the form $m/2^n$. It is readily shown that both f and its inverse are uniformly continuous, and hence both may be extended to their limit points, thus giving a homeomorphism between C and the interval (0, 1).

(2) It will be noted that mapping theorems are used in the proof of (5.1) only in showing, under (ii), that the set S there defined is closed in R. This can be done independent of mapping theorems as follows. Beginning with the sentence "If not let $K = K_\sigma(y) \cdots$" we proceed as follows: If not let $K_1 = K_0(y)$ and, using property S, express K as the sum of a finite number of locally connected continua $[G']$ each of diameter less than 1 and less than $\rho(y, C)$. Let G_1 be the sum of all these continua not intersecting X_n and let Y_1 be the component of G_1 containing y. There exists at least one continuum K_2 of $[G']$ such that $K_2 \cdot X_n \neq 0 \neq K_2 \cdot Y_1$. Again using property S express K_2 as the sum of a finite number of locally connected continua $[G'']$ each of diameter less than 1/2 and less than $\rho(Y_1, C)$. Let G_2 be the sum of all these not intersecting X_n and let Y_2 be the component of G_2 containing Y_1, and so on. Continuing in this way we obtain an infinite sequence of locally connected continua $y \subset Y_1 \subset Y_2 \subset Y_3 \subset \cdots \subset K_0 - K_0 \cdot X_n$ such that if $Y = \overline{\sum Y_i}$, clearly $Y = \sum Y_i +$ a single point of X_n. Thus Y is a locally connected continuum in K_0 and $X_1, X_2, \cdots, X_n, Y$ is a regular ϵ-chain of locally connected continua in S from a to y. Accordingly S is closed in R.

(3) Finally, by leaning heavily on the mapping theorem (4.1), we get the following simple and direct proof of (5.1) due to J. L. Kelley:

Let $f(I) = M$ be a continuous mapping of the unit interval I onto M. Let us say that a closed subset F of I has property P provided $f(F) \supset a + b$ and if (xy) is any maximal segment of $I - F$, then $f(x) = f(y)$.

Then property P is inducible. For let F_i be closed and have property P and $F_1 \supset F_2 \supset F_3 \supset \cdots$. Let $F = \prod F_i$ and let (xy) be a maximal open segment of $I - F$. Then there exists a sequence $(x_1y_1), (x_2y_2), \cdots$ of maximal segments complementary to $F_1, F_2, \cdots$ such that $(xy) = \sum (x_iy_i)$, $x_i \to x$, $y_i \to y$. Since $f(x_i) = f(y_i)$, clearly this gives $f(x) = f(y)$.

Thus by the Brouwer Reduction Theorem [I, (11.1)], since I has property P, there exists a closed subset A of I which has property P irreducibly. We shall show that $f(A) = T$ is a simple arc in M from a to b.

First, T is a continuum. For if there were a separation $T = T_1 + T_2$, since $f^{-1}(T_1)$ and $f^{-1}(T_2)$ are disjoint compact sets on I there would be an open segment (xy) in $I - A$ with $f(x) \in T_1$, $f(y) \in T_2$ contrary to $f(x) = f(y)$. Now let $p \in T - (a + b)$, $P' = A \cdot f^{-1}(p)$ and $a_0 \in A \cdot f^{-1}(a)$, $b_0 \in A \cdot f^{-1}(b)$. Then by the irreducibility of A, the interval a_0b_0 of I must contain A and must break up into intervals: $a_0b_0 = a_0p_1 + p_1p_2 + p_2b_0$ where $A \cdot p_1p_2 = P' = p_1 + p_2$ (p_1 may $= p_2$). Thus $A = (A_1 + P') + (A_2 + P')$, where $(A_1 + P')$ and $(A_2 + P')$ are compact and intersect in just P' and $A \cdot ap_1 - p_1 = A_1 \supset a_0$, $A \cdot p_2b - p_2 = A_2 \supset b_0$. Thus if $T_1 = f(A_1)$, $T_2 = f(A_2)$ we have

$$(*) \qquad T - p = T_1 + T_2 \quad \text{where} \quad T_1 \supset a,\ T_2 \supset b.$$

Now this must be a separation. For if $T_1 \cdot T_2 \supset q$, we could find $q_1 \in A_1$ and $q_2 \in A_2$ such that $f(q_1) = f(q_2) = q$. But clearly if we omit from A the points on the open segment (q_1q_2) of I, we obtain a closed proper subset of A having property P, contrary to the irreducibility of A. Thus $T_1 \cdot T_2 = 0$; and since $T_1 + p$ and $T_2 + p$ are compact, (*) is a separation. Accordingly A is a simple arc in M from a to b.

REFERENCES

§2. (2.2) Hausdorff [1].
(2.4) Alexandroff [2], Hausdorff [2].
§3. Whyburn [3].
§4. (4.1) Hahn [2], Mazurkiewicz [1].
(4.2) Whyburn [3].
(4.3), (4.7) Alexandroff [7] and Urysohn. (New methods of proof.)
(4.8) Harrold [1].
§5. Moore [1, 6], Mazurkiewicz [1], Tietze [1], Kaluzsay, [1], Kuratowski [3], Menger [2]. For treatment given here, Whyburn [3].
In connection with §§4, 5 see also Aronszajn [1].

CHAPTER III

CUT POINTS. NON-SEPARATED CUTTINGS

1. **Fundamental preliminaries.** If M is a connected set and p is a point of M such that the set $M - p$ is not connected, then p will be called a *cut point* of M. More generally, any set X such that $M - X$ is not connected is called a *cutting* of M. A cut point of a connected set then, is a point whose complement in the set is not connected. Thus if X is a cut point, or in fact any cutting of the connected set M, then $M - X$ is the sum of two mutually separated sets $M_1(X)$ and $M_2(X)$ and we write

$$\text{(i)} \qquad M - X = M_1(X) + M_2(X).$$

However, this separation of $M - X$ into sets $M_1(X)$ and $M_2(X)$ may not be unique, or, in other words in some cases the sets $M_1(X)$ and $M_2(X)$ may be chosen in more than one way. For example, if M is the sum of three straight line intervals ap, bp, and cp each two having only p in common, then we can choose any one of the sets $[ap - p]$, $[(ap - p) + (bp - p)]$, $[(ap - p) + (cp - p)]$, for $M_1(p)$. It is only in case $M - p$ is the sum of two *connected* mutually separated sets that the decomposition (i) is unique. It is of interest to note that it will be proved below that this is the case for all save a countable number of the cut points of any connected set.

A cut point, or in general a cutting X, of M will be said to *separate* two given points or point sets A and B in M if the two sets $M_1(X)$ and $M_2(X)$ in (i) can be so chosen that $M_1(X) \supset A$ and $M_2(X) \supset B$; and X is said to separate a single set N in M when $M_1(X)$ and $M_2(X)$ can be so chosen that $N \cdot M_1(X) \neq 0 \neq N \cdot M_2(X)$. If a and b are any two points of M, let us denote by $E(a, b)$ the set of all those points (if any) which separate a and b in M. We shall show presently that every such set $E(a, b)$ possesses a *natural order*.

A set or collection of sets E is said to be *ordered* or to possess an order, provided a definition of "preceding" (indicated $<$) can be given so that (1) to any two elements x and y of E, either x precedes y (indicated $x < y$) or y precedes x ($y < x$ or $x > y$), (2) if $x < y$ then y does not precede x, and (3) if $x < y$ and $y < z$, then $x < z$. In other words, the relation "preceding" or "$<$" must be *defined* for each pair of points, and be *asymmetric* and *transitive*. If, in addition, for each $x \in E$, when P_x denotes the sum of all elements of E preceding x and F_x the sum of all those following x (i.e., preceded by x) we have

$$\bar{P}_x \cdot F_x = P_x \cdot \bar{F}_x = 0$$

we shall say that the order in E is a *natural order*.

Now let X and Y be any two cuttings of M and set

(i) $$M - X = M_1(X) + M_2(X),$$

(ii) $$M - Y = M_1(Y) + M_2(Y),$$

representing decompositions of $M - X$ and $M - Y$, respectively, into mutually separated sets. Then if i, j, r, and s are positive integers such that $i + j = 3 = r + s$, it follows immediately that the following equation is valid:

(1.1) $$M = M_i(X) + M_r(Y) + M_j(X) \cdot M_s(Y) + X + Y.$$

With the aid of this equation, we deduce at once the result

(1.2) *If neither of the sets X and Y separates the other and $X \cdot Y = 0$, we may choose the indices i and r such that*

(a) $$X \subset M_r(Y) \quad \textit{and} \quad Y \subset M_i(X);$$

and these relations imply also the relations

(b) $$M_j(X) \cdot M_s(Y) = 0, \qquad M_j(X) + X \subset M_r(Y),$$
$$\textit{and} \quad M_s(Y) + Y \subset M_i(X).$$

Clearly this is the case, because by virtue of the relations (a) we may omit the last two terms in equation (1.1); and since M is connected, the term $M_j(X) \cdot M_s(Y)$ must vanish. This fact gives at once the remaining two relations (b).

DEFINITION. *A collection G of disjoint subsets of M will be called non-separated provided that no element of G separates in M two points belonging to any other single element of G.*

Now let E be any non-separated collection of sets in M each of which separates two given points a and b in M. Let $x, y \in E$ and let the indices in (i) and (ii) above be chosen so that

(iii) $$M_1(x) \cdot M_1(y) \supset a \quad \text{and} \quad M_2(x) \cdot M_2(y) \supset b.$$

The element x of E will be said to *precede* the element y and this fact is indicated by the notation $x < y$, provided that for at least one set of decompositions satisfying (i), (ii) and (iii) it is true that $x \subset M_1(y)$. We shall now show that this definition gives a natural order to the elements of E.

First, for any two elements x and y of E, at least one of the relations $x < y$ and $y < x$ must be valid. For if x does not precede y, then by (1.2) (a), $r = 2$ and hence $s = 1$. By (b) and (iii) it follows that $j = 2$ and hence $i = 1$. Therefore by (a), $y \subset M_1(x)$, which means $y < x$.

Second, only one of the relations $x < y$ and $y < x$ can be valid. For if $x < y$, then for some decomposition satisfying (ii) we have $x \subset M_1(y)$ so that $r = 1$.

Similarly if $y < x$, for some decomposition (i) we have $y \subset M_1(x)$ so that $i = 1$. But then $i = r, j = s$, which is impossible by (b) and (iii).*

Finally, for any three elements z, x, and y of E, the relations $z < x$, $x < y$ imply that $z < y$. For then $z \subset M_1(x)$ and $x \subset M_1(y)$. Hence in (1.2), $r = 1$ and $s = 2$. By (b) and (iii) it follows that $j = 1$. Therefore by the second relation in (b), $z \subset M_1(x) + x \subset M_1(y)$, which gives $z < y$.

Thus we have proved the following result:

(1.3) *If each element of the non-separated collection E of disjoint subsets of M separates the two points a and b in M, then the collection E possesses a natural order.*

In particular, since any collection of single points obviously is non-separated, we have

(1.31) *The set $E(a, b)$ of all points of M separating two given points, $ab \in M$ possesses a natural order.*

(1.4) *If X is any cutting of M giving a separation*

$$M - X = M_1 + M_2,$$

there exists a closed subset X' of X and a separation

$$M - X' = M_1' + M_2',$$

$$\text{where} \quad M_1' \supset M_1, \; M_2' \supset M_2.$$

To prove this, let M_1' be the set of all $x \in M$ such that

$$\rho(x, M_1) < \rho(x, M_2)$$

and let $M_2' = M - \bar{M}_1'$, $X' = \bar{M}_1' - M_1'$. Then obviously $M_1 \subset M_1' \subset M_1 + X$ and M_1' and M_2' are open. Thus X' is closed. Since $X' \subset \bar{M}_1'$ we have $X' \subset X$. Accordingly $M_2 \subset M_2'$ and obviously $M - X' = M_1' + M_2'$ is a separation.

Definition. *A subset X of M is said to separate M irreducibly or to be an irreducible cutting of M provided X separates M (i.e., $M - X$ is not connected) but no proper subset of X separates M. Similarly, X is said to separate M irreducibly between two points, a, b of M provided X separates M between a and b but no proper subset of X separates M between a and b.*

The result just proved yields at once

(1.41) Corollary. *Any irreducible cutting of M, or indeed any irreducible cutting of M between two points of M, is a closed set in M.*

(1.5) Theorem. *If G is any uncountable collection of disjoint cuttings of a connected set M and G^* denotes the sum of all the sets of G, then some element X of G separates in M a pair of points belonging to $G^* - X$.*

Proof. For if not, then for $X \in G$, we have a separation

* It results from this that if $x < y$, then *for any decompositions whatever satisfying* (i), (ii), *and* (iii), *we have* $x \subset M_1(y)$, $y \subset M_2(x)$. For if $x \subset M_2(y)$ for some decomposition, we have $r = 2 = j$, $s = 1 = i$ since by (b) and (iii) $j \neq s$. But this would give $y \subset M_1(x)$ or $y < x$.

$$M - X = Q_x + R_x,$$

where $Q_x \supset G^* - X$.

Now by (1.4) we may suppose the elements X of G are all closed sets, so that Q_x and R_x are open sets. But now for $X, Y \in G, X \neq Y$, we have

$$M = [Q_x + Q_y] + R_x \cdot R_y$$

since $Q_x \supset Y$ and $Q_y \supset X$. Thus $R_x \cdot R_y = 0$, because M is connected and the sets here exhibited are open and disjoint. But then the sets $[R_x]$ are disjoint and open, which is impossible since there are uncountably many of them whereas each of them would have to contain a neighborhood of the fundamental sequence.

(1.51) COROLLARY. *No continuum of convergence K of a connected set M contains any uncountable collection of mutually exclusive cuttings of M. Indeed, if $a, b \in K$, no subset of K separates a and b in M.*

For if $K = \lim [K_n]$ and K contains such a collection G, by (1.5) we have for some $X \in G$ a separation $M - X = R_1 + R_2$, where $R_1 \cdot G^* \neq 0 \neq R_2 \cdot G^*$. Either R_1 or R_2, say R_1, contains K_{n_i} for infinitely many n_i's, since $K_n \cdot X = 0$ for every n. But then we have that every point of $R_2 \cdot G^* \subset K$ is a limit point of $\sum_i K_{n_i} \subset R_1$, which is impossible since R_1 and R_2 are separated.

(1.52) COROLLARY. *If G is any uncountable set of cut points of a connected set M, then some two points of G are separated in M by a third point of G.*

(1.53) COROLLARY. *All save possibly a countable number of the cut points of a connected set M which belong to a connected subset C of M are cut points also of C.*

For if G denotes the set of all cut points of M which lie in C but are not cut points of C, clearly no point of G separates any two points of G in M. Thus by (1.5), G must be countable.

(1.54) COROLLARY. *If C is any connected subset of a connected set M, $\bar{C} - C$ contains at most a countable number of cut points of M.*

2. Non-separated cuttings. If (G) is any collection of sets, a point p will be called a *condensation point of the collection* $[G]$ provided every neighborhood of p contains points from uncountably many elements of G.

(2.1) LEMMA. *Let $[G]$ be any naturally ordered collection of disjoint sets in a separable metric space. At most a countable number of elements G of $[G]$ can contain a point which is not a condensation point both of the collection $[P_g]$ of all predecessors of G in $[G]$ and the collection $[F_g]$ of all successors of G in $[G]$.*

Proof. Let $[G_1]$ be the collection of all $G_1 \in [G]$ such that G_1 contains a point g_1 which is not a condensation point of $[P_{g_1}]$. For each G_1 there exists a neighborhood R_n of the fundamental sequence containing g_1 which intersects only a countable number of the sets $[P_{g_1}]$. For a given n, R_n can be chosen only a countable number of times for such sets G_1. For if not, R_n would contain an uncountable set E of the points $[g_1]$ and hence would contain two distinct condensation points g_1' and g_1'' of E which belong to E, where we may suppose the notation chosen so that $G_1' < G_1''$. But then since the order in $[G]$ is natural, g_1' is not a limit point of the sum of all the sets in $[F_{g_1''}]$, and thus uncountably many of the sets of $[G_1]$ intersecting R_n must precede G_1'' and thus belong to $[P_{g_1''}]$, contrary to the choice of R_n to correspond to G_1''.

Since thus each R_n can be chosen only a countable number of times to correspond to sets $G_1 \in [G_1]$, it follows that the collection $[G_1]$ is countable. Similarly, the collection $[G_2]$ of all elements G_2 of $[G]$ which contain points g_2 such that g_2 is not a condensation point of $[F_{g_2}]$ is countable. Hence the sum of the collections $[G_1]$ and $[G_2]$ is countable and our lemma is established.

(2.11) Corollary. *If $[H]$ denotes the collection remaining after omitting from the collection $[G]$ in (2.1) all elements G which contain a point that is not a condensation point both of the predecessors and of the successors of G, then for any two elements H_1 and H_2 of $[H]$ there are uncountably many elements of $[H]$ between H_1 and H_2, i.e., elements H with $H_1 < H < H_2$ (assuming $H_1 < H_2$).*

For if $p \in H_1$, then since p is not a limit point of the sum of all the successor^S of H_2 in $[G]$ but is a condensation point of the collection of successors of H^1 in $[G]$, there exist uncountably many elements of $[G]$ between H_1 and H_2. Since $[H]$ contains all but a countable number of the elements of $[G]$, our result follows.

Definitions. *A non-separated collection $[G]$ of subsets of M will be said to be saturated provided that if $G \in [G]$ and p is any point of $M - G$, there exists at least one element G' of $[G]$ which separates p and G in M.*

If Q is any collection of subsets of M (possibly each consisting of a single point of M), and n is an integer, a closed subset X of M is said to have potential order less than or equal to n in M relative to Q provided there exists a monotone decreasing sequence of open sets in M: $U_1, U_2, \cdots$ such that

$$\bar{U}_1 \supset U_1 \supset \bar{U}_2 \supset U_2 \supset \bar{U}_3 \supset \cdots;$$

such that $X = \prod U_i$ and for each i, $F(U_i)$ is contained in the sum of at most n elements of Q. If X is of potential order less than or equal to n but not of potential order less than or equal to $n - 1$ in M relative to Q, X is said to be of potential order n in M relative to Q.

(2.2) Theorem. *Any non-separated collection $[G]$ of cuttings of a connected set M contains a saturated subcollection $[Q]$ differing from $[G]$ by only a countable number of elements and such that each element Q of $[Q]$ is closed in M, irreducibly separates M into just two components and has potential order 2 in M relative to $[Q]$.*

Proof. Let $P = \sum_1^\infty p_i$ be a countable subset of M which is dense in M and omit from $[G]$ all elements intersecting P, retaining the notation $[G]$ for the reduced collection. For each i and j, let $[G_{ij}]$ be the collection of all elements of $[G]$ which separate p_i and p_j. Then since each element G of $[G]$ gives a separation

$$M - G = M_1 + M_2$$

and clearly both M_1 and M_2 must intersect P, it follows that for some i and j, G belongs to $[G_{ij}]$. For each i, j, $[G_{ij}]$ is naturally ordered by (1.3). Hence, using (2.1), it follows that if we omit from $[G_{ij}]$ all elements which contain a point not a condensation point both of its predecessors and of its successors in $[G_{ij}]$ and call $[H_{ij}]$ the resulting collection, then $[H_{ij}]$ differs from $[G_{ij}]$ by only a countable number of elements.

Now for each i, j, let us omit from $[H_{ij}]$ all elements H such that for some $p_h \epsilon M - H$ no element of $[H_{ij}]$ separates p_h and H in M and call the resulting collection $[Q_{ij}]$. Then for each i, j only a countable number of such sets H are omitted from $[H_{ij}]$ so that $[Q_{ij}]$ differs from $[H_{ij}]$ by a countable set of elements. For consider the collection of sets $[H + p_h]$. Let us order this collection by assigning to its elements the same precedence relations as have the sets H in the collection $[H_{ij}]$, i.e., $(H_1 + p_{h_1}) < (H_2 + p_{h_2})$ provided $H_1 < H_2$. This ordering is natural. For let $X = (H + p_h) \epsilon [H + p_h]$. Since we have a separation

$$M - H = M_1 + M_2, \qquad \text{where } M_1 \supset p_i, M_2 \supset p_j$$

and M_1 contains all elements of $[H_{ij}]$ preceding H and M_2 contains all such elements following H, and since H cannot separate any set H_1 from its corresponding point p_{h_1}, it follows that M_1 contains all the sets $(H_1 + p_{h_1})$ preceding X and M_2 contains all the ones following X. Accordingly, the ordering is natural. Also, since for any two elements H_1 and H_2 of $[H_{ij}]$, by (2.11), there exists an element H_3 of $[H_{ij}]$ which separates H_1 and H_2 in M and hence separates also the points p_{h_1} and p_{h_2} when they exist, it follows that the sets of the collection $[H + p_h]$ are disjoint. But then by (2.1), there exists an element $H + p_h$ of this collection such that for any separation

$$M - H = M_1 + M_2, \qquad M_1 \supset p_i, M_2 \supset p_j,$$

p_h is a limit point both of M_1 and of M_2, and this is absurd since $p_h \subset M_1 + M_2$. Therefore $[H_{ij}]$ differs from $[Q_{ij}]$ by only a countable number of elements.

Thus if $[Q]$ denotes the collection of all sets $Q_{ij} \epsilon [Q_{ij}]$ for all i and j, $[Q]$ is a subcollection of $[G]$ differing from $[G]$ by only a countable number of elements. We shall show that $[Q]$ satisfies also the remaining requirements in our theorem.

This collection is saturated. For let $Q \epsilon [Q]$ and $p \epsilon M - Q$. Then $Q \epsilon [Q_{ij}]$ for some i, j, and there exists an $H \epsilon [H_{ij}]$ separating p and Q. We may suppose $H < Q$ in the ordering of $[H_{ij}]$. By (2.11) it follows that there exists an element Q_1 of $[Q]$ which separates H and Q in M. Thus if we take separations

$$M - H = M_p + M_Q, \qquad M_p \supset p, M_Q \supset Q + Q_1,$$

$$M - Q_1 = M_1 + M_2, \qquad M_1 \supset H, M_2 \supset Q,$$

it follows by (1.2) that $M_p \subset M_1$. Thus Q_1 separates p and Q.

Every element Q of $[Q]$ irreducibly separates M. For let

$$(*) \qquad M - Q = M_1 + M_2$$

be any separation. Choose $p_i \in M_1$, $p_j \in M_2$. Then $Q \in [Q_{ij}]$ and since every point of Q is a condensation point both of the predecessors and of the successors of Q in the ordering of $[Q_{ij}]$ it follows that

$$\bar{M}_1 \cdot \bar{M}_2 \supset Q.$$

It follows from this that every $Q \in [Q]$ is closed in M. Furthermore, the sets M_1 and M_2 in (*) must be connected. For if we had a separation, say,

$$M_2 = N_1 + N_2 \qquad \text{where } p_j \in N_2,$$

then for any $H \in [H_{ij}]$ we have $H \subset M_1 + N_2$. Thus if $p \in N_1$, no element of $[H_{ij}]$ can separate p and Q in M, contrary to $Q \in [Q_{ij}]$. Accordingly, each set Q of $[Q]$ separates M irreducibly into just two components.

It remains only to show that each element Q of $[Q]$ has potential order 2 in M relative to $[Q]$. To this end let

$$M - Q = M_1(Q) + M_2(Q)$$

be a separation. Let n_1 be the least integer such that $p_{n_1} \subset M_1(Q)$. Since $[Q]$ is saturated there exists an element X_1 of $[Q]$ and a separation

$$M - X_1 = M_1(X_1) + M_2(X_1), \qquad M_1(X_1) \supset p_{n_1}, \; M_2(X_1) \supset Q.$$

Let n_2 be the least integer such that $p_{n_2} \subset M_2(X_1) \cdot M_1(Q)$. Similarly, there exists $X_2 \in [Q]$ and a separation

$$M - X_2 = M_1(X_2) + M_2(X_2), \qquad M_1(X_2) \supset p_{n_2}, \; M_2(X_2) \supset Q.$$

Let n_3 be the least integer such that $p_{n_3} \subset M_2(X_2) \cdot M_1(Q)$, and so on. Continuing in this way we obtain a sequence (X_i) of elements of Q such that for each i

$$\text{(i)} \qquad M_1(X_i) + X_i \subset M_1(X_{i+1}), \qquad M_2(X_i) \supset Q + M_2(Q)$$

and such that for any $p_k \in P \cdot M_1(Q)$ an i exists so that $p_k \subset M_1(X_i)$. Similarly we find a sequence (Y_i) of elements of Q such that for each i

$$\text{(ii)} \qquad M_2(Y_i) + Y_i \subset M_2(Y_{i+1}), \qquad M_1(Y_i) \supset Q + M_1(Q)$$

and such that for any $p_m \in P \cdot M_2(Q)$ an i exists so that $p_m \subset M_2(Y_i)$. Now for each i, let

$$U_i = M_1(Y_i) \cdot M_2(X_i).$$

Then for each i, U_i is open (since it is the intersection of two open sets) and by (i) and (ii) we have

$$U_1 \supset \bar{U}_2 \supset U_2 \supset \bar{U}_3 \supset U_3 \cdots \supset Q,$$

$$F(U_i) \subset X_i + Y_i.$$

Further, since $P \subset \sum [M_1(X_i) + M_2(Y_i)]$ and $[Q]$ is saturated, we have $Q = \prod U_i$. For if $p \in M - Q$ we have, say, $p \in M_1(Q)$; and some $Q' \in [Q]$ separates p and Q in M. Since P is dense in M, clearly this gives $p \in M_1(X_k)$ for some k. Accordingly Q has potential order 2 in M relative to (Q).

(2.21) COROLLARY. *Referring to the notation used in the proof of (2.2), for any element Q of $[Q]$, any neighborhood V of Q and any integer n, we have either*

$$\bar{U}_n \subset V \quad \textit{or} \quad \bar{U}_n \cdot F(V) \neq 0.$$

For suppose the contrary. Then $E = (M - V) \cdot U_n \neq 0$, and E is open in U_n, since $\bar{U}_n \cdot F(V) = 0$. Furthermore the set

$$K = M_1(X_n) + X_n + E + Y_n + M_2(Y_n)$$

must be connected. For if not we have a separation

$$K = K_1 + K_2 \qquad \text{where } K_1 \supset M_1(X_n) + X_n,\ K_2 \supset M_2(Y_n) + Y_n$$

since the sets $M_1(X_n) + X_n$ and $M_2(Y_n) + Y_n$ are connected. Either $E \cdot K_1$ or $E \cdot K_2$, say $E \cdot K_2$, is nonvacuous; but this gives the separation

$$M_1(Y_n) = E \cdot K_2 + [M_1(Y_n) - E \cdot K_2],$$

contrary to the fact that $M_1(Y_n)$ is connected. Thus K is connected. Clearly this is impossible since $K \supset X_n + Y_n$ whereas $K \cdot Q = 0$. This establishes our corollary.

Note. Essentially the same argument applies for any open set V with $V \cdot Q \neq 0$.

DEFINITION. *If G is any collection of subsets of M, and n is any integer, a subset X of M is said to be of order less than or equal to n in M relative to G provided that if V is any neighborhood of X in M there exists a neighborhood U of X in M with $U \subset V$ and such that $F(U)$ is contained in the sum of at most n elements of G. Clearly this amounts to saying that for any $\epsilon > 0$ a neighborhood U of X exists with $U \subset V_\epsilon(X)$ and $F(U) \subset$ sum of at most n of the elements of G.*

If X is of order less than or equal to n in M relative to G but not of order less than or equal to $n - 1$ in M relative to G, X is said to be of *order n in M relative to G.*

(2.22) COROLLARY. *If M is a generalized continuum and the elements of the non-separated collection $[G]$ of cuttings of M in (2.2) are compact, each element Q of the collection $[Q]$ in (2.2) is of order 2 in M relative to $[Q]$.*

For let V be any neighborhood of Q. Since Q is compact and M is locally compact, by the Borel Theorem we may suppose that V is chosen so that $\bar{V}$ and hence $F(V)$ is compact. Now by (2.21), either there exists an integer i such that $\bar{U}_i \subset V$, in which case our result follows, or else for every n we have

$$\bar{U}_n \cdot F(V) \neq 0.$$

But now since $F(V)$ is compact, this latter relation would give

$$0 \neq \prod [\overline{U}_n \cdot F(V)] = F(V) \cdot \prod \overline{U}_n = F(V) \cdot Q,$$

which is absurd since $Q \subset V$. Thus for some i, $\overline{U}_i \subset V$ and our corollary is proved.

(2.3) Existence theorem. *Let M be a connected and locally connected set, let $f(x)$ be any continuous mapping of M onto the unit interval* (0, 1) *and let $a \in f^{-1}(0)$, $b \in f^{-1}(1)$. For each x, $0 < x < 1$, we can choose a closed subset X of $f^{-1}(x)$ such that the collection $[X]$ is a non-separated collection of cuttings of M between a and b.*

Proof. For each x, $0 < x < 1$, let $R_a(x)$ be the component of $M - f^{-1}(x)$ containing a, let $R_b(x)$ be the component of $M - \overline{R_a(x)}$ containing b and let $X = F[R_b(x)]$. Then clearly

$$X = F[R_a(x)] \cdot F[R_b(x)] \subset f^{-1}(x).$$

Further, if $0 < x_1 < x_2 < 1$, we have

$$X_1 \subset \overline{R_a(x_1)} \subset R_a(x_2),\ X_2 \subset \overline{R_b(x_2)} \subset R_b(x_1).$$

Hence the collection $[X]$ is non-separated.

Since for any two disjoint closed sets A and B in a connected set M, there exists a continuous mapping $f(x)$ of M onto (0, 1) so that $A \subset f^{-1}(0)$, $B \subset f^{-1}(1)$. We have

(2.31) *If A and B are disjoint closed and connected subsets of a connected and locally connected set M, there exists an uncountable non-separated collection of cuttings of M between A and B.*

An application of (2.2) yields at once

(2.4) Divisibility theorem. *If A and B are disjoint closed and connected subsets of a connected and locally connected space M, there exists a division of M into three disioint sets R, F and G where R and G are regions containing A and B, respectively, and F is the boundary of both R and G, i.e., $M = R + F + G$, $F = F(R) = F(G)$.*

3. **Cut points. Order theorems.** We turn now to some applications of the results in the preceding section to the special case of a collection of cut points of a connected set. Clearly any collection of cut points (or of disjoint connected cuttings) of a connected set M is non-separated.

Thus we have

(3.1) Theorem. *Any set G of cut points of a connected set M contains a subset Q such that $G - Q$ is countable and any point q of Q is of potential order 2 in M relative to Q and cuts M into just two components. If M is a generalized continuum, each point of Q is of order 2 in M relative to Q.*

In particular if we take G to be the set of all cut points of the connected set M, we get

(3.2) Cut point-order theorem. *All save possibly a countable number of the cut points of a* $\left\{\begin{matrix}\textit{connected set}\\ \textit{generalized continuum}\end{matrix}\right\}$ *M are of* $\left\{\begin{matrix}\textit{potential order}\\ \textit{order}\end{matrix}\right\}$ *2 in M.*

Note. A point $p \in M$ which is of order or potential order n in M *relative to* M is said to be merely of order or potential order n in M.

4. **Properties of the set** $E(a, b) + a + b$. For any two points a and b of a connected set M we have denoted by $E(a, b)$ the set of all points of M which separate a and b in M. We assume throughout this section that M is a connected space and a and b are any two distinct points of M.

(4.1) *If M is semi-locally-connected at $x \in M$ and x does not separate two points a and b in M, i.e., x non $\in E(a, b) + a + b$, there exists a closed and connected subset N of M with*

$$a + b \subset N \subset M - x.$$

Proof. Let $V_1, V_2, V_3, \cdots$ be a sequence of neighborhoods of x in M such that $\delta(V_i) < 1/i$, $M - V_i$ has just a finite number of components, $V_i \supset \bar{V}_{i+1}$, and $a \in M - V_1$. For each i, let R_i be the component of $M - V_i$ containing a, let $R = \sum_1^\infty R_i$. Since there are only a finite number of components of $M - V_{i+1}$, every point of R_i is an interior point of R_{i+1} and thus of R. Hence R is open in M. But also, since for any i there exists a k such that for $m, n > k$, $R_m \cdot (M - V_i) = R_n \cdot (M - V_i)$, R must be closed in $M - x$. Hence b belongs to R and hence to R_n for some n, because otherwise $M - x = R + (M - x - R)$ would be a separation of $M - x$ between a and b. Taking $N = R_n$, our conclusion is satisfied.

(4.11) *For any connected space M and any $a, b \in M$, $E(a, b) + a + b$ contains every one of its limit points at which M is semi-locally-connected or regular.*

For if x is such a point, then x non $\in E(a, b) + a + b$ gives $E(a, b) \subset N \subset M - x$ so that x is not a limit point of $E(a, b)$.

(4.12) *If M is semi-locally-connected or regular, $E(a, b) + a + b$ is closed.*

As a consequence of the proof of (4.1) we have

(4.13) *If the connected set M is semi-locally-connected at $x \in M$ and $M - x$ has k components, there exists an arbitrarily small neighborhood U of x in M such that $M - U$ has exactly k components.*

Thus in particular

(4.14) *If x is a non-cut point of the connected set M and M is semi-locally-connected at x, there exists an arbitrarily small neighborhood of x in M whose complement in M is connected.*

This gives at once

(4.15) *If x is a non-cut point of a locally connected generalized continuum M, there exists an arbitrarily small region in M containing x whose complement is connected.*

For by [I, (13.21)], M is semi-locally-connected at x. Hence by (4.14) there exists a neighborhood V of x of diameter less than ϵ (preassigned $\epsilon > 0$)

so that $M - V$ is connected. Let R be the component of V containing x. Then R is open in M, $\delta(R) < \epsilon$ and it follows at once that $M - R$ is connected.

If M is locally connected we have a much stronger result as follows

(4.2) *If M is locally connected, every set $E(a, b) + a + b$ is compact.*

First, this set is closed. For let p be any limit point of $E(a, b)$ distinct from a and from b. Then if p does not belong to $E(a, b)$, the component R of $M - p$ which contains a also contains b, for otherwise $R + [M - R - p]$ is a separation of $M - p$ between a and b. But then covering R by regions whose closures lie in R and applying the chain lemma [II, (3.1)] to R we obtain a finite connected chain C in R from a to b such that $\bar{C} \cdot p = 0$; and since C is connected, we have $E(a, b) \subset C$, which is impossible since $\overline{E(a, b)} \supset p$.

Also $E(a, b) + a + b$ is compact. For if, on the contrary, $E(a, b)$ contains an infinite sequence $P = (p_1, p_2, \cdots)$ having no limit point, then by covering M by regions each of which contains at most one point of P and applying [II, (3.1)], we obtain a connected chain C from a to b containing only a finite number of points of P. But this is impossible since C must contain $E(a, b)$.

(4.3) *In any continuum M if G is the set of all cut points and N is any subcontinuum irreducible between a and b, then $N \cdot G - E(a, b)$ is countable.*

For if not, then by (1.52), some point z of this set separates two other points x and y of this set in M and hence also in N. Thus $N - z = N_x + N_y$, a separation. But since $N_x + z$ and $N_y + z$ are continua, neither can contain $a + b$ and hence we have $z \subset E(a, b)$.

(4.31) Corollary. *If M is a locally connected continuum and ab is any arc in M, then $ab \cdot G = E(a, b) +$ a countable set.*

Exercise. Show that conversely, if G is any subset of an arc ab such that $G + a + b$ is the sum of a closed set K containing $a + b$ and a countable set L, then there exists a locally connected continuum M such that in M, $E(a, b) + a + b = K$ and G is the set of all cut points of M.

(4.4) Theorem. *Any compact naturally ordered set K has a first point.*

For suppose K has no first point. Let $P = \sum p_i$ be a countable set such that $\bar{P} = K$, and set $p_1 = p_{n_1}$. For any $x \in K$, denote by ax the set consisting of x together with all points of K preceding x. Now $ap_{n_1} - p_{n_1}$ contains points of K and hence points of P since P is dense in K; let p_{n_2} be the point of lowest index of P in $ap_{n_1} - p_{n_1}$. Similarly let p_{n_3} be the point of least index of P in $ap_{n_2} - p_{n_2}$, and so on. In general, let p_{n_i} be the point of least index of P in $ap_{n_{i-1}} - p_{n_{i-1}}$. Now we have the order $\cdots < p_{n_i} < p_{n_{i-1}} < \cdots < p_{n_1}$, and this set of points has a limit point p which must belong to K. Furthermore p must be the first point of K; for clearly $p < \cdots < p_{n_i} < p_{n_{i-1}} < \cdots < p_{n_1}$ and if $ap - p$ contained a point x of K it would contain a point p_k of P and we would have

$$p_k < p < \cdots < p_{n_i} < p_{n_{i-1}} < \cdots < p_1 .$$

This is impossible by the method of choice of the points p_{n_i}, since after at most k steps, p_k would have been chosen, as $k \leqq n_i$. Thus K has a first point.

5. Borel set classification of the cut points. Let the letters F and G denote closed and open sets, respectively. Any set which can be expressed as the sum of a countable number of sets F (i.e., closed sets) is said to be an F_σ; any set which is the product of a countable number of sets G (i.e., open sets) is called a G_δ. For example, the set of rational points of a line is an F_σ and the set of irrational points a G_δ. Similarly a set which is the product of a countable number of F_σ sets is an $F_{\sigma\delta}$, and a set which is the sum of a countable number of G_δ sets is a $G_{\delta\sigma}$. Similarly we define the sets $F_{\sigma\delta\sigma}$, $F_{\sigma\delta\sigma\delta}$, $\cdots$ and $G_{\delta\sigma\delta}$, $G_{\delta\sigma\delta\sigma}$, $\cdots$ by alternating the σ- and δ-operations. The sets defined in this manner are known as the *Borel sets*. It is clear that in order to obtain actually new types of sets in the iteration we must alternate the σ- and δ-operations because $F_{\sigma\sigma} = F_\sigma$ and $G_{\delta\delta} = G_\delta$.

The Borel sets are generalized closed and open sets and it is seen at once that they have many properties of these sets. Just as closed and open sets are complementary, so are also the sets F_σ and G_δ, the sets $F_{\sigma\delta}$ and $G_{\delta\sigma}$, and so on, i.e., the complement of every F_σ is a G_δ, and conversely; the complement of every $F_{\sigma\delta}$ is a $G_{\delta\sigma}$, and conversely; and so on. Also all closed or open sets are at the same time F_σ's and G_δ's. Similarly all F_σ and G_δ sets are at the same time $F_{\sigma\delta}$'s and $G_{\delta\sigma}$'s and so on.

Just as the notions of closed and open sets are *relative* in the sense that they depend on the space in which the sets are imbedded, so also are the notions of the Borel sets in general. Thus any set X whatever is both an F_σ and a G_δ, just as it is both closed and open, when we consider X alone as the space. However if a set X is a Borel set of a certain class, say G_δ, in every metric space which contains it, then X is said to be an *absolute* Borel set of this class (in this case an *absolute* G_δ). It is seen at once that the Borel sets in any compact or in any separable locally compact space are all absolute.

We shall now show that the set of all cut points of any continuum is a Borel set of class $G_{\delta\sigma}$ in general and of class F_σ in case the continuum is locally connected.

(5.1) *If M is any generalized continuum, every set $E(a, b)$ is the sum of a G_δ-set and a countable set.*

For let G denote the set of all points of M which are of order less than or equal to 2 relative to $E(a, b)$, i.e., all points x of M such that x lies in an arbitrarily small neighborhood whose boundary consists of not more than 2 points of $E(a, b)$. Then G is a G_δ-set. To see this we have only to let, for each n, G_n be the sum of all open sets in M of diameter less than $1/n$ whose boundaries consist of not more than 2 points of $E(a, b)$ and note that G_n is open and $G = \prod_1^\infty G_n$. Now by (4.11) we have $E(a, b) + a + b \supset G$. Since $a + b$ is closed, we have that $G \cdot E(a, b)$ is a G_δ-set. And by (3.1) we have that $E(a, b) - G \cdot E(a, b)$ is countable, which completes the proof.

(5.2) THEOREM. *The set G of all cut points of any generalized continuum M is a $G_{\delta\sigma}$-set, i.e., the sum of a countable number of G_δ-sets.*

This is an immediate consequence of the theorem just proved and of the fact that G is the sum of a countable number of the sets $E(a, b)$.

(5.3) THEOREM. *If M is any connected and locally connected space, every set $E(a, b)$ in M is both a G_δ and an F_σ and the set G of all cut points of M is both an F_σ and a $G_{\delta\sigma}$.*

For by (4.2), every $E(a, b) + a + b$ is compact and hence $E(a, b)$ is both an F_σ and a G_δ. And since G is the sum of a countable number of sets $E(a, b)$, we have that G is both an F_σ and a $G_{\delta\sigma}$.

DEFINITION. *A set K is said to be perfect provided it is compact and dense in itself.*

For example, any continuum is perfect, as is also the Cantor discontinuum [see II, §4]. Now it is seen at once that *every compact set K is the sum of a perfect set P and a countable set C.* For we have only to take P as the set of all condensation points of K, i.e., the points x such that every neighborhood V of x contains uncountably many points of K, and take $C = K - P$. It follows at once from [I, (3.8)] that C is countable, and that P is perfect. (The set P may, of course, be vacuous, and it will be in case K is countable.)

(5.4) *If the set of all cut points of a generalized continuum M is uncountable it contains a perfect set.*

Since every uncountable set G contains a dense in itself subset, namely, the set of all condensation points of G which belong to G, this theorem follows at once from either (5.1) or (5.2) and

(5.5) THEOREM OF W. H. YOUNG. *Every G_δ-set G, in a locally compact separable space, which contains a dense in itself subset H contains a perfect set.*

Proof. We have $G = \prod_1^\infty G_n$ where each G_n is open. Let us select two points of H which we designate by (.0) and (.1) and then take disjoint neighborhoods $G(.0)$ and $G(.1)$, respectively, of these such that the closure of each of these neighborhoods is compact and contained in G_1 and each is of diameter less than 1. Now take a point (.01) of H in $G(.0)$ distinct from (.0), [(.0) we now call (.00)] and take disjoint 1/2-neighborhoods $G(.00)$ and $G(.01)$ of (.00) and (.01), respectively, whose closures lie wholly in G_2 and also in $G(.0)$. Similarly take a point (.11) of H in $G(.1)$ distinct from (.1), [(.1) is now called (.10)], and 1/2-neighborhoods $G(.10)$ and $G(.11)$ with closures lying in $G_2 \cdot G(.1)$. Let us continue this process indefinitely. For each n, let K_n denote the sum of the $2^n 1/n$-neighborhoods $G(.a_1a_2 \cdots a_n)$, $(a_i = 0, 1)$, and let $K = \prod_1^\infty K_n$. Then since $\bar{K}_n \subset K_{n-1} \subset G_{n-1}$ we have $K = \prod_1^\infty \bar{K}_n$, so that K is compact and $K \subset G$. And since, for each n, every point of K is contained in a $1/n$-neighborhood $G(.a_1a_2 \cdots a_n)$ which contains infinitely many of the points $(.a_1a_2 \cdots a_k)$ all of which belong to K, then K is dense in itself and therefore perfect. Thus our theorem is proved.

Now it will be noted that each point $x \in K$ can be represented in one and only one way as the product of a sequence of the neighborhoods chosen, so we have $x = G(.a_1) \cdot G(.a_1a_2) \cdots$, $(a_i = 0, 1)$. Then if we set $x = (.a_1a_2a_3 \cdots)$ and regard the indices $.a_1a_2a_3 \cdots$ as numbers in the dyadic system we have established a (1-1) correspondence between the set K and the number interval

(0, 1). Thus K is uncountable and indeed every point of K is a condensation point of K. This is a property of perfect sets in general. Since every perfect set is a G_δ which is dense in itself we have

(5.51) COROLLARY. *No countable set is perfect. Thus every compact countable set has at least one isolated point.*

6. Non-cut points. Properties of simple arcs. We begin with the

(6.1) NON-CUT POINT EXISTENCE THEOREM. *Every continuum has at least two non-cut points.*

For suppose, on the contrary, that every point of some continuum M save possibly one is a cut point. Then if x is a cut point of M and $M - x = A + B$ is a separation of $M - x$ into mutually separated sets, at least one of the sets A and B, say A, contains no non-cut point of M. Let $P = \sum p_i$ be a countable set of points dense in M. Let p_{n_1} be the first point of P belonging to A. Then since every point of A is a cut point of M, $M - p_{n_1} = E_1 + F_1$ where E_1 and F_1 are mutually separated and E_1 contains x. Let p_{n_2} be the first point of P belonging to F_1 and let us write $M - p_{n_2} = E_2 + F_2$ where E_2 and F_2 are mutually separated and $E_2 \supset p_{n_1}$. In general, for each $i > 1$, let p_{n_i} be the first point of P belonging to F_{i-1} and write $M - p_{n_i} = E_i + F_i$, where E_i and F_i are mutually separated and $E_i \supset p_{n_{i-1}}$.

Let $F = \prod_1^\infty F_i$. Then since $F_i \supset (F_{i+1} + p_{n_{i+1}}) = \bar{F}_{i+1}$ for each i, it follows that F is nonvacuous and contains no point of $E = \sum_1^\infty (E_i + p_{n_i})$. But since for each i, $E_i + p_{n_i}$ is connected and is a subset of E_{i+1} (for it does not contain $p_{n_{i+1}}$), it follows that E is connected. And since for each i, $E_{i+1} \supset \sum_1^{n_i} p_n$, therefore $E \supset P$ and thus $\bar{E} = M$. Thus every point of F is a limit point of E and is a non-cut point of M, contrary to supposition.

(6.2) THEOREM. *In order that a continuum M should be a simple continuous arc it is necessary and sufficient that it have at most two non-cut points.*

The necessity is an immediate consequence of the definition of an arc. To prove the sufficiency, let a and b be the two non-cut points of M and let x be any other point of M. Then $M - x = A + B$ where A and B are mutually separated and $A \supset a$. Furthermore B must contain b; for if not, then $a + b \subset A$ and since, by (6.1), the continuum $B + x$ has at least two non-cut points, there must exist a point y of B so that $(B + x) - y$ is connected. But then $(A + x) + (B + x) - y = M - y$ is connected, and hence y is a non-cut point of M, contrary to the fact that a and b are the only non-cut points of M. Therefore $A \supset a$ and $B \supset b$, and thus every point x of $M - (a + b)$ separates a and b in M. Consequently M is an arc from a to b.

We now establish some elementary properties of arcs which result from the definition.

(i) *The end points of an arc ab are the non-cut points of it.*

(ii) *If T is an arc ab then $T - (a + b)$ is connected.*

For suppose $T - (a + b) = T_1 + T_2$, mutually separated sets. Then setting $T_1 + a + b = A$, $T_2 + a + b = B$, we see that since $T - a$ is connected, $T_1 + b$ and $T_2 + b$ are connected. Thus since a is a limit point of at least one of these sets, either A or B, say A, must be connected. But then clearly no point of T_2 can separate a and b in T, contrary to the definition of T.

(iii) *Every simple arc T is locally connected.*

For by the Cut Point-Order Theorem it follows that all save a countable number of points of T are points of order 2 and hence T is locally connected at all save a countable number of points. But no continuum can fail to be locally connected at just a countable number of points [I, (12.3)]. Thus T is locally connected everywhere.

(iv) *Any arc T from a to b is naturally ordered.*

This follows immediately from (1.3), since in T we have $E(a, b) = T - (a + b)$.

(v) *If x is any interior point of an arc ab, then $ab = ax + xb$ where ax and xb are arcs from a to x and from x to b and $ax \cdot xb = x$.*

For by definition of arc, $ab - x = T_a + T_b$, where $T_a \supset a$, $T_b \supset b$. Set $T_a + x = ax$, $T_b + x = xb$. Then ax is an arc from a to x. For by [I, (8.3)], it is a continuum, and if y is any point of $ax - (a + x)$, $ax - y$ is not connected. For if it were, then $(ax - y) + xb = ab - y$ would be connected, contrary to the fact that ab is an arc from a to b. Similarly, xb is an arc from x to b, and clearly $ax \cdot xb = x$.

We remark here that by the way order was defined in the sets $E(a, b)$, it follows that order can be so defined on an arc ab that a point y precedes a point x if and only if y is on the arc ax of ab.

(vi) *Any closed subset K of an arc ab has a first point in the order a, b.*

This is a corollary to (4.4). However the following independent proof is of interest on account of its simplicity.

Proof for (vi). We may suppose a non ϵ K. Let C_a be the component of $ab - K$ containing a. By [I, (10.2)] K contains at least one limit point p of C_a. Then p must be the first point of K in the order a, b. For let $x \epsilon ap - (a + p)$. Then since $ab = ax + xb$ and $p \epsilon xb - x$, $p \epsilon \bar{C}_a$, we have $C_a \cdot (ax - x) \neq 0 \neq C_a \cdot (xb - x)$. Accordingly we must have $x \epsilon C_a$, since C_a is connected. Whence x non ϵ K, so that p is the first point of K.

(vii) *On any arc ab, for each $\epsilon > 0$, there exists a monotone increasing chain of points $a = x_0, x_1, x_2, \cdots, x_n = b$ such that the diameters of all the arcs $x_i x_{i+1}$ $(i = 0, 1, \cdots, n - 1)$ are less than ϵ.*

For by (iii), ab is locally connected and hence has property S. Thus $ab = \sum_1^n C_i$, where each C_i is a continuum of diameter less than ϵ. Let C_{n_1} be one of these sets containing a, set $x_0 = a$ and let x_1 be the last point of C_{n_1}. Now x_1 and the points immediately following it belong to some set C_{n_2} different from C_{n_1}. Let x_2 be the last point of C_{n_2}. Likewise x_2 and the points immediately following belong to a set C_{n_3} different from C_{n_1} and C_{n_2}. Let x_3 be the last point of C_{n_3}, and so on. Since the sets $C_{n_1}, C_{n_2}, \cdots$ thus obtained are all distinct whereas there are only n of the sets C_i altogether, it follows that after a finite number m of steps we arrive at $x_m = b$. Now since, for each i, $(i = 0, 1, \cdots, n-1)$, we have $x_i + x_{i+1} \subset C_{n_{i+1}}$, it follows that $x_i x_{i+1} \subset C_{n_{i+1}}$ and hence $\delta(x_i x_{i+1}) < \epsilon$.

Finally we prove

(6.3) THEOREM. *Any simple continuous arc ab is homeomorphic with the unit interval* (0, 1).

By (vii), there exists in ab a monotone increasing chain C^1 which we may suppose consists of exactly $2^{v_1} + 1$ points

$$a = x_0^1, x_1^1, x_2^1, \cdots, x_{2^{v_1}}^1 = b,$$

such that the diameter of the arc in ab joining any two successive points is less than 1. Likewise for each i, $(0 \leqq i \leqq 2^{v_1})$, there exists in the arc $x_i^1 x_{i+1}^1$ a monotone increasing chain C_i^1 which we may suppose in each case contains exactly $2^{v_2} + 1$ points

$$x_i^1 = x_{i \cdot 2^{v_2}}^2, x_{i \cdot 2^{v_2}+1}^2, \cdots, x_{(i+1)2^{v_2}}^2 = x_{i+1}^1$$

such that the diameter of the arc in ab joining any two successive points is less than 1/2. Continue this process indefinitely

Now to the values

$$t = \frac{i}{2^{v_1+v_2+\cdots+v_k}}, \qquad 0 \leqq i \leqq 2^{v_1+v_2+\cdots+v_k},$$

we make correspond the points x_i^k and write $f(t) = x_i^k$. Thus we define f over the set D of all dyadic rational numbers on (0, 1). Next to prove f uniformly continuous over D we have only, for a given $\epsilon > 0$, to take k so large that $1/k < \epsilon/4$ and then take $\delta = 1/2^{v_1+v_2+\cdots+v_k}$, [see II, §3]. Likewise to prove f^{-1} uniformly continuous over $f(D)$, for a given $\epsilon > 0$, we take k so that $1/2^{v_1+v_2+\ldots+v_k} < \epsilon/4$ and then take δ less than the least distance from any arc $x_m^k x_{m+1}^k$ to the set $ax_{m-1}^k + x_{m+2}^k b$. Now since f and f^{-1} are single-valued and uniformly continuous, we have only to extend f and f^{-1} to the limit points of D and $f(D)$, respectively, and the desired homeomorphism between ab and (0, 1) is obtained. [See II, (2.3).]

Note. Another proof of this theorem may be obtained by using the fact established in [II, §5] that any two points of a locally connected continuum can be joined by an arc T in that continuum which is homeomorphic with (0, 1). For by (iii) any arc ab is locally connected. Hence ab contains an arc T from a to b which is homeomorphic with (0, 1). But clearly $T \equiv ab$ by the definition of arc.

By a similar method of proof we obtain the following much more general result

(6.4) THEOREM. *If K is any compact naturally ordered set, there exists an order-preserving homeomorphism mapping K onto a subset of the unit interval* (0, 1).

Proof. Let a and b be the first and last points, respectively, of K. We first show that for any $\epsilon > 0$ there exists a finite monotone sequence of points of K

$$a = x_0, x_1, x_2, \cdots, x_n = b$$

so that any interval (x_i, x_{i+1}) of K which contains interior points is of diameter less than ϵ and contains infinitely many points and, in case K is infinite, n may be any sufficiently large integer. This will be called a "subdivision" of K.

To prove this, let K_a denote the set of all points x of K such that the "interval" (a, x) of K can be so subdivided and let $K_b = K - K_a$. Then K_a is closed. For if $y \in \bar{K}_a - K_a$ there would exist a point x of K_a with $\rho(x, y) < \epsilon$ and, indeed, with δ [interval (x, y)] $< \epsilon$. Hence if we add y to a subdivision of (a, x) we get a subdivision of (a, y). Thus $y \in K_a$ and K_a is closed. Also K_b is closed. For if $x \in K_a \cdot \bar{K}_b$ we could find a point $y \in K_b$ with $\delta[(x, y)] < \epsilon$, and if we add y to a subdivision of (a, x) we get a subdivision of (a, y), contrary to $y \in K_b$. Thus if $K_b \neq 0$ there exists a first point y of K_b and a last point x of K_a. But clearly adding y to a subdivision of (a, x) gives a subdivision of (a, y). Hence $K_b = 0$ so that $K = K_a$. Thus there exists a subdivision of $K = (a, b)$.

Now let (x_i, x_{i+1}) be an interval of this subdivision which contains infinitely many points. If there is a first point after x_i or a last point before x_{i+1} let y be such a point. If neither of these exist, let y be any interior point of (x_i, x_{i+1}). Then if we add y to the subdivision, clearly we get a subdivision of K having one more point and yet retaining all properties of the previous subdivision. Thus our statement is proved.

Now we may assume K infinite. Let

$$a = x_0^1, x_1^1, \cdots, x_{2^{n_1}}^1 = b$$

be a subdivision of K as above for $\epsilon = 1$. Let us define

$$f(x_i^1) = i/2^{n_1}, \qquad i = 0, 1, \cdots, 2^{n_1}.$$

Now take similar subdivisions of all intervals (x_i^1, x_{i+1}^1) which contain interior points, using $\epsilon = 1/2$ and adjusting so that each such interval contains exactly $2^{n_2} + 1$ points

$$x_i^1 = x_{i \cdot 2^{n_2}}^2, x_{i \cdot 2^{n_2}+1}^2, \cdots, x_{(i+1)2^{n_2}}^2 = x_{i+1}^1.$$

Let us define

$$f(x_{i \cdot 2^{n_2}+j}^2) = f(x_i^1) + \frac{j}{2^{n_1+n_2}}, \qquad j = 0, 1, \cdots, 2^{n_2}.$$

Continue this process indefinitely. In view of the argument given for (6.3) it is clear that in this way we obtain an order-preserving (1-1) mapping $f(x)$ of a subset D of K which is dense in K onto a subset of $(0, 1)$ so that both f and f^{-1} are uniformly continuous. Hence f can be extended to yield the required order-preserving homeomorphism of K into a subset of $(0, 1)$.

7. Simple closed curves. A set of points J which is the sum of two simple arcs axb and ayb having just their end points in common is called a *simple closed curve*. The following properties result at once from the above properties of arcs.

(i) *If α and β are any two points of a simple closed curve J, then $J = \alpha x\beta + \alpha y\beta$, where $\alpha x\beta$ and $\alpha y\beta$ are arcs and $\alpha x\beta \cdot \alpha y\beta = \alpha + \beta$.*

(ii) *No point cuts any simple closed curve.*

(iii) *Any simple closed curve is homeomorphic with a circle.*

We turn now to the following characterization of simple closed curves:

THEOREM. *In order that a continuum be a simple closed curve it is necessary and sufficient that it be disconnected by the omission of any two of its points.*

The necessity of the condition results from (i) above. We shall prove the sufficiency of the condition. Let C be a continuum satisfying the conditions. Then C can have no cut point. For if $C - x = C_1 + C_2$, (C_1 and C_2 mutually separated), by the Non-Cut Point Existence Theorem the continua $C_1 + x$, $C_2 + x$ have non-cut points x_1 and $x_2 \neq x$, respectively. Thus $C_1 + x - x_1$ and $C_2 + x - x_2$ are connected and so is their sum $C_1 + C_2 + x - (x_1 + x_2) = C - (x_1 + x_2)$, contrary to our condition. Thus if x is any point of C, then $C - x$ is connected and every point of $C - x$ is a cut point of $C - x$. Since $C - x$ is locally compact, it follows by the Cut Point-Order Theorem that all save a countable number of points $C - x$ are points of order 2 of $C - x$ and hence also of C. Thus C is regular and consequently locally connected.

Now C contains a simple closed curve J; for let a and b be two points of C and axb be an arc in C from a to b. Since $C - x$ is connected, it contains an arc aob. On xa and xb, respectively, of axb let u and v denote the first points belonging to aob. Then clearly uxv of $axb + uv$ of aob is a simple closed curve J in C.

Now $J = C$. For if $C - J$ contains a point x, $C - x$ is connected and every point of $C - x$ is a cut point of $C - x$. But this is impossible because $C - x \supset J$ and at most a countable number of points of J can be cut points of $C - x$ by (1.53).

8. **Separating points.** A point p of a set M will be called a *separating point* of M provided p separates M between some two points a and b of the component C of M containing p, i.e., $M - p = M_a + M_b$, where M_a and M_b are mutually separated and contain a and b, respectively.

It is obvious that any such separating point p of M is a cut point of C which separates a and b in C. Thus if M is connected, its cut points and its separating points are identical. However, if M is not connected, a cut point of a component of M is not necessarily a separating point of M. For example, let M consist of an interval ab together with an infinite sequence of disjoint intervals converging to ab; then every point of $ab - (a + b)$ cuts the component ab of M but no such point is a separating point of M.

(8.1) *Only a countable number of components of any separable metric space M can contain separating points of M.*

Let us suppose, on the contrary, that there exists an uncountable collection G of components X of M each of which contains a separating point x of M. Then x separates M between some pair of points a_x and b_x of X. Now since G, and therefore the collection of pairs (a_x, b_x), is uncountable, there must exist at least one of these pairs a_x, b_x which is the limit of a convergent sequence (a_{x_i}, b_{x_i}) of the others—indeed so that $a_{x_i} \to a_x$ and $b_{x_i} \to b_x$. For if this were not so and if $R_1, R_2, \cdots$ denotes a fundamental sequence of neighborhoods in M, then for each $X \epsilon G$ there would exist a pair of neighborhoods R_i, R_j such that $R_i \supset a_x$ and $R_j \supset b_x$ but such that no other pair a_y, b_y is contained in the pair R_i, R_j in the sense that $a_y \subset R_i$ and $b_y \subset R_j$. But since the collection of all possible pairs R_i, R_j is countable whereas there are uncountably many pairs a_x, b_x, it follows that some single pair R_i, R_j of neighborhoods must be chosen for two distinct pairs of points a_x, b_x and a_y, b_y. Thus we have $a_x + a_y \subset R_i$ and $b_x + b_y \subset R_j$, contrary to the method of selecting the pairs R_i, R_j.

Thus there exists a pair a_x, b_x and a sequence (a_{x_i}, b_{x_i}) such that $a_{x_i} \to a_x$ and $b_{x_i} \to b_x$. But clearly this is impossible, since we have the separation $M - x = M_{a_x} + M_{b_x}$ and either M_{a_x} or M_{b_x}, say M_{a_x}, contains infinitely many of the sets X_i so that b_x would have to be a limit point of M_{a_x} as $b_{x_i} \to b_x$.

(8.2) *If U is any uncountable set of separating points of a separable metric space M, there exists a component Q of M, a pair of points a, b of Q and an uncountable subset H of U such that each point of H separates a and b in M.*

For, by (8.1), there exists a component Q of M such that $Q \cdot U$ is uncountable. Let $P = \sum p_i$ be dense in Q. Now each point of $Q \cdot U$ separates some pair of points of Q in M and hence must separate also some pair p_i, p_j of points of P in M, since P is dense in Q. Since there are only a countable number of such pairs possible whereas $Q \cdot U$ is uncountable, it follows that there exists at least one such pair a, b and an uncountable subset H of $Q \cdot U$ each point of which separates a and b in M.

(8.3) *If p is any point of a component C of a compact metric space K, if R is any neighborhood of p such that there exists no convergent sequence of components of K whose limit contains points of both $C \cdot R$ and $C \cdot (K - \bar{R})$, then for each $\epsilon > 0$ there exists a neighborhood V such that $R \cdot C \subset V$, $V \subset V_\epsilon(R)$, and $F(V) \subset C \cdot F(R)$.*

Proof. For each integer $n > 0$ let $V_n = V_{\epsilon/2n}[F(R)]$, $R_n = R - R \cdot V_n$, and $U_n = K - R - V_n$. From our hypothesis it follows that for each $n > 0$ there exists a positive number $d_n < 1/n$ such that no component of K except C intersects each of the sets $\overline{V_{d_n}(C)}$, R_n and U_n. Let $V_{d_0}(C) = K$ and, for each $n > 0$, let K_n be the sum of all components X of K such that

$$\text{(i)} \qquad X \cdot F(R) \neq 0 \neq X \cdot [\overline{V_{d_{n-1}}(C)} - V_{d_n}(C)].$$

For each $n > 0$ and for each component X of K_n there exists a $1/n$-neighborhood G_x of X with $G_x \cdot C = 0$ and $F(G_x) = 0$ and, when $n > 1$, such that G_x does not intersect both R_{n-1} and U_{n-1}; and since K_n is closed, it is covered by a finite number of these neighborhoods. Let us select such a finite covering of K_n and let G_n be the sum of all neighborhoods G_x in this covering such that $G_x \subset R + V_{n-1}$ and let H_n be the sum of the remaining ones. Then for each n we have

(ii) $$G_n \subset R + V_{n-1}, \quad F(G_n) = F(H_n) = 0, \quad K_n \subset G_n + H_n,$$

and

$$H_n \subset U_{n-1} + V_{n-1}, \qquad \text{for } n > 1.$$

Now set $G = \sum G_n$, $H = \sum H_n$, $V = R + G - H \cdot (R + G)$. Then

$$F(G) \subset \bar{R} \cdot C, \qquad F(H) \subset (K - R) \cdot C.$$

Then since $C \cdot H = 0$ we have $R \cdot C \subset V$ and, since $V \subset R + G \subset R + V_1$, we have $V \subset V_\epsilon(R)$. Further, since $H \cdot (R + G) = \bar{H} \cdot (R + G)$, V is an open set. Thus we have only to show that $F(V) \subset C \cdot F(R)$. Now from (ii) we have that $F(V) \subset \bar{R}$ and, since every point of $F(R) - C \cdot F(R)$ belongs to some K_n and hence either to V or to H and thus not to $F(V)$, we have only to prove that no point x of R belongs to $F(V)$. Clearly this is impossible since any such point x would have to belong to $F(H)$ and (ii) gives $F(H) \subset K - R$. Therefore $F(V) \subset C \cdot F(R)$ and our theorem is proved.

(8.31) COROLLARY. *If a closed set X separates C into two separated sets C_a and C_b then in order that X separate C_a and C_b in K it is necessary and sufficient that the limit of no convergent sequence of components of K should intersect both C_a and C_b.*

(8.32) COROLLARY. *In order that a point p of a component C of K be a separating point of K it is necessary and sufficient that there exist a separation of $C - p = C_a + C_b$ such that the limit of no convergent sequence of components of K intersects both C_a and C_b.*

(8.4) SEPARATING POINT-ORDER THEOREM. *All save a countable number of the separating points of any compact metric space K are points of order 2 of K.*

Proof. Suppose, on the contrary, that there exists an uncountable set U of separating points of K no one of which is of order 2 in K. By (8.2) there exists a component Q of K and an uncountable subset H of U each point of which separates two fixed points a and b of Q in K. Now since Q is a continuum and every point of H is a cut point of Q separating a and b in Q it follows by the Cut Point-Order Theorem for continua, (3.2), that there exists a countable subset H_0 of H such that if $N = H - H_0$, then every point of N is a point of order 2 in Q relative to N. Let $p \in N$ and let ϵ be any positive number. Then since N is ordered as in (1.31), there exist points x and y of N in the order a, x, p, y, b such that, in the notation of §1, $\delta[Q_2(x) \cdot Q_1(y)] < \epsilon/4$, where $Q_1(x) \cdot Q_1(y) \supset a$ and $Q_2(x) \cdot Q_2(y) \supset b$ so that $Q_2(x) \cdot Q_1(y) \supset p$. Now if for each $z \in Q_2(x) \cdot Q_1(y)$ we take a neighborhood R_z of z of diameter less than $\epsilon/4$ and also less than $1/2\, \rho[z, Q_1(x) + Q_2(y)]$ and let $R = \sum R_z$, then R is a neighborhood of p in K with $\delta(R) < \epsilon/2$, $R \cdot Q = Q_2(x) \cdot Q_1(y)$ and $F(R) \cdot Q = x + y$.

Now the limit of no convergent sequence of distinct components of K can intersect both $R \cdot Q$ and $Q \cdot (K - \bar{R}) = Q_1(x) + Q_2(y)$; for if so, this limit would have to intersect either both $Q_1(x)$ and $Q_2(x)$ or both $Q_1(y)$ and $Q_2(y)$; and each of these is impossible by (8.31), since x and y, respectively, separate these pairs of sets in K. Therefore, by (8.3), there exists a neighborhood V of p in K such that $\delta(V) < \delta(R) + \epsilon/2 < \epsilon/2 + \epsilon/2 = \epsilon$ and $F(V) \subset Q \cdot F(R) = x + y$. But this proves that p is a point of order 2 in K, contrary to supposition, and thus our theorem is proved.

(8.41) Corollary. *All save a countable number of the separating points of any locally compact separable metric space L are points of order 2 of L.*

For we can write $L = \sum L_i$, where each L_i is open and $\bar{L}_i$ is compact. Then each separating point of L lies in some L_i and is a separating point of $\bar{L}_i$. Thus by (8.4) every uncountable set of separating points of L contains at least one point which is a point of order 2 of some L_i and thus also of L since L_i is open.

(8.42) Corollary. *Every uncountable set H of separating points of a locally compact separable metric space L contains a countable subset H_0 such that if $N = H - H_0$, then every point of N is a point of order 2 in L relative to N.*

Exercises. (1) Prove that the set of all points which separate two fixed points a and b of a component C in a space L as above is the sum of a G_δ-set and a countable set.

(2) Show that the set of all separating points of any such space L is a $G_{\delta\sigma}$.

9. Local separating points.

Definition. *A point p of a locally compact separable metric space L will be called a local separating point of L provided p is a separating point of some open set in L.*

(9.1) *If p is a separating point of an open set R in L then it is also of every open set U such that $p \subset U \subset R$.*

We have a separation $R - p = R_a + R_b$, where a and b belong to the component Q of R containing p. Let $U \cdot R_a = U_1$ and $U \cdot R_b = U_2$. Then $U - p = U_1 + U_2$ and since $Q \cdot R_a + p$ and $Q \cdot R_b + p$ are generalized continua, it follows by [I, (10.2)] that the components V_1 and V_2 of $U \cdot (Q \cdot R_a + p)$ and $U \cdot (Q \cdot R_b + p)$, respectively, containing p are nondegenerate. Let $a' \in V_1 - p$ and $b' \in V_2 - p$. Then since $V_1 \subset U_1$, $V_2 \subset U_2$, we have a separation of $U - p$ between a' and b' and $a' + b' \subset V_1 + V_2 + p$, the component of U containing p.

(9.2) Local separating point-order theorem. *All save a countable number of the local separating points of any locally compact separable metric space L are points of order 2 of L.*

For let $R_1, R_2, \cdots$ be a fundamental sequence of neighborhoods in L, let G be the set of all local separating points of L and, for each i, let G_i be the set of all separating points of R_i. Now by definition of local separating point we have $\sum G_i \subset G$. But also every point p of G must belong to some G_i; for p is a separating point of some open set T and there exists an i such that $p \subset R_i \subset R$, so that by (9.1) $p \in G_i$. Thus $G \subset \sum G_i$, so therefore $G \equiv \sum G_i$. Since each R_i is locally compact, by (8.41) all save a countable number of points of G_i are points of order 2 of R_i and hence also of order 2 in L, it follows that all save a countable number of the points $G = \sum G_i$ are points of order 2 of L.

(9.21) COROLLARY. *Every uncountable set G of local separating points of L contains a countable subset G_0 such that if $Q = G - G_0$, then every point of Q is a point of order 2 in L relative to Q.*

(9.3) THEOREM. *In order that a point p of a generalized continuum M should be a local separating point of M it is necessary and sufficient that it be an isolated point of some closed cutting X of M and be a limit point of both M_1 and M_2, where M_1 and M_2 are some two mutually separated sets whose sum is $M - X$.*

To show the sufficiency, we have only to take conditionally compact neighborhoods R and G such that $p \subset G \subset \bar{G} \subset R$ and $R \cdot X = p$ and set $R_1 = R \cdot M_1$ and $R_2 = R \cdot M_2$. Then $R - p = R_1 + R_2$ is a separation of R. Also if N is the component of R containing p, then $N \cdot R_1 \neq 0 \neq N \cdot R_2$. For since p is a limit point of M_1 and hence of R_1 then either p is a limit point of $N \cdot R_1$ or there exists a sequence $K_1, K_2, \cdots$ of components of $\bar{G} \cdot R_1$ converging to a limit continuum K containing p. Then, in the latter case, $K - p \subset R_1$ and $K \subset N$ so that $K - p \subset R_1 \cdot N$, and $K - p \neq 0$ since it contains at least one point of $F(G)$. Similarly $N \cdot R_2 \neq 0$.

To prove the necessity, let p be a local separating point of M. Then p is a separating point of some open set R in M so that we have a separation $R - p = R_1 + R_2$, where both R_1 and R_2 intersect the component C of R containing p. Now if we set $X = F(R) + p$, $M_1 = R_1$ and $M_2 = M - X - M_1 = R_2 + M - \bar{R}$, we have a separation $M - X = M_1 + M_2$ and p is an isolated point of X and is a limit point of both M_1 and M_2 since it is a limit point of both R_1 and R_2.

(9.4) *If a cutting X of a generalized continuum M contains no perfect set, then X contains at least one local separating point of M.*

Proof. Let $M - X = N_1 + N_2$ be a separation of $M - X$ and let $\bar{N}_1 \cdot \bar{N}_2 = K$. Then K is closed and $K \subset X$; and if $M_1 = \bar{N}_1 \cdot (M - K)$ and $M_2 = \bar{N}_2 \cdot (M - K)$, then M_1 and M_2 are mutually separated and $M_1 + M_2 = M - K$ since X contains no perfect set. Also every point of K is a limit point of both M_1 and M_2 since $N_1 \subset M_1$ and $N_2 \subset M_2$. Now since (§5) every compact set is the sum of a perfect set and a countable set and since K is locally compact it follows that K is countable. Hence, by (5.51), K contains at least one isolated point p and, by (9.3), p is a local separating point of M.

(9.41) COROLLARY. *Every countable cutting of a generalized continuum M contains at least one local separating point of M.*

(9.42) Corollary. *Every point of any finite irreducible cutting of a generalized continuum M is a local separating point of M.*

Note. A cutting X of M is said to be irreducible provided no proper subset of X is a cutting of M.

(9.43) Corollary. *In any regular curve or any rational curve the local separating points are everywhere dense.*

Note. A continuum M is called a regular curve or a rational curve (following Menger and Urysohn) provided each point of M is contained in arbitrarily small neighborhoods whose boundaries (rel. M) are finite or countable, respectively. See Chapter V below.

(9.5) *A point p of a locally connected generalized continuum M is a local separating point of M if and only if it is a cut point of some region in M.*

This follows from (9.1) together with the fact that in a locally connected space, every neighborhood of a point p contains a region containing p.

Exercise. Show that the set of all local separating points of any generalized continuum M is a $G_{\delta\sigma}$ and that if M is locally connected this set is an F_σ.

References

§1. Moore [7], Whyburn [4], Zarankiewicz [1, 2].
§§2, 3. Whyburn [4, 5, 6]. See also Kuratowski [9].
§4. Whyburn [1].
(4.15) Gehman [1], Ayres [1].
(4.2) Whyburn [7, 8]. For the proof given, see Wilder [2].
§5. (5.1)–(5.4) Whyburn [9]. (5.3) See also Zarankiewicz [1].
(5.5) Young [1].
§§6, 7. Moore [8].
In connection with (6.3) and (6.4) see Whyburn [3].
§9. Whyburn [10].

CHAPTER IV

CYCLIC ELEMENT THEORY

1. Conjugate points. Simple links. E_0-sets. Two points a and b of a connected set M will be said to be *conjugate* provided no point separates a and b in M, i.e., provided $E(a, b) = 0$. A point p of M which is of order one in M will be called an *end point* of M, i.e., p is an end point of M provided there exist arbitrarily small neighborhoods of p in M the boundary of each of which consists of a single point.

(1.1) *If a point p of a generalized continuum M is neither a cut point nor an end point of M, there exists a point of M other than p which is conjugate to p.*

Proof. Let $p \in M$ and suppose p is not a cut point and not conjugate to any other point of M. Then p can belong to no continuum of convergence of M and accordingly M is semi-locally-connected at p.

Hence for any $\epsilon > 0$ there exists a conditionally compact neighborhood V of p of diameter less than ϵ such that $M - V$ is connected. Let C be the component of V containing p and let q be a point of $M - V$ which is a limit point of C. Since q is not conjugate to p, there exists a point x and a separation

$$M - x = M_p + M_q .$$

Obviously $x \in C$. Hence $M - V \subset M_q$. Accordingly $M_p \subset V$. Thus $\delta(M_p) < \epsilon$, $F(M_p) = p$. Hence p is an end point.

DEFINITIONS. *If p is neither a cut point nor an end point of a connected set M, $p \in M$, the set consisting of p together with all points of M conjugate to p will be called a simple link of M. A nondegenerate connected subset of M which has no cut point and is maximal with respect to the property of being a connected subset of M without cut points will be called an E_0-set in M.*

It results at once from these definitions that any two distinct points of a simple link or of an E_0-set are conjugate to each other.

By (1.1) we have

(1.2) *Any simple link of a generalized continuum M is nondegenerate. Every point of M is either a cut point, an end point or a point of a single simple link of M.*

(1.3) *If a and b are points of the closure of a simple link L of a connected set M, L contains every irreducible continuum in M from a to b. Thus if M is a continuum, every simple link of M is a continuum.*

Proof. Let N be a continuum in M irreducible from a to b. Suppose there is a point p of N which is not in L. Then there exists a point x of M and a separation

$$M - x = M_1 + M_2 ,$$

where $M_1 + x \supset L$, $M_2 \supset p$. But then

$$a + b \subset \bar{L} \subset M_1 + x$$

and $N \cdot M_1 + x$ is a proper subcontinuum of N containing $a + b$ since

$$N - x = N \cdot M_1 + N \cdot M_2$$

is a separation. Accordingly $N \subset L$.

Now if M is a continuum it contains an irreducible continuum between any two of its points. Thus any simple link of it is a continuum.

If M is not compact, a simple link of M may fail to be connected even though M is a semi-locally-connected generalized continuum. For let C_n be the circle with center (0, 0) and radius $1 - 1/n$, let I be the open interval from (0, 0) to (1, 0), let α and β be disjoint open arcs on $x^2 + y^2 = 1$ neither of which contains the point (1, 0) and let $M = \alpha + \beta + I + \sum_1^\infty C_n$. Then M is a semi-locally-connected generalized continuum and $\alpha + \beta$ is a simple link of M.

However, we do have

(1.4) *Any simple link or E_0-set in a connected set M is closed in M. Any two different simple links or E_0-sets in M can intersect in at most one point. Any point common to two simple links is a cut point of M.*

Proof. Let L be a simple link and suppose $p \in \bar{L} - L$. Then there exists a point x separating p from $L - x$. But then x separates p from $\bar{L} - x$ so that p non $\in \bar{L}$. Hence L is closed. Now clearly an E_0 must be closed since otherwise it could be increased without losing the property of being connected and having no cut point.

Any two different E_0-sets, say E_0^1 and E_0^2, can have at most one point in common, because otherwise $E_0^1 + E_0^2$ would be connected and have no cut point.

Finally, let L_1 and L_2 be distinct simple links of M. There is a point in one of these sets which is not in the other, say $p \in L_1 - L_1 \cdot L_2$. Since p non $\in L_2$, there exists a point x and a separation

$$M - x = M_1 + M_2, \qquad \text{where } p \in M_1,\ M_2 \cdot L_2 \neq 0.$$

This gives

$$L_1 \subset M_1 + x, \qquad L_2 \subset M_2 + x.$$

Accordingly

$$L_1 \cdot L_2 \subset x.$$

Since x is a cut point, our assertion follows.

From [III, (1.52)] we have at once

(1.5) *Any simple link or E_0-set in a connected set M contains at most a countable number of cut points of M.*

Indeed, it follows from [III, (3.1)] that there can exist at most a countable number of cut points a of a connected set M such that a is conjugate to any other point whatever of M. Hence we have the stronger result

(1.6) *If S denotes the sum of all the simple links* (*or* E_0-sets) *of a connected set M, S contains at most a countable number of cut points of M.*

It follows that any E_0-set in a connected set M contains non-cut points of M and, since clearly no point of an E_0 or of a simple link can be an end point, *any* E_0 *is contained in a simple link of M.* The converse is not true in general even for continua.

However we have

(1.7) *If M is a semi-locally-connected continuum, no simple link of M can have a cut point of itself. Thus the simple links and the E_0-sets in M are identical.*

Proof. Suppose, on the contrary, that a simple link L of M has a cut point x. Let

$$L - x = L_a + L_b$$

be a separation. Let $a \in L_a$, $b \in L_b$. Since x cannot separate a and b in M (for otherwise it would separate one of them from the defining point of L), by [III, (4.1)] there exists a continuum N such that

$$a + b \subset N \subset M - x$$

and we may suppose N is irreducible between a and b. But by (1.3), $N \subset L$, which is absurd.

Thus L has no cut point. Since clearly L must be maximal with respect to the property of being connected and having no cut point, it follows that every simple link of M is an E_0-set. To see that every E_0-set E_0 is a simple link we have only to take a simple link L of M containing E_0 and note that L must be equal to E_0.

2. Cyclic elements. Let M be a semi-locally-connected continuum. By a *cyclic element* of M will be meant a cut point of M, an end point of M or a simple link or E_0-set in M. (Since the simple links and E_0-sets are identical, obviously it is immaterial which is used here.)

The cut points and end points of M will be called *degenerate* cyclic elements and the simple links or E_0-sets nondegenerate or *true* cyclic elements of M.

It has already been demonstrated that every point of M belongs to at least one cyclic element. Further, a non-cut point can belong to only one cyclic element. Indeed only a countable number of points of M can belong to more than one cyclic element and all these points must be cut points.

(2.1) *If E is a true cyclic element of a continuum M and R is a component of* $M - E$, *there exists a point x of E such that R is a component of* $M - x$.

Proof. Let $a \in R$ and let b be a non-cut point of M belonging to E. By [III, (4.12)], $E(a, b) + a + b$ is closed and hence compact. Since $E(a, b) + a \subset \bar{R}$, either there exists a last point x of $E(a, b)$ in the order (a, b) or else b is a limit point of $E(a, b)$ and hence of R. But since b is a non-cut point of M, for any $c \in E - b$ there exists a continuum N with $a + c \subset N \subset M - b$; and since $N \supset E(a, b)$, b cannot be a limit point of $E(a, b)$. Thus there exists a last point x in $E(a, b)$. Now we must have $x \in E$. For if not, there is a point y in $E(x, b)$ and clearly $y \in E(a, b)$. Thus $x \in E$; whence, $R \subset M - x$; and since x separates R from $E - x$, it follows that R is a component of $M - x$.

3. **A-sets.** For any metric space M we shall call a subset A of M an *A-set* provided that $M - A = \sum G_i$ where G_i is open, $G_i \cdot G_j = 0$ for $i \neq j$, $F(G_i)$ contains at most one point, and $\delta(G_i) \to 0$, i.e., $M - A$ is the sum of a finite number or of a null sequence of disjoint open sets each having at most one boundary point.

Note. A sequence (collection, family) of sets is called a *null sequence* (*null collection* or *family*) provided that for any $\epsilon > 0$ at most a finite number of its elements are of diameter greater than ϵ.

(3.1) *For any A-set $A \subset M$ and any connected set $Z \subset M$, $A \cdot Z$ is either vacuous or connected.*

Proof. Supposing the contrary, we have a separation

$$A \cdot Z = A_1 + A_2 .$$

For any i, let $B_i^1 = B_i^2 = 0$ if $Z \cdot G_i = 0$; let $B_i^1 = Z \cdot G_i$ and $B_i^2 = 0$ if $Z \cdot G_i \neq 0$ and $F(G_i) \subset A_1$; and let $B_i^1 = 0$ and $B_i^2 = Z \cdot G_i$ if $Z \cdot G_i \neq 0$ and $F(G_i) \subset A_2$. Let

$$Z_1 = A_1 + \sum B_i^1 , \qquad Z_2 = A_2 + \sum B_i^2 .$$

Then clearly $Z = Z_1 + Z_2$ and $Z_1 \cdot Z_2 = 0$. If Z_1 contains a limit point x of Z_2, clearly $x \in A$, since each B_i^1 is open in Z. But since x is not a limit point of any single set B_i^2 and $\delta(B_i^2) \to 0$, there would exist a sequence of sets $B_{i_j}^2 \to x$. This gives $F(B_{i_j}^2) \to x$, which is impossible since by definition $F(B_{i_j}^2) \subset A_2$.

(3.11) *If M is connected, so also is any A-set in M.*

The same method of proof suffices to establish the following somewhat more general result for the case of a continuum M.

(3.2) *If $A = H + K$ is any division of an A-set A in a continuum M into $\left\{\begin{matrix}\text{sets}\\ \text{closed sets}\\ \text{continua}\end{matrix}\right\}$ H and K, there exists a division $M = L + N$ of M into $\left\{\begin{matrix}\text{sets}\\ \text{closed sets}\\ \text{continua}\end{matrix}\right\}$ L and N such that $L \cdot N = H \cdot K$, $L \supset H$, $N \supset K$.*

To see that the sets obtained in the last part of this theorem are continua it is only necessary to note that the sets $G_i + F(G_i)$ are continua. We may remark that only in this part of the theorem is it essential that M be a continuum, for the other two parts hold for an arbitrary metric set M.

As an immediate consequence, we get

(3.22) *If two points x and y of an A-set A in a metric space M are separated in A by a subset X of A, then x and y are also separated in M by X.*

(3.3) THEOREM. *If M is a semi-locally-connected continuum, the following four properties for subcontinua A of M are equivalent:*

(a) *If A contains more than one point of a cyclic element E_0 of M, we have* $A \supset E_0$.

(b) *For any $p \epsilon A$ we have either $p = A$ or $p \subset$ some cyclic element $E_0 \subset A$.*

(c) *The boundary* of each component of $M - A$ is a single point.*

(d) *A is an A-set.*

Proof. (a) *implies* (b). For let $p \epsilon A$. If p is either a cut point or an end point the result is obvious. If not, there exists a unique, nondegenerate cyclic element E_0 containing p. If $E_0 \cdot A \neq p$, we have $E_0 \subset A$ by (a). Hence we may suppose $E_0 \cdot A = p$. Let $x \epsilon E_0 - p$, $y \epsilon A - p$. Since p is not a cut point, there exists a continuum N with $x + y \subset N \subset M - p$. The component Q of $N - N \cdot A$ containing x has a limit point q in A, $q \neq p$. Since $A \supset p + q \subset E_0 + Q + q$ and $A \cdot (E_0 + Q + q) = p + q$, clearly p and q are not separated in M by any single point. Whence, $p + q \subset E_0 \subset A$ by (a).

(b) *implies* (a). For let $E_0 \cdot A \supset p + q$. Let N be an irreducible continuum in A from p to q. By (1.3), $N \subset E_0$. Hence $E_0 \cdot A$ contains a non-cut point x of M. Whence $x \subset E_0 \subset A$.

(a) *implies* (c). For suppose the boundary of a component R of $M - A$ contains two distinct points p and q. Then since $(R + p + q) \cdot A = p + q$, there exists a cyclic element E_0 with $p + q \subset E_0 \subset A$. Whence, $R \subset M - E_0$. But this is impossible since the component of $M - E_0$ containing R can have only one boundary point.

(c) *implies* (a). For let $E_0 \cdot A \supset p + q$. Suppose there exists $x \epsilon E_0 - E_0 \cdot A$. Let R be the component of $M - A$ containing x and let $y = \bar{R} - R$. Since $E_0 - y$ is connected and contains x and at least one of the points p and q, say p, there is a continuum N with $p + x \subset N \subset M - y$. But clearly N contains a point of $\bar{R} - R$, contrary to $y \cdot N = 0$.

(c) *implies* (d). In the first place, each component R of $M - A$ is open. For let $y = \bar{R} - R$. Then R is identical with the component D of $M - y$ containing R. If $D \cdot A \neq 0$ we could find a continuum N in D intersecting both R and A, by [III, (4.1)]; and this is impossible since clearly $N \cdot A$ would contain a limit point of R, contrary to $\bar{R} - R = y$. Hence $D \cdot A = 0$, which gives $D = R$. Now D, and hence R, is open since for any $x \epsilon D$ there exists a neighborhood V of y with $\bar{V} \cdot x = 0$ and such that $M - V$ has only a finite number of components.

Thus it remains to show that the components of $M - A$ form at most a null sequence. Suppose on the contrary, that there exists a sequence of components $R_1, R_2, \cdots$ of $M - A$ converging to a nondegenerate limit continuum K. By what was just shown, $K \subset A$. Now if $p_i = F(R_i)$, we may suppose the sequence $[R_i]$ chosen so that $p_i \rightarrow p \epsilon A$. But it is seen at once that this contradicts the semi-local-connectedness of M at p.

Finally, that (d) *implies* (c) is at once evident.

Note. An A-set A in a s.l.c. continuum M will be called a *true A-set* pro-

* "Boundary" here means the set $\bar{R} - R$, where R is the component of $M - A$.

vided that for each $p \in A$ we have $p \subset$ some cyclic element $\subset A$. Clearly any A-set is a true A-set except when it reduces to a single non-cut non-end point.

(3.31) *If A is an A-set in a semi-locally-connected continuum M, the components of $M - A$ are open sets each bounded by a single point of A and they form a null sequence at most.*

(3.32) *If p is any point of a semi-locally-connected continuum M, and S is a component of $M - p$, $S + p$ is an A-set.*

(3.4) *A subcontinuum A of a locally connected continuum M is an A-set if and only if it has the property*: (e) *for any simple arc ab in M, $a, b \in A$ implies $ab \subset A$.*

Proof. Obviously any A-set has property (e), since by (3.1) the intersection of an A-set with any simple arc in M is either vacuous or connected. To prove the converse, let A be a continuum in M having property (e). If A is not an A-set, then by property (c) of (3.3) there exists a component R of $M - A$ and two distinct points a and b of $F(R)$. Let R_a and R_b be disjoint regions in M containing a and b, respectively, and let $x \in R_a \cdot R$, $y \in R_b \cdot R$. There exist arcs ax, xy, yb in R_a, R and R_b, respectively. The set $ax + xy + yb$ contains an arc ab; but since $ax \cdot yb = 0$, $ab \cdot R \neq 0$. Hence ab is not contained in A, contrary to property (e). Accordingly, A is an A-set.

(3.5) *If M is a semi-locally-connected continuum,* (i) *every cyclic element of M is an A-set and* (ii) *the product of any number whatever of A-sets is an A-set.*

Both of these conclusions result at once from property (b) of (3.3).

(3.6) *If Z is a* $\left\{\begin{matrix}\textit{locally connected}\\ \textit{semi-locally-connected}\end{matrix}\right\}$ *subset of a connected set M and A is any A-set in M, then $A \cdot Z$ is either vacuous or* $\left\{\begin{matrix}\textit{locally connected}\\ \textit{semi-locally-connected}\end{matrix}\right\}$.

The first result is an immediate consequence of (3.1).

To prove the second one, let Z be semi-locally-connected, let $x \in A \cdot Z$ and let $\epsilon > 0$ be given. Let V be a neighborhood of x in Z of diameter less than ϵ such that $Z - V$ has only a finite number of components, say $Z_1, Z_2, \cdots, Z_n$. Then since, by (3.1), $A \cdot Z_i$ is connected, $A \cdot Z - A \cdot V$ has only the components $A \cdot Z_1, A \cdot Z_2, \cdots, A \cdot Z_n$. Accordingly, $A \cdot Z$ is semi-locally-connected at x, because $A \cdot V$ is a neighborhood of x in $A \cdot Z$ of diameter less than ϵ.

Now in case M is $\left\{\begin{matrix}\text{locally connected}\\ \text{semi-locally-connected}\end{matrix}\right\}$ we may take $Z = M$ in (3.6) and thus we get $A \cdot Z = A$. Whence

(3.61) *Any A-set in a* $\left\{\begin{matrix}\textit{locally connected}\\ \textit{semi-locally-connected}\end{matrix}\right\}$ *and connected set M is itself* $\left\{\begin{matrix}\textit{locally connected}\\ \textit{semi-locally-connected}\end{matrix}\right\}$.

(3.7) *If A is any A-set in a semi-locally-connected continuum M, any cyclic element of A is a cyclic element of M. Thus any A-set in A is an A-set in M.*

Proof. Any cut point of A is a cut point of M by (3.22). Any end point x

of A which is not an end point of M must be a cut point of M; because if not, x would be in a true cyclic element E of M with $E \subset A$, which is impossible since x is an end point of A. Finally any true cyclic element of A is a true cyclic element of M, because no point y of $M - A$ can be conjugate to more than one point of A, namely the boundary point of the component of $M - A$ containing y.

DEFINITIONS. *If $f(X) = Y$ is a continuous transformation such that $Y \subset X$ and for each $y \in Y$, $f(y) = y$, f is said to be a retracting transformation or a retraction of X into Y. A continuous transformation $f(A) = B$ is said to be monotone provided that for each $b \in B$, $f^{-1}(b)$ is a continuum and is said to be A-set reversing provided $f^{-1}(b)$, $b \in B$, is an A-set in A.*

(3.8) *If M is a continuum and $f(M) = A$ is retracting and such that $[f^{-1}(x)]_{x \in A}$ is a null collection or has only a countable number of nondegenerate elements, then f is monotone and each component C of $M - A$ has just the one limit point $f(C)$ in A. Furthermore, f is uniquely determined by A.*

Proof. Let $x \in A$ be any limit point of C. Since $f(C)$ is connected it must reduce to a single point, because otherwise it would contain a point y such that $y = f^{-1}(y)$ which manifestly is absurd. Since $\bar{C} \supset x$ and $f(x) = x$, this gives $f(C) = x$ by continuity. Hence x is the only limit point of C in A.

Now f must be monotone and is uniquely determined, since for each $x \in A$, $f^{-1}(x)$ consists of x together with all components C of $M - A$ having x as their limit point in A.

(3.81) COROLLARY. *If M is semi-locally-connected, A is an A-set and f is A-set reversing.*

(3.9) THEOREM. *In order that a subset A of a semi-locally-connected continuum M be an A-set it is necessary and sufficient that there exist a retraction $f(M) = A$ such that $[f^{-1}(x)]_{x \in A}$ has only a countable number of nondegenerate elements. When such a retraction exists it is uniquely determined and is actually A-set reversing and hence monotone and the collection $[f^{-1}(x)]_{x \in A}$ is a null collection.*

4. Continua of convergence.

(4.1) THEOREM. *If $[K_n]$ is an infinite sequence of disjoint continua in a semi-locally-connected continuum M converging to a nondegenerate continuum K, there exists a cyclic element E containing K and such that $K = \lim (E \cdot K_n)$.*

Proof. As any two points of K are conjugate, clearly there exists a cyclic element E containing K. Also, $\limsup (E \cdot K_n) \subset K$. We have to show $\liminf (E \cdot K_n) \supset K$. Suppose, on the contrary, that there is a point x of K not belonging to $\liminf (E \cdot K_n)$. Then there exists an $\epsilon > 0$ and an infinite sequence of integers n_i such that $K_{n_i} \cdot V_\epsilon(x) \neq 0 = E \cdot K_{n_i} \cdot V_{3\epsilon}(x)$, for each i. Now for each i, $K_{n_i} \cdot V_{3\epsilon}(x)$ contains a continuum N_i of diameter greater than ϵ. Since $N_i \subset M - E$, it follows by (3.31) that there exists a component R of $M - E$ containing infinitely many of the sets N_i. But clearly this is impossible, since $F(R)$ is a single point, whereas lim sup $N_i \subset E \subset M - R$ and $\delta(\text{lim sup } N_i) \geqq \epsilon$.

(4.11) *Any continuum of convergence of a semi-locally-connected continuum M is a continuum of convergence of some single cyclic element of M.*

(4.2) *The true cyclic elements of a semi-locally-connected continuum form a null sequence at most.*

For otherwise there would exist a sequence of distinct cyclic elements $[E_i]$ with $\delta(E_i) > d > 0$ $(i = 1, 2, \cdots)$, which converges to a continuum K. Now for each i, $E_i \cdot E_j$ can be different from 0 for only a finite number of integers j, because by (3.31) only a finite number of components R of $M - E_i$ can be such that $\bar{R}$ contains an E_j and clearly no such set R can contain more than one E_j which intersects E_i. Hence we may assume without loss of generality that the sets E_i are disjoint. But then (3.1) would give a cyclic element E such that $E_i \cdot E$ is nondegenerate for almost all i, which clearly is absurd.

5. **Cyclic chains.** For any two distinct points a and b of a semi-locally-connected continuum M, the product of all A-sets in M containing $a + b$ will be called the *cyclic chain* from a to b and is denoted by $C(a, b)$.

By (3.5) we have that every cyclic chain is an A-set.

(5.1) *If a and b are conjugate points of a semi-locally-connected continuum M, $C(a, b)$ is a cyclic element of M.*

For since $C(a, b)$ is an A-set and therefore a continuum, it contains an irreducible continuum N from a to b. Since a and b are conjugate, by [III, (4.3)] N can contain at most a countable number of cut points of M. Let p be a non-cut point of M belonging to N. Since the cyclic element E of M containing p contains both a and b and is an A-set, we have $E = C(a, b)$.

(5.2) THEOREM. *For any two distinct points a and b of a semi-locally-connected continuum M, $C(a, b) = E(a, b) + a + b +$ all cyclic elements E such that $E \cdot [E(a, b) + a + b]$ consists of exactly two points.*

For let X denote the set on the right. Then X is an A-set. For since $E(a, b) + a + b$ is compact, X is compact. Also X is connected. For if not we have a separation $X = X_a + X_b$. Since $E(a, b) + a + b = K$ is linearly ordered and compact, there is a last point x of K in X_a and a first point y of K in X_b following x. Now x and y must be conjugate; for if z separates x and y it readily follows that $z \in E(a, b)$, contrary to the fact that there is no point of $E(a, b)$ "between" x and y. Hence there exists a cyclic element N, with $x + y \subset N$. But since clearly N cannot contain three points of K, we have $N \cdot K = x + y$ so that $N \subset X$, contrary to $x \in X_a$, $y \in X_b$. Thus X is connected. Now for any $x \in X$, there exists a cyclic element E_0 with $x \subset E_0 \subset X$. This follows by the definition of X except when $x = a$ or $x = b$. Suppose $x = a$. Now if a is a cut point or an end point of M, again our result follows. If not, let E_0 be the (nondegenerate) cyclic element containing a. Let R be the component of $M - E_0$ containing b and let $z = F(R)$. Then since $a \neq z$, we have $z \in E(a, b)$

$+ b$. Consequently $a + z = E_0 \cdot K$ and $E_0 \subset X$. A similar proof works for the case $x = b$. Thus X is an A-set.

Now since any A-set containing $a + b$ clearly must contain X, we have $C(a, b) \supset X$; and since X is an A-set containing $a + b$, we have $X \supset C(a, b)$. Whence $X = C(a, b)$. (See Note 8, p. 280.)

6. ***H*-sets.** Any connected subset H of a semi-locally-connected continuum M having the property that each point of H is contained in a cyclic element of M lying wholly in H will be called an *H-set* in M. [Compare with property (b) of (3.3).] Thus an H-set is any connected set which can be obtained by adding together cyclic elements of M. It follows of course that any single cyclic element is an H-set, as is also any nondegenerate A-set.

Throughout this section M will denote a semi-locally-connected continuum and H will stand for an arbitrary H-set in M.

From the definition we have

(6.1) *If the sum of any number whatever of H-sets is connected, it is likewise an H-set.*

(6.2) *If the set $H \cdot C$ contains either two distinct points or one non-cut point of M, where C is any true cyclic element, then $C \subset H$.*

For, by (3.1), $H \cdot C$ is connected and hence, by (1.5), contains at least one non-cut point of the space. Since each non-cut point belongs to only one cyclic element, we have $C \subset H$.

(6.3) *If a and b are any two points of an H-set H, then $C(a, b) \subset H$.*

For since H is connected and contains $a + b$, we evidently have $E(a, b) + a + b \subset H$. Thus every cyclic element which contains two points x and y of $E(a, b) + a + b$ contains two points in common with H; and consequently, by (6.2), $C \subset H$. Thus, by virtue of the definition of the chain X, it follows that $C(a, b) = X \subset H$.

(6.4) *Every H-set has the property that every arc or irreducible continuum ab in M joining two points a and b of H is contained in H.*

For, by (6.3), we have $ab \subset C(a, b) \subset H$.

(6.5) *If Z is any connected set, $H \cdot Z$ is either vacuous or connected.*

For let a and b be any two points of $H \cdot Z$. Then by (3.1) the set $Z \cdot C(a, b)$ is connected and, by (6.3), we have

$$a + b \subset Z \cdot C(a, b) \subset Z \cdot H.$$

From (6.5) we have immediately

(6.6) *Every H-set is semi-locally-connected. If M is locally connected, so also is every H-set in M.*

(6.7) *The product of any number whatever of H-sets is an H-set.*

To prove this, let G be any collection of H-sets and let P denote the product of all the elements of this collection. Then P is a cyclic-element-sum. For let p be any point of P. If p is cut point, it is itself a cyclic element; and if it is a non-cut point, there exists one and only one cyclic element C which contains p; and since p is contained in all the H-sets of G, it follows that $C \subset P$.

Now P is also connected. For let a and b be any two points of P. Then by (6.3) the cyclic chain $C(a, b)$ is contained in all the H-sets of G and, consequently, in their product P. It follows that P is connected and thus it is an H-set.

(6.8) *Every point of the set* $\bar{H} - H$ *is a cyclic element (thus either a cut point or an end point of* M*). Hence every set* H_0 *such that* $H \subset H_0 \subset \bar{H}$ *is likewise an* H*-set.*

For let p be any point of $\bar{H} - H$, and suppose, contrary to what we wish to show, that p is neither a cut point nor an end point of M. Then there exists a true cyclic element C containing p. Now $H \cdot C \neq 0$; for, if we had $H \cdot C = 0$, the boundary $F(R)$ of the component R of $S - C$ containing H would be a cut point of M and consequently would be different from p; and it would follow that p was not a limit point of H, contrary to hypothesis. Thus the set $H \cdot C$ is not vacuous; and since, by (3.1), it is connected and cannot contain C, it follows from (6.2), that it reduces to a single point q. Consequently $(H + p) \cdot C = p + q$ which is impossible, because $(H + p) \cdot C$ is connected by (3.1).

(6.81) Corollary. $\bar{H}$ *is an* A*-set.*

(6.9) *If* R *is any component of the complement of an* H*-set* H*, the set* $\bar{R} \cdot \bar{H}$ *reduces to a single point.*

Suppose, on the contrary, that $\bar{R} \cdot \bar{H}$ contains two distinct points a and b. By (6.8), the set $H_0 = H + a + b$ is an H-set. But then we have

$$H_0 \cdot (R + a + b) = (H + a + b)(R + a + b) = a + b.$$

This is impossible, by (6.5), because the set $R + a + b$ is connected.

7. Cyclic chain development. Imbedding Theorem.

(7.1) Cyclic chain approximation theorem. *Let* S *be a semi-locally-connected continuum. There exist two sequences (finite or infinite) of points* (p_i) *and* (q_i) $(i = 1, 2, 3, \cdots)$, *such that the cyclic chains* $C(p_i, q_i)$ *have the following properties:*

(a) *For each* i, $C(p_i, q_i) \cdot \sum_{j=1}^{i-1} C(p_j, q_j) = q_i$.

(b) *Setting* $H_n = \sum_1^n C(p_i, q_i)$ *and* $H = \sum_1^\infty H_n = \sum_1^\infty C(p_n, q_n)$, *every point of the set* $S - H$ *is an end point of* S.

(c) *The diameter of the components of* $S - H_n$ *tends to* 0 *uniformly with* $1/n$.

(d) $$\lim_{n \to \infty} \delta[C(p_n, q_n)] = 0.$$

Proof. Since S is separable, there exists a countable set $D = r_1 + r_2 + \cdots$ such that $\bar{D} = S$. Set $r_1 = p_1$ and $q_1 = r_2$. Let n_2 be the smallest integer such that the point r_{n_2} is not contained in the cyclic chain $C(p_1, q_1)$. [If there exists no such integer, evidently $C(p_1, q_1) = S$.] Let q_2 be the point which is the boundary of the component S_2 of $S - C(p_1, q_1)$ containing the point $p_2 = r_{n_2}$.

Let n_3 be the smallest integer such that r_{n_3} is not contained in the set $H_2 = C(p_1, q_1) + C(p_2, q_2)$, and let q_3 be the boundary of the component S_3 of $S - H_2$ which contains the point $p_3 = r_{n_3}$, and so on. In general if the set D is not contained in the set $H_{n-1} = C(p_1, q_1) + \cdots + C(p_{n-1}, q_{n-1})$, (in which case the theorem is evident), we take for n_n the smallest integer such that r_{n_n} is not contained in H_{n-1} and for q_n the boundary point of the component S_n of $S - H_{n-1}$ which contains the point $p_n = r_{n_n}$. We shall now prove that the conditions (a)–(d) are satisfied.

Since, by (3.32), each set $S_i + q_i$ is an A-set and hence an H-set, it follows in view of (6.3) that we have $C(p_i, q_i) \subset S_i + q_i$, and consequently (a) is satisfied.

Now let x be any point of $S - H$. It follows from the definition of the points p_n that, for each n, $\sum_1^n r_i \subset H_n$; and consequently, $D \subset H$. Thus $\bar{D} = S = \bar{H}$; and by virtue of (6.8) it follows that the point x is either a cut point or an end point of the space. Now since H is connected and $\bar{H} = S$, it follows that x is not a cut point. Therefore x is an end point and (b) is proved.

Now suppose, contrary to (c), that for some $\epsilon > 0$ there exists, for every n, at least one component R_n of $S - H_n$ of diameter greater than ϵ. Let $[R_{n_i}]$ be a convergent subsequence of $[R_n]$ and let L be its limit set. Then L is a continuum and $\delta(L) \geqq \epsilon$. Hence there are at least two points a and b of L which are not end points of S. Hence, by (b), there exists an integer k such that $a + b \subset H_k$. Accordingly, $L \cdot H_k$ is a nondegenerate continuum. But $R_{n_i} \subset S - H_k$ for $n_i > k$ and only a finite number of the components of $S - H_k$ can intersect $\sum_{n_i > k} R_{n_i}$, and hence $L \cdot H_k$ must be a finite set. Thus our supposition that (c) does not hold leads to a contradiction.

Property (d) follows immediately from (c). For, given any $\epsilon > 0$, let n be chosen so that each component of $S - H_n$ is of diameter less than ϵ. Then since $C(p_i, q_i) - q_i$, $(i > n)$ is contained in a single component of $S - H_n$, it follows that $\delta[C(p_i, q_i)] < \epsilon$ for $i > n$.

If the set D is chosen so as to contain no end point of S, no p_i or q_i will be an end point and hence H will contain no end point. Whence

(7.11) COROLLARY. *The points p_i and q_i can be so chosen that $S - H$ is exactly the set of all end points of S.*

DEFINITION. *A set X will be said to be topologically contained in a set Y provided Y contains a set homeomorphic with X. In other words, there exists a topological mapping $f(X) = X'$ where $X' \subset Y$.*

(7.2) IMBEDDING THEOREM. *Any semi-locally-connected continuum M is topologically contained in a locally connected continuum L such that for every cyclic element C of L, $M \cdot C$ is a cyclic element of M. Furthermore, L can be chosen in the space Q_ω so that every true cyclic element of L is a topological "sphere" whose interior contains no point of M.*

Proof. In case M is a cyclic chain $C(p, q)$ the theorem is proved as follows. Since the set $E(p, q) + p + q$, which we shall call K, is compact and naturally ordered, by [III, (6.4)], there exists a topological, order-preserving transformation $f(x)$ mapping this set into a subset of a linear interval ab of Q_ω so that $f(p) = a$, $f(q) = b$. Now each true cyclic element E_i of M contains exactly two points x_i and y_i of $E(p, q) + p + q$. For each such E_i, let S be the sphere in Q_ω having the interval $f(x_i)f(y_i)$ of ab as a diameter and, using [II, (1.7)], let us extend f to E_i in such a way that $f(E_i) \subset S_i$ and f is topological. Then clearly the extended f maps M topologically into a subset of $L = f(K) + \sum S_i$; and this latter set satisfies all the requirements of our theorem, since its cyclic elements are the S_i together with the individual points $f(K)$.

In the general case we make use of the Cyclic Chain Approximation Theorem (7.1). We take $M = S$ and use the notation of (7.1). Clearly by extending the case just treated we can obtain, for each n, a topological mapping

$$f(H_n) \subset H'_n$$

in such a way that H'_n satisfies all the conditions imposed on L in our theorem and so that if $H' = \sum H_n$, $\bar{H}' = L$, then L is a locally connected continuum having the sets H_n as A-sets, every point of $L - H'$ is an end point of L, and the diameters of the components of $L - H'_n$ converge uniformly to 0 with $1/n$. (Note: a slight distortion of the "spheres" employed will be necessary when three or more cyclic elements have a common point.)

This gives a topological mapping $f(H) = K \subset H'$. We proceed to show that this mapping is uniformly continuous on H. Let $\epsilon > 0$ be given. Let k be an integer such that every component of $S - H_k$ or of $L - H'_k$ is of diameter less than $\epsilon/6$. Since the mapping $f(H_k) \subset H'$ is topological and H_k is compact, there exists a $\delta > 0$, $\delta < \epsilon/3$, such that if $x, y \in H_k$ and $\rho(x, y) < 3\delta$, then $\rho[f(x), f(y)] < \epsilon/3$. Since there are only a finite number of components of $S - H_k$ of diameter greater than $\delta/2$, there exists a number $d > 0$ such that $d < \delta/2$ and if $x, y \in S - H_k$ and C_x and C_y are the components of $S - H_k$ containing x and y, respectively, and $\rho(x, y) < d$, then either $F(C_x) = F(C_y)$ or $\delta(C_x) < \delta > \delta(C_y)$. We may suppose d also chosen so that any two points of M or of L at a distance apart less than d lie together in a connected subset of M or L of diameter less than δ.

Now let $x, y \in H$, $\rho(x, y) < d$. If $x, y \in H_k$ we have $\rho[f(x), f(y)] < \epsilon$ since $d < \delta$. If $x \in H_k$, $y \in M - H_k$, let C_y be the component of $M - H_k$ containing y and let $p_y = F(C_y)$. Then $\rho(y, p_y) < \delta$ since x and y lie in a connected subset of M of diameter less than δ. Thus $\rho(p_y, x) < 3\delta$; whence

$$\rho[f(x), f(p_y)] < \epsilon/3,$$

and

$$\rho[f(p_y), f(y)] < \epsilon/6$$

because the diameter of the component of $L - H'_k$ containing $f(y)$ is less than $\epsilon/6$ and $f(p_y)$ is the boundary of this component. Accordingly

$$\rho[f(x), f(y)] < \epsilon.$$

Finally, if $x, y \in M - H_k$ let C_x and C_y be the components of $M - H_k$ containing x and y, respectively. Let $p_x = F(C_x)$, $p_y = F(C_y)$. If $p_x = p_y$, then $\rho[f(x), f(y)] < \epsilon$ because the components C'_x and C'_y of $L - H'_k$ containing $f(x)$ and $f(y)$, respectively, are of diameter less than $\epsilon/3$ and they have the same boundary, i.e., $F(C'_x) = F(C'_y) = f(p_x)$. If, on the other hand, $p_x \neq p_y$, then by the choice of d we have

$$\delta(C_x) < \delta > \delta(C_y).$$

Hence

$$\rho(x, p_x) < \delta > \rho(y, p_y).$$

Whence

$$\rho(p_x, p_y) < 3\delta.$$

Accordingly

$$\rho[f(p_x), f(p_y)] < \epsilon/3.$$

Also

$$\rho[f(x), f(p_x)] < \epsilon/6,$$
$$\rho[f(y), f(p_y)] < \epsilon/6,$$

because

$$\delta(C'_x) < \epsilon/6 > \delta(C'_y).$$

Therefore

$$\rho[f(x), f(y)] < \epsilon.$$

Thus we have shown that f is uniformly continuous on H. In exactly the same manner we show that the mapping

$$f^{-1}(K) = H$$

is uniformly continuous. Hence these may be extended to yield a topological mapping

$$f(M) = \bar{K} \subset L$$

satisfying all conditions in our theorem.

DEFINITION. *For any integer $n > 0$ a locally connected continuum C each true cyclic element of which is homeomorphic with an m-dimensional sphere (i.e., the set in R^{m+1} satisfying $x_1^2 + x_2^2 + \cdots + x_{m+1}^2 = 1$) will be called an m-dimensional cactoid. For $m = 1$ we shall use the term boundary curve and for $m = 2$ the term cactoid.*

As a corollary to (7.2) and its proof we have

(7.3) *If for some integer m each true cyclic element of a semi-locally-connected continuum M is homeomorphic with a subset of an m-dimensional sphere, then M can be topologically imbedded in an m-dimensional cactoid C in R^{m+1} which is the* *boundary of the unbounded component of its complement in R^{m+1} and so that the intersection of M with each cyclic element of C is a cyclic element of M.*

As a special case we have

(7.31) *Any m-dimensional cactoid which is not a dendrite is homeomorphic with the boundary of a bounded connected region in R^{m+1}.*

Clearly the region may be taken bounded.

(7.32) *Any dendrite is homeomorphic with a dendrite on a plane or on a sphere.*

8. **Nodal sets. Nodes.** A closed subset N of a set M will be called a *nodal set* in M provided $N \cdot \overline{M - N}$ contains at most one point. It will be noted that if N is nodal, so also is the set $\overline{M - N}$. Also in case M is connected and p is a cut point of M, then for any separation $M - p = M_1 + M_2$, the sets $M_1 + p$ and $M_2 + p$ are nodal sets.

If M is a semi-locally-connected continuum, the true cyclic elements of M which are nodal sets and the end points of M will be called *nodes* of M. Thus a node of M is either an end point of M or a true cyclic element of M which contains at most one cut point of M. If M has no cut points clearly M is a node of itself; and indeed, the only nodal subsets of M are M itself together with the individual points of M.

(8.1) THEOREM. *Any nondegenerate nodal subset N of a semi-locally-connected continuum S contains at least one node of S.*

Proof. Let us take a cyclic chain development of S as given by (7.1). Let n_1 be the least integer such that p_{n_1} belongs to $N - N \cdot \overline{S - N}$. If the cyclic element of S in H_{n_1} which contains p_{n_1} is not a node of S there exists a component C_1 of $S - H_{n_1}$ which lies in N and is such that its boundary point x_1 in H_{n_1} is different from the point $N \cdot \overline{S - N}$. Let $N_1 = C_1 + x_1$. Then N_1 is a nodal subset. Let n_2 be the least integer such that $p_{n_2} \subset C_1$. If the cyclic element of S in H_{n_2} which contains p_{n_2} is not a node of S there exists a component C_2 of $S - H_{n_2}$ which lies in C_1 and is such that its boundary point x_2 is different from x_1. Let $N_2 = C_2 + x_2$. Continue this process. Either we find a node of S in N or else the process continues indefinitely. In the latter case we obtain an infinite decreasing sequence $N \supset N_1 \supset N_2 \supset \cdots$ where $N_i = C_i + x_i$, $\delta(N_i) \to 0$, and N_i does not contain x_{i-1}. Let $p = \prod_1^\infty N_i$. Then p is an end point of S, because $p = \prod C_i$ (since $p \neq x_i$ for every i), and $\delta(C_i) \to 0$, $F(C_i) = x_i$. Hence p is a node and thus in any case N contains a node of S.

(8.2) THEOREM. *Any semi-locally-connected continuum which has a cut point has at least two nodes.*

9. **Cyclic connectedness.** A set of points X is said to be *cyclicly connected* provided every two points of X lie together on some simple closed curve in X.

We begin with two lemmas concerning locally connected continua having no cut points. Let C denote such a continuum.

(9.1) LEMMA. *If A and B are nondegenerate, closed, and mutually exclusive subsets of C, there exist two mutually exclusive arcs in C joining A and B.*

There exists an arc ab in C where $ab \cdot A = a$, $ab \cdot B = b$; and if p is a point of $A - a$, clearly there exist points x which can be joined in C to p by an arc px containing no point of ab. Thus there exist points x such that mutually exclusive arcs ab and px exist in C so that

$$(1) \qquad ab \cdot A \supset a, \qquad ab \cdot B \supset b, \quad \text{and} \quad px \cdot A \supset p.$$

Let G denote the set of all such points x. We shall show that $G = C$. Suppose this is not so. Then since C is connected and clearly G is open in C, it follows that at least one point y of $C - G$ is a limit point of G. There exists an arc $a'b'$ in $C - y$ such that $a'b' \cdot A = a'$ and $a'b' \cdot B = b'$. Let R be a region containing y but having no point or boundary point in $a'b' + A$. Then R contains a point x of G, and there exist arcs ab and px satisfying (1). Since y does not belong to G, it follows at once that $ab \cdot R \neq 0$. The arcs ab and px contain arcs aw and pr, respectively, such that $aw \cdot \bar{R} = w$ and $pr \cdot \bar{R} = r$. Let $H = A + aw + pr$. Then $a'b'$ contains an arc uv such that $uv \cdot H = u$ and $uv \cdot B = v$. Let T denote one of the sets aw and pr which does not contain u and let Z denote the other one of these sets. Let Q be a region containing the point $T \cdot \bar{R}$ and containing no point of $Z + uv$. Then $Z + uv$ contains an arc mn and $T + Q + R$ contains an arc qy such that $mn \cdot A \supset m$, $mn \cdot B \supset n$, $qy \cdot A \supset q$, and $mn \cdot qy = 0$. But this is impossible since y does not belong to G. Therefore $G = C$. Accordingly G contains a point x of B, and thus there exist two mutually exclusive arcs ab and px joining A and B.

(9.2) LEMMA. *Every poini x of C is an interior point of some arc axb in C.*

This is obvious if x is a cut point of some region R in C; for then it is only necessary to take a and b in different components of $R - x$ and any arc ab in R will contain x. Thus we may suppose that x is a cut point of no region in C. Now let a and b be any two distinct points of $C - x$. Let R_1 be a region containing x of diameter less than 1 so that $\bar{R}_1 \cdot (a + b) = 0$. By [I, §15], there exists a locally connected subcontinuum E_1 of C of diameter less than 1 which contains $\bar{R}_1$ but does not contain either a or b. Since x is not a cut point of R_1, it cannot be a cut point of E_1; and since it is not an end point of E_1 it therefore lies in some nondegenerate cyclic element C_1 of E_1. By (9.1) there exist in C two mutually exclusive arcs aa_1 and bb_1 where $aa_1 \cdot C_1 = a_1$ and $bb_1 \cdot C_1 = b_1$. Obviously, we may suppose $a_1 \neq x \neq b_1$. Let R_2 be a region in C_1 containing x of diameter less than 1/2, such that $\bar{R}_2 \cdot (a_1 + b_1) = 0$. There exists a locally connected subcontinuum E_2 of C_1 of diameter less than 1/2 which contains $\bar{R}_2$ but does not contain either a_1 or b_1. Again we may suppose that x cuts no region in C_1; and it follows just as in the case of E_1 that x lies in some nondegenerate cyclic element C_2 of E_2. By (9.1) there exist in C_1 two mutually exclusive arcs a_1a_2 and b_1b_2 such that $a_1a_2 \cdot C_2 = a_2$ and $b_1b_2 \cdot C_2 = b_2$. Let R_3 be a region in C_2 of diameter less than 1/3, and so on. If we continue this process indefinitely, it is clear that the point set

$$x + aa_1 + a_1a_2 + a_2a_3 + \cdots + bb_1 + b_1b_2 + \cdots$$

so obtained is an arc axb in C.

(9.3) Cyclic connectedness theorem. *In order that a locally connected continuum S be cyclicly connected it is necessary and sufficient that S have no cut point.*

The necessity is immediate. For if S has a cut point p, and if a and b are points of S which are separated by p, then no simple closed curve in S can contain $a + b$, since each of the two arcs in such a curve would necessarily contain p.

To prove the sufficiency, let C be any such S having no cut point and let a and b be any two points of C. Now every point x of C lies on some simple closed curve in C; for by (9.2) there exists an arc pxq in C and there exists an arc pyq in $C - x$, and clearly $pxq + pyq$ contains a simple closed curve containing x. Thus there exist simple closed curves C_a and C_b in C containing a and b, respectively. Now if $C_a \cdot C_b = 0$, then by (9.1) there exist mutually exclusive arcs mn and uv in C where $mn \cdot C_a = m$, and $mn \cdot C_b = n$, $uv \cdot C_a = u$, and $uv \cdot C_b = v$; and in this case clearly the set mn + arc nbv of $C_b + uv$ + arc nam of C_a is a simple closed curve in C containing $a + b$. If $C_a \cdot C_b = p$, a single point, then $C - p$ contains an arc uv so that $uv \cdot C_a = u$ and $uv \cdot C_b = v$; and in this case uv + arc vbp of C_b + arc pau of C_a is a simple closed curve in C containing $a + b$. Finally, if $C_a \cdot C_b$ contains more than one point, then C_b contains an arc pbq, where $pbq \cdot C_a = p + q$; and in this case pbq + arc paq of C_a is a simple closed curve in C containing $a + b$. Thus the cyclic connectivity theorem is established.

Corollary 1. *Every true cyclic element of any locally connected continuum S is cyclicly connected.*

Corollary 2. *Any two conjugate points a and b of any locally connected continuum S lie together on a simple closed curve in S.*

(9.4) In any locally connected continuum S let us call any nondegenerate cyclicly connected subset C of S which is not a proper subset of any other cyclicly connected subset of S a *maximal cyclic* subset of S. Then by (1.2) and (9.3) we have at once that the maximal cyclic subsets of S are identical with the true cyclic elements of S. Thus the cyclic elements of S could be defined (as they actually were first defined) as (1) the cut points of S, (2) the end points of S, and (3) the maximal cyclic subsets of S.

(9.5) Three point theorem (Ayres). *In order that a locally connected continuum S be cyclicly connected it is necessary and sufficient that for each triple of points p_1, p_2, p_3 of S taken in any order p_i, p_j, p_k there exists an arc $p_ip_jp_k$ in S.*

To prove the necessity of the condition, let us suppose S is cyclicly connected. Then there exists a simple closed curve J in S containing $p_i + p_j$. If $p_k \subset J$, then J contains an arc $p_ip_jp_k$. If not, then by virtue of the cyclic connectivity of S there exists an arc p_kp_j in S not containing p_i. On this arc in the order p_kp_j let x be the first point of J, and let xp_k be the subarc of p_kp_j from x to p_k. Now J contains an arc p_ip_jx and $xp_k + p_ip_jx$ is an arc $p_ip_jp_k$.

The sufficiency of the condition results at once from (9.3). For if S has the three point property, it can have no cut point, since if z separates x and y in S, clearly S can contain no arc xyz.

However, a direct proof for the sufficiency can be given as follows. Let a and b be any two points of a space S having the three point property and let c be any third point of S. Then S contains arcs abc and bac. Let t denote the arc ac of bac. On the arcs ba and bc of abc let u and v, respectively, denote the first points of t. Then arc ubv of abc + arc uv of $t = J$ is a simple closed curve containing b. If it also contains a our proof is finished. If not, then on the arcs ab and au of bac let x and y be the first points of J. Then arc xay of bac + arc xby of J is a simple closed curve in S containing $a + b$.

10. Some equivalences. In the preceding section we saw that for locally connected continua with the aid of cyclic connectedness a different though equivalent definition for *cyclic element* could be given. In this section we shall show that similar equivalences exist for the H-sets.

DEFINITION. *For any subset K of a locally connected continuum S, let the set $M(K)$ of all points x of S such that x is on some arc of S joining two points of K be called the arc-curve of K [$M(K) = K$ in case K consists of a single point].*

We shall now prove that the following four classes of subsets of any locally connected continuum S are identical:

(I) the H-sets (adopting here the convention that any set consisting of a single point is an H-set);

(II) the sets Q having the property that their product $Q \cdot Z$ with any connected set Z is connected (or vacuous);

(III) the sets (closed or not) having property (e) of (3.4);

(IV) the arc-curves.

We shall demonstrate the identity of these four classes by establishing the following cyclic inclusion for the classes (I)–(IV):

$$(I) \subset (II) \subset (III) \subset (IV) \subset (I).$$

The inclusion (I) $\subset$ (II) means simply that every H-set H has the property that $H \cdot Z$ is connected (or vacuous) for every connected subset Z of S, and this is identical with (6.5) which has already been proved.

To prove (II) $\subset$ (III), let Q be any set of class (II). We have to show that Q has property (e); but this is immediate. For if ab is any arc with $a + b \subset Q$, then since $ab \cdot Q$ is connected we must have $ab \subset Q$.

We shall prove (III) $\subset$ (IV) by showing that if K is any set having property (e), then $M(K) = K$ and hence K is an arc-curve. Clearly this must be so, for if ab is any arc with $a + b \subset K$, we have $ab \subset K$ by property (e). This proves $M(K) \subset K$, and obviously $K \subset M(K)$ by definition of arc-curve.

Finally, to prove (IV) $\subset$ (I), let $M(K)$ be the arc-curve of any set K. We have to show that $M(K)$ is an H-set. Obviously $M(K)$ is connected, and we may suppose that $M(K)$ contains more than one point since the other case is trivial. Let C be any true cyclic element such that $C \cdot M(K)$ contains at least one non-cut point of S. Then $C \cdot M(K)$ contains at least two distinct points a and b, each of which either belongs to K or is the boundary point of a component S_a or S_b of $S - C$ containing points a' and b', respectively, of K. If $a \subset K$, let $aa' = a$; if not, let aa' be an arc in $S_a + a$ from a to a'; and similarly let $bb' = b$ if $b \subset K$ and otherwise let bb' be an arc in $S_b + b$ from b to b'. Now by the Three Point Theorem, any point x of C lies on an arc axb in C; and $a'a + axb + bb'$ is an arc $a'xb'$, so that $x \subset M(K)$. Thus $C \subset M(K)$ and $M(K)$ is an H-set.

11. **Applications: Cyclicly extensible and cyclicly reducible properties.** The usefulness of the notion of cyclic elements is due in general to the so-called *acyclic* or *dendritic* structure of the space relative to its cyclic elements. It is seen with the aid of the theorems already established that there is a great deal of resemblance between the structure of an arbitrary locally connected or semi-locally-connected continuum S with respect to its cyclic elements and that of a dendrite (i.e., a locally connected continuum S which contains no simple closed curve, or in other words, whose cyclic elements all reduce to single points).

There are a large number of properties which are possessed by the whole semi-locally-connected continuum S provided only that these same properties are possessed by the cyclic elements of S. A property of this type will be called a *cyclicly extensible property*. More precisely, a property is cyclicly extensible provided that when each cyclic element of S has this property, the space S itself likewise has this property. On the other hand, a property is said to be *cyclicly reducible* provided that when S has this property so also does every cyclic element of S.

We give now a number of such properties. Throughout this section, S denotes a semi-locally-connected continuum.

(11.1) *The property of containing no continuum of convergence is cyclicly extensible and reducible.*

The reducibility is trivial. Furthermore, the extensibility results immediately from (4.11). For if no cyclic element of S has a continuum of convergence, then neither can S by (4.11).

In the next chapter we shall see that the property of having no continuum of convergence is equivalent to the property of being hereditarily locally connected, where a continuum is said to be hereditarily locally connected provided every one of its subcontinua is locally connected.

(11.2) *The property of containing no connected subset which is not arcwise connected is cyclicly extensible and reducible.*

Again the reducibility is obvious. To prove the extensibility, we suppose that every connected subset of any cyclic element of S is arcwise connected. Then let Z be any connected subset of S and let a and b be any two points of Z. Now, by (5.2), $C(a, b) = K + a + b + \sum C_i$, where each C_i is a true cyclic element containing just two points x_i and y_i of $K + a + b$ and where $K = E(a, b)$. Clearly $K + a + b \subset Z$. For each i, $Z \cdot C_i$ is connected and hence arcwise connected, and therefore it contains an arc x_iy_i. But it is seen at once that $K + a + b + \sum x_iy_i$ is an arc in Z from a to b. Thus Z is arcwise connected and our result is established.

(11.21) COROLLARY. *Every connected subset of a dendrite or of a boundary curve (i.e., a locally connected continuum every true cyclic element of which is a simple closed curve) is arcwise connected.*

For, in the case of a dendrite, each cyclic element is a single point and hence has the property in question. For the same reason we have the following result for dendrites:

(11.3) *Any cyclicly extensible property P which is possessed by every set consisting of a single point is possessed by all dendrites.*

DEFINITION. *A continuum K is said to be unicoherent provided that however K be expressed as the sum of two continua M and N, the product set $M \cdot N$ is a continuum.*

For example, an arc or a dendrite or a circle plus its interior is unicoherent; a circle or a torus is not unicoherent.

(11.4) *The property of being unicoherent is cyclicly extensible and reducible.*

To prove the extensibility let us suppose each true cyclic element of S is unicoherent. Let $S = M + N$, where M and N are continua. Then $M \cdot N$ must be connected. For let $a, b \in M \cdot N$. Then, by (5.2), $C(a, b) = K + a + b + \sum C_i$, where K is the set of all points separating a and b and each C_i contains exactly two points of $K + a + b$. Now since $a + b \subset M \cdot N$ clearly $K \subset M \cdot N$. But since for each i, $C_i = C_i \cdot M + C_i \cdot N$, and $C_i \cdot M$ and $C_i \cdot N$ are continua and C_i is unicoherent, it follows that $C_i \cdot M \cdot N$ is connected. Then clearly $K + a + b + \sum C_i \cdot M \cdot N$ is connected, and since it is contained in $M \cdot N$ and contains $a + b$, it follows that $M \cdot N$ is connected. Thus S is unicoherent.

The property of unicoherence is also reducible. For let C be any true cyclic element of a unicoherent s.l.c. continuum S, and let $C = M + N$ where M and N are continua. Let $M_0 = M +$ all components R of $S - C$ such that $F(R) \subset M$, and let $N_0 = N +$ all components of $S - C$ which are not in M_0. Then $M_0 + N_0 = S$ and $M_0 \cdot N_0 = M \cdot N$; and since S is unicoherent the set $M_0 \cdot N_0 = M \cdot N$ is connected. Thus C is unicoherent.

We shall now deduce from a general criterion for extensibility that the properties of being (1) a regular curve, (2) a rational curve, and (3) of dimension not exceeding n, are cyclicly extensible. A continuum M is said to be (1) *regular*, (2) *rational*, (3) *of dimension not exceeding n* in the Menger-Urysohn sense provided each point of M is contained in arbitrarily small neighborhoods (rel. M) whose boundaries are (1) finite, (2) countable, (3) of dimension not exceeding $n - 1$, respectively, where the null set and only the null set is of dimension -1.

We give now the following general criterion:

(11.5) THEOREM. *Let G be any system of closed subsets of a s.l.c. continuum S such that every set consisting of a single point belongs to G. The property P of a set to be separated between each pair of its points by a set of the system G is cyclicly extensible.*

Proof. Let us suppose, in accordance with our hypothesis, that any two points of a cyclic element C are separated in C by some set of the system G. Let a and b be any two points of the whole space S. Now if a and b lie together in a cyclic element C, then by hypothesis there exists a set X of G which separates a and b in C; and by (3.22), X likewise separates a and b in S. If a and b do not lie together in a cyclic element, then they are not conjugate; and consequently there exists a point x which separates them in S. Since each set consisting of a single point belongs to the system G, it follows that in this case also a and b are separated in S by a set of G. Thus S has property P.

(11.6) *In order to apply this theorem,* let G_f be the system of all finite subsets of S, let G_c be the system of all closed and countable subsets of S and, for each n ($n = 0, 1, 2, 3, \cdots$), let G_n be the system of all closed sets of dimension less than or equal to n. For each α ($\alpha = f, c, 0, 1, 2, 3, \cdots$), let P_α be the property of a set to be separated between any two of its points by a set of the system G_α. It follows from the preceding theorem that all the properties P_α ($\alpha = f, c, 0, 1, 2, 3, \cdots$) are cyclicly extensible.

Now it is easily shown with the aid of the Borel Theorem (see Chapter V), since the systems G are additive; that every space S which has property P_α ($\alpha = f, c, 0, 1, 2, 3, \cdots$) also has the property that each of its points is contained in arbitrarily small neighborhoods whose boundaries are sets belonging to the system G_α. In other words, *the properties of being a regular curve, of being a rational curve, or of having a dimension less than or equal to n ($n = 1, 2, 3, \cdots$) are identical, respectively, with the properties P_f, P_c, P_{n-1}; and consequently these properties are cyclicly extensible.*

12. **Degree of multicoherence.**

In this section we consider a number which measures the degree of multicoherence, or non-unicoherence, and establish its cyclic additivity in locally connected continua.

DEFINITION (Eilenberg). *For a continuum X and a decomposition*

$$(\dagger) \qquad X = X_1 + X_2, \ (X_1, X_2 \text{ continua})$$

let $r(X_1, X_2)$ = the number of components of $X_1 \cdot X_2$ less one and define

$$r(X) = \text{l.u.b. } r(X_1, X_2)$$

for all decompositions (†).

(12.1) LEMMA. *In a locally connected continuum let A be an A-set containing true cyclic elements* $C_1, C_2, \cdots$; let

$$A = H + K$$

be a decomposition of A into continua such that $H \cdot K$ *intersects each node of A. For each i let* $H_i = H \cdot C_i$, $K_i = K \cdot C_i$. *Then*

$$(*) \qquad r(H, K) = \sum r(H_i, K_i).$$

Proof. If $r(H, K)$ is finite, the number of pairs of components of $H \cdot K$ is finite; and since there can be at most one C_i intersecting each set of a given pair of such components, there can be only a finite number of integers i such that

$$\text{(i)} \qquad r(H_i, K_i) > 0.$$

Thus in any case we may suppose (i) holds for only a finite number of integers i and these may be taken to be $1 \leqq i \leqq k$.

If $k = 1$, (*) follows because each component of $H \cdot K$ must contain a component of $H_1 \cdot K_1$ and no component Q of $H \cdot K$ can contain two components of $H_1 \cdot K_1$ since $Q \cdot C_1$ is connected.

Suppose we have demonstrated (*) for all values of $k < n$ and let us consider the case $k = n > 1$. There exists a division

$$A = A_1 + A_2$$

where A_1 and A_2 are A-sets and $A_1 \cdot A_2$ reduces to a single point p and where both A_1 and A_2 contain sets C_i satisfying (i). Since then both A_1 and A_2 contain less than n such sets C_i satisfying (i) it follows by our assumption that

$$\text{(ii)} \qquad \begin{aligned} r(H \cdot A_1, K \cdot A_1) &= \sum_{1}^{j} r(H_i, K_i), \qquad j < n, \\ r(H \cdot A_2, K \cdot A_2) &= \sum_{j+1}^{n} r(H_i, K_i), \end{aligned}$$

where for simplicity we have chosen the notation so that $C_i \subset A_1$ for $i \leqq j$, $C_i \subset A_2$ for $j < i \leqq n$.

Now since $p = H \cdot K \cdot A_1 \cdot A_2$ it results that the number of components in $H \cdot K \cdot A_1$ not containing p plus the number of components in $H \cdot K \cdot A_2$ not containing p equals the number of components in $H \cdot K$ not containing p. That is,

$$\text{(iii)} \qquad r(H \cdot A_1, K \cdot A_1) + r(H \cdot A_2, K \cdot A_2) = r(H, K).$$

Thus from (ii) and (iii) we get on adding equations (ii)

$$r(H, K) = \sum_{i=1}^{n} r(H_i, K_i)$$

which is (*) for the case $k = n$. Hence by complete induction we have established (*) in any case in the sense that if either side of (*) is finite the other is finite and the two are equal.

(12.2) THEOREM. *If M is a locally connected continuum with true cyclic elements $E_1, E_2, \cdots$, we have*

$$r(M) = \sum_i r(E_i).$$

Proof. Let m be any integer and suppose $\sum_i r(E_i) \geqq m$. There exists an integer k such that

$$\sum_1^k r(E_i) \geqq m.$$

Let A be the smallest A-set in M containing $\sum_1^k E_i$. Then every node of A is one of the sets E_i $(i \leqq k)$. Hence each E_i $(i \leqq k)$ contains only a finite set F_i of cut points of A.

For each $i \leqq k$, there exists a decomposition

(i) $$E_i = H_i + K_i, \ (H_i, K_i \text{ continua}),$$

such that

$$r(H_i, K_i) \geqq \min\,[m, r(E_i)].$$

Let $F_i = p_1 + p_2 + \cdots + p_s$. If p_1 does not belong to $H_i \cdot K_i$, take an arc p_1q_1 in E_i so that $p_1q_1 \cdot H_i \cdot K_i = q_1$ and add p_1q_1 to both H_i and K_i. Clearly this does not change $r(H_i, K_i)$. Similarly if p_2 does not belong to the now modified $H_i \cdot K_i$, add in an arc p_2q_2 to H_i and K_i so that $p_2q_2 \cdot H_iK_i = q_2$ and so on. Clearly, then we may suppose the decomposition (i) chosen so that

(ii) $$F_i \subset H_i \cdot K_i.$$

Now let

$$F = A - \sum_1^k E_i,$$

$$H = F + \sum_1^k H_i,$$

$$K = F + \sum_1^k K_i.$$

Then

$$A = H + K.$$

Also, since $F + \sum F_i$ is closed it results from (ii) that both H and K are compact; and since each H_i or K_i is connected it follows at once that H and K are continua. Furthermore every node of A intersects both H and K and for each i, $H_i = H \cdot E_i$, $K_i = K \cdot E_i$ $(i \leqq k)$.

Hence by the lemma,

$$r(H, K) = \sum_{j=1}^k r(H_i, K_i).$$

Thus by (i)

(iii) $$r(H, K) \geqq \min\left[km, \sum_{1}^{k} r(E_i)\right] \geqq m.$$

Now if we let

$$H_0 = H + \text{all components } Q \text{ of } M - A \text{ with } F(Q) \subset H,$$

$$K_0 = K + \text{all components } Q \text{ of } M - A \text{ with } F(Q) \subset K - H \cdot K,$$

we have

$$M = H_0 + K_0,$$

$$H_0 \cdot K_0 = H \cdot K.$$

Whence by (iii),

$$r(H_0, K_0) = r(H, K) \geqq m.$$

Accordingly we have shown that

$$r(M) \geqq \sum_i r(E_i).$$

To establish the reverse inequality, let m be any integer and suppose

$$r(M) \geqq m.$$

Let

$$M = H_0 + K_0$$

be a decomposition of M into continua so that

$$r(H_0, K_0) \geqq m.$$

Let A be the smallest A-set in M containing $H_0 \cdot K_0$ and let

$$H = H_0 \cdot A, \qquad K = K_0 \cdot A.$$

Then

$$A = H + K,$$

and every node of A intersects $H \cdot K$ since $H \cdot K = H_0 \cdot K_0$ and A is the least A-set containing $H_0 \cdot K_0$.

Hence by the lemma, if $C_1, C_2, \cdots$ are the sets E_i lying in A and $H_i = C_i \cdot H$, $K_i = C_i \cdot K$,

$$r(H, K) = \sum r(H_i, K_i).$$

Whence

$$m \leqq r(H_0, K_0) = r(H, K) = \sum r(H_i, K_i) \leqq \sum r(C_i) \leqq \sum r(E_i).$$

References

§1. Whyburn [11, 1], Moore [1], Kelley [1].

§§2–8. Whyburn [1, 11, 12, 13, 14], Kuratowski [10] and Whyburn. In connection with (3.8), (3.9) on retraction, see Borsuk [1]. In connection with (7.3) on cactoids for $m = 2$ see Moore [9]. For the case $m = 1$ of (7.31) see Ayres [2]. For (7.32) see Ważewski [1] and Gehman [2]. For related imbedding results, see Kuratowski [7], Whitney [1], Claytor [1], Adkisson [1] and MacLane.

§9. Whyburn [12, 7, 15], Ayres [3, 4]. For extensions see Rutt [1], Nöbeling [1], Zippin [3].

§10. For notion of arc-curve and related material, see Ayres [5, 6].

§11. Same references as for §§2–8.

§12. Whyburn [16]. For definitions of $r(X)$ and related material see Eilenberg [1, 2, 3]. Also see Chapters VIII and XI. For closely related material to the cyclic element theory, see also Hall [1], MacLane [1], Radó [1] and Reichelderfer, Wheeler [1], Youngs [1, 2], Harry [1], and Jones [1].

CHAPTER V

SPECIAL TYPES OF CONTINUA

1. Dendrites. A continuum M is said to be a *dendrite* or an *acyclic curve* provided it is locally connected and contains no simple closed curve. In other words, a dendrite is a locally connected continuum S every cyclic element of which reduces to a single point.

We first consider a number of properties each of which characterizes the dendrites among the continua.

(1.1) *In order that a continuum M should be a dendrite, each one of the following conditions is both necessary and sufficient:*

(i) *that every two points of M be separated by a third point;*

(ii) *that every point of M be either a cut point or an end point;*

(iii) *that every subcontinuum of M contain uncountably many cut points of M;*

(iv) *that the order and the component number* for every point of M be equal whenever either of these is finite;*

(v) *that the common part of any two connected subsets be connected.*

Proof. (i) Sufficiency: By [III, (1.51)], M is locally connected; and if M contained a simple closed curve J, then no two points of J could be separated by any single point; and thus M is a dendrite. Necessity: Let a and b be any two points of a dendrite D. By [IV, (5.2)] we have that, in D, $K + a + b = C(a, b)$, where K is the set of all points separating a and b. Thus $K \neq 0$.

(ii) Sufficiency: By the Cut Point-Order Theorem [III, (3.2)], it follows that M is regular and hence locally connected, at all save a countable number of points and, hence is locally connected. It can contain no simple closed curve, because no point of a simple closed curve is an end point and by [III, (1.53)] only a countable number of such points can be cut points. Necessity: The cyclic elements of a dendrite D are all degenerate and hence each point of D is either a cut point or an end point.

(iii) Sufficiency: M is locally connected, since by [III, (1.51)] it can have no continuum of convergence; and it can contain no simple closed curve J, because J could contain only a countable number of cut points of M. Necessity: As in the proof of (i) we have $C(a, b) = K + a + b$ in any dendrite D. Thus K is uncountable and if N is any connected subset of D and $a, b \in N$ we have $K \subset N$.

(iv) Sufficiency: Since the component number of every non-cut point of M is one, then every non-cut point of M is an end point; and hence, by (ii), M is a dendrite. Necessity: Let p be any point of a dendrite D. If the component number $\alpha(p)$ of p is finite and equal to k, then $D - p$ has k components

* The component number $\alpha(p)$ of any point p of a connected set M is the number (finite or infinite) of components of $M - p$.

$C_1, C_2, \cdots, C_k$. Since p is not a cut point of $C_i + p$ $(1 \leqq i \leqq k)$ it is an end point of this set for each i; and it follows immediately that the order of p in D is k. On the other hand, if the order of p in D is finite and equal to n, then clearly the component number $\alpha(p)$ is finite and does not exceed n. Hence by what we have just shown, $\alpha(p) = n$.

(v) Sufficiency: M is locally connected. For if not, it has a continuum of convergence K; and if G is any uncountable collection of disjoint cuttings of K, by [III, (1.51)] there exists an $X \in G$ such that $M - X$ is connected. But then $(M - X) \cdot K = K - X$ and $K - X$ is not connected, contrary to hypothesis. Clearly M contains no simple closed curve J, for we would have $J = axb + ayb$ where $axb \cdot ayb = a + b$. Necessity: Let H_1 and H_2 be arbitrary connected subsets of a dendrite D such that $H_1 \cdot H_2 \neq 0$. Now H_1 and H_2 are H-sets and hence, by [IV, (6.7)] so also is $H_1 \cdot H_2$.

Among the locally connected continua the dendrites obviously may be characterized as follows:

(1.2) *In order that a locally connected continuum be a dendrite either of the following conditions is both necessary and sufficient*:

(i) *that each cyclic element reduce to a single point* (*i.e., there are no true cyclic elements*);

(ii) *that one and only one arc exists between any two points.*

We mention in conclusion the following additional properties of dendrites:

(1.3) *For any dendrite D*:

(i) *every subcontinuum is a dendrite,*

(ii) *every connected subset is arcwise connected,*

(iii) $D = E + \sum_{i=1,2,\cdots} p_i q_i$, *where every point of E is an end point of D and the sets $p_i q_i$ are simple arcs such that* $\lim_{i \to \infty} \delta(p_i q_i) = 0$ *and* $p_n q_n \cdot \sum_1^{n-1} p_i q_i = q_n$,

(iv) *the branch points* (*i.e., points of order greater than* 2) *are countable.*

To prove (i) we have only to note that every subcontinuum of a dendrite is an A-set. Now (ii) has already been proved in [IV, (11.21)]; however it follows from [IV, (6.3)] together with the fact that every connected subset of a dendrite is an H-set. Also (iii) is a corollary to the Cyclic Chain Approximation Theorem [IV, (7.1)] since in a dendrite every cyclic chain $C(p_i q_i)$ is an arc $p_i q_i$. Finally, (iv) results from the facts that every branch point of a dendrite is a cut point and that only a countable number of the cut points of any continuum can be branch points (Cut Point-Order Theorem).

2. Hereditarily locally connected continua. A continuum M is said to be *hereditarily locally connected* provided every subcontinuum of M is locally connected.

In this section we shall develop a number of properties of such continua, most of which characterize them among continua in general.

(2.1) *In order that a continuum H should be hereditarily locally connected it is necessary and sufficient that it have no continuum of convergence.*

The sufficiency has been proved in [I, (12.2)]. To prove the necessity, suppose that a continuum H has a continuum of convergence K. Let $a, b \in K$ and let $K_1, K_2, K_3, \cdots$ be a sequence of disjoint continua in H converging to K. Now we wish to show that some subcontinuum of H is not locally connected, and hence we may suppose that H itself is locally connected. Let R be a region in H containing a but such that $\bar{R} \cdot b = 0$. There exists an integer k such that if $n \geqq k$, $K_n \cdot R \neq 0$. Let N denote the continuum $K + \bar{R} + \sum_{n=k}^{\infty} K_n$. Then N is not locally connected. For if it were, then there would exist a region V in N containing b and such that $\bar{V} \cdot \bar{R} = 0$. But then we have $\bar{V} = \bar{V} \cdot K + \bar{V} \cdot K_k + \bar{V} \cdot K_{k+1} + \cdots$, a decomposition of the continuum $\bar{V}$ into a countable number (>1) of disjoint closed sets; and this is impossible by [I, (10.3)].

DEFINITION. *A sequence $X_1, X_2, X_3, \cdots$ of sets is called a null sequence provided that* $\lim_{n \to \infty} \delta(X_n) = 0$. *Likewise a collection G of sets will be called a null collection or a null family provided that for each $\epsilon > 0$ not more than a finite number of the set G are of diameter greater than ϵ* [see IV, §3].

(2.2) COROLLARY. *In a hereditarily locally connected continuum, the components of any open set form a null sequence. Thus any collection of disjoint regions is a null collection.*

DEFINITION. *If M is any set and p is any point of M, then by the quasi-component of M containing p is meant the set consisting of p together with all points x of M such that M is not separated between p and x, i.e., there exists no separation $M = M_p + M_x$ where $M_p \supset p$ and $M_x \supset x$.*

It follows from the definition that the quasi-components of any set M are disjoint and closed in M. Also each component of M is contained in a single quasi-component. In general the quasi-components of a set may be different from the components of this set. However, in [I, (9.3)], it was shown that for compact sets, the components and quasi-components are identical; and we shall prove presently that the same is true of *any subset whatever* of a hereditarily locally connected continuum.

(2.3) *Let p be any point of a quasi-component C of an arbitrary subset N of a hereditarily locally connected continuum H, and let R be any neighborhood of p. Then there exists a neighborhood V of p such that $R \cdot C \subset V \subset R$ and $F(V) \cdot N \subset F(R) \cdot C$.*

Proof. Let E denote the collection of all quasi-components X of N different from C and such that $X \cdot F(R) \neq 0$. For each $X \in E$ there exists a separation $N = N_c + N_x$ between C and X, and if G_x denotes the set of all points of H which are nearer to N_x than to N_c, then G_x is open, $G_x \supset X$, and

(i) $$F(G_x) \cdot N = 0.$$

Since the collection $[G_x]$ covers E, by the Lindelöf Theorem there exists a countable sequence $G_1, G_2, G_3, \cdots$ of the sets G_x which also covers E. Now set $G_1 = U_1, G_2 - G_2 \cdot \bar{G}_1 = U_2, G_3 - G_3 \cdot (\bar{G}_1 + \bar{G}_2) = U_3, \cdots, G_n - G_n \cdot \sum_1^{n-1} \bar{G}_i =$

$U_n, \cdots$. Then $U_1, U_2, U_3, \cdots$ is a sequence of disjoint open sets which, by virtue of (i), covers E. Furthermore, for each n, we have

$$\text{(ii)}\qquad F(U_n) = F\left[G_n - G_n\cdot\sum_1^{n-1}\bar{G}_i\right] \subset F(G_n) + \sum_1^{n-1} F(G_i) = \sum_1^{n} F(G_i).$$

Hence, by (i), we have, for each n,

$$\text{(iii)}\qquad F(U_n)\cdot N = 0.$$

Now the set $U = \sum U_n$ is open and hence by (2.2) it follows that if $W_1, W_2, W_3, \cdots$ denote the components of U each of which contains points of E, then $[W_k]$ is a null sequence covering E. Furthermore, since for each k there exists an n such that $W_k \subset U_n$ and $F(W_k) \subset F(U_n)$, it follows by (iii) that $F(W_k)\cdot N = 0$; and hence if W denotes the set $\sum_1^\infty W_k$ we have

$$\text{(iv)}\qquad F(W)\cdot N \subset N\cdot\sum F(W_k) + N\cdot F(R) \subset F(R)\cdot C$$

because $E \subset W$. Thus if we set $V = R - R\cdot\bar{W}$, we have $R\cdot C \subset V \subset R$, since $C\cdot W = 0$, and also $F(V)\cdot N \subset F(R)\cdot C$, by (iv), since $F(V) \subset F(W) + F(R)$ and $F(R)\cdot N - F(R)\cdot C \subset E \subset W$. This completes the proof.

(2.4) *In order that a continuum H be hereditarily locally connected it is necessary and sufficient that the components and the quasi-components of any subset of H be identical.*

To prove the necessity, it suffices to show that any quasi-component of any subset N of a hereditarily locally connected continuum H is connected. Suppose, on the contrary, that some quasi-component C of such a set N is not connected. Then we have a separation $C = C_1 + C_2$, and if R denotes the set of all points of H which are nearer to C_1 than to C_2, then R is open, $R \supset C_1$, $\bar{R}\cdot C_2 = 0$, and $F(R)\cdot C = 0$. But then, applying (2.3), we obtain a neighborhood V such that $C_1 = R\cdot C \subset V \subset R$ and $F(V)\cdot N \subset F(R)\cdot C = 0$; and hence we have the separation $N = N\cdot V + N - N\cdot\bar{V}$ between C_1 and C_2, contrary to the fact that C is a quasi-component of N.

The sufficiency results at once from (2.1). For if H is not hereditarily locally connected, it has a continuum of convergence K; and if $a, b \in K$ and $K_1, K_2, K_3, \cdots$ is a sequence of disjoint continua in H converging to K and $N = a + b + \sum K_n$, clearly a and b belong to the same quasi-component but to distinct components of N.

(2.5) *In order that a continuum H be hereditarily locally connected it is necessary and sufficient that every connected subset of H be locally connected.*

The sufficiency is obvious. To prove the necessity, let Z be any connected subset of H, where H is hereditarily locally connected, let p be any point of Z and U be any neighborhood of p. Let R be a neighborhood of p such that $\bar{R} \subset U$ and let C be the component of $\bar{R}\cdot Z$ containing p. By (2.4) and (2.3) there exists a neighborhood V of p such that $R\cdot C \subset V \subset R$ and $F(V)\cdot\bar{R}\cdot Z = F(V)\cdot Z \subset F(R)\cdot C$. Then we must have $V\cdot Z \subset C$. For if not, there would exist a component C_0 of $\bar{R}\cdot Z$ different from C and which is contained wholly in V; and then applying (2.3) using the neighborhood $V = R_0$ of C_0 we would obtain a neighborhood V_0 of C_0 such that $F(V_0) \cdot Z \subset F(V) \cdot C_0 = 0$ and $V_0 \cdot C = 0$, and clearly this is impossible since Z is connected. Thus every point of $V \cdot Z$ lies together with p in the connected subset C of U and hence Z is locally connected.

(2.6) *In order that the continuum H should be hereditarily locally connected it is necessary and sufficient that the components of any subset of H form a null family.*

The condition is necessary. For let H be any hereditarily locally connected continuum and suppose, contrary to our theorem, that there exists an infinite sequence $K_1, K_2, K_3, \cdots$ of distinct components of some subset of H all of which are of diameter greater than some given positive number ϵ. Set $K = \sum K_n$. Then for each n, K_n is a component of K. By (2.4), K_n is also a quasi-component of K, for each n. Thus there exists a separation of K into two mutually separated sets H_1 and H_2 containing K_1 and K_2, respectively, and hence there exist two mutually exclusive open sets G_1 and Q_1 containing H_1 and H_2, respectively, and such that the boundary of neither of these sets contains a point of K. One of these sets, say Q_1, contains infinitely many of the sets $[K_n]$. Likewise there exists a separation of $K \cdot Q_1$ into two nonvacuous mutually separated sets L_1 and L_2, and hence there exist two mutually exclusive open subsets G_2 and Q_2 of Q_1 containing L_1 and L_2, respectively, the boundary of neither of which contains a point of K. One of these sets, say Q_2, contains infinitely many of the set $[K_n]$. There exists a separation of $K \cdot Q_2$, and so on. Continuing this process indefinitely, we obtain an infinite sequence of disjoint open sets $G_1, G_2, G_3, \cdots$, such that for each i, G_i contains some component K_{n_i} of K. But since $\delta(K_{n_i}) > \epsilon$, we have $\delta(G_i) > \epsilon$ for every i, and this is impossible by (2.2).

The sufficiency of the condition is an immediate consequence of the existence of a continuum of convergence $K = \lim [K_n]$ in any continuum which is not hereditarily locally connected, because for each i, K_i is a component of $\sum K_n$ and $\delta(K_i) >$ some $\epsilon > 0$, which contradicts our condition.

(2.7) *Every connected subset Q of an hereditarily locally connected continuum H has property S.*

Let $\epsilon > 0$ be given and for each $x \in Q$ let Q_x be the component of $Q \cdot V_{\epsilon/4}(x)$ containing x. Let $x_1 \in Q$. If there exists a point $x \in Q$, such that $Q_x \cdot Q_{x_1} = 0$, let x_2 be such a point. Then if there exists a point $x \in Q$ such that $Q_x \cdot (Q_{x_1} + Q_{x_2}) = 0$, let x_3 be such a point, and so on. Since for each i, Q_{x_i} is a component of $\sum Q_{x_i}$ and $\delta(Q_{x_i}) > \epsilon/5$, it follows by (2.6) that this process must terminate after a finite number, say n, of steps. Thus each set Q_x intersects some set Q_{x_i}, $i \leqq n$. Therefore if for each i, $(i \leqq n)$ Q_i denotes the sum of all sets Q_x such that $Q_x \cdot Q_{x_i} \neq 0$, then Q_i is connected, $\delta(Q_i) < \epsilon$, and $Q = \sum_1^n Q_i$.

(2.8) *If H is an hereditarily locally connected continuum and N is any subset whatever of H, then*

(a) *if C is any component of N and R is any open set, there exists an open set V such that $R \cdot C \subset V \subset R$ and $F(V) \cdot N \subset F(R) \cdot C$;*

(b) *if for any $p \in N$ the component of N containing p reduces to just p, then p is contained in arbitrarily small neighborhoods whose boundaries do not intersect N. (In other words, N is of dimension 0 at p.)*

Result (a) is an immediate consequence of (2.3) and (2.4) and (b) follows at once from (a).

3. **Rationality of the hereditarily locally connected continua.** In this section we shall prove that any hereditarily locally connected continuum is a rational curve. First, however, we establish two lemmas.

(3.1) LEMMA. *Let M be any subset of a separable metric space and let $R_1, R_2, \cdots$ be a fundamental sequence of neighborhoods in this space. Then there exists a countable subset D of M such that every pair of points p and q of M which belong to a neighborhood R_i and are separated in $M \cdot R_i$ by some countable set are separated also in $M \cdot R_i$ by some subset of D.*

Proof. If for an arbitrary triple of integers i, j, k, for which

$$M \cdot R_j + M \cdot R_k \subset M \cdot R_i,$$

there exists some countable set which separates the set $M \cdot R_j$ and $M \cdot R_k$ in $M \cdot R_i$, let us select some such countable set and denote it by D_{ijk}; if no such set exists, we let D_{ijk} be the null set. We shall now show that the countable set $D = \sum_{i,j,k=1}^{\infty} D_{ijk}$ satisfies the conditions of our lemma. To this end let p and q be any two points of $M \cdot R_i$ which are separated in $M \cdot R_i$ by some countable set E. We may suppose that E is closed in $M \cdot R_i$ since it contains a closed (rel. $M \cdot R_i$) subset which separates p and q in R_i. Now we have a separation $M \cdot R_i - E = M_p + M_q$, where $M_p \supset p$ and $M_q \supset q$. Then there exist neighborhoods R_j and R_k of the fundamental sequence containing p and q, respectively, and such that $M \cdot R_j \subset M_p$ and $M \cdot R_k \subset M_q$. Thus the sets $M \cdot R_j$ and $M \cdot R_k$ likewise are separated in $M \cdot R_i$ by the countable set E, and hence by definition of the set D_{ijk}, they are separated in $M \cdot R_i$ by the subset D_{ijk} of D.

DEFINITION. *A region R in a locally connected space is said to join two point sets A and B provided R contains at least one point of A and at least one point of B. Two regions R_1 and R_2 are said to be strongly separated provided that $\bar{R}_1 \cdot \bar{R}_2 = 0$.*

(3.2) LEMMA. *If N is any connected and locally connected metric space having no cut point and A and B are any two mutually exclusive nondegenerate subsets of N, then there exist two strongly separated regions in N joining A and B.*

Proof. We may suppose that A and B are closed, for obviously they contain nondegenerate closed subsets. Let p denote some point of A. Since $N - p$ is connected, there exists a region R_{ab} joining A and B and such that $p \cdot \bar{R}_{ab} = 0$;

for it is only necessary to cover $N - p$ with a set of regions no one of which contains p or has p on its boundary and then take R_{ab} = the sum of the elements of a finite connected chain of these regions joining some point a of A and some point b of B. [See II, (3.1).] Also there exists a region R_{ax} containing p and such that $\bar{R}_{ax} \cdot \bar{R}_{ab} = 0$. Thus there exist points x of N and two strongly separated regions R_{ab} and R_{ax} exist joining A and B and A and x, respectively. Let S denote the set of all such points x of N. We shall show that $S = N$. Suppose this is not so. Then since obviously S is open in N and N is connected, it follows that some point y of $N - S$ is a limit point of S. Since $N - y$ is connected, it follows just as above in the case of p that there exist two strongly separated regions G_{ab} and G such that G_{ab} joins A and B and G contains y. The region G contains a point x of S; and hence there exist two strongly separated regions R_{ab} and R_{ax} joining A and B and A and x, respectively.

Now let R_1 and R_2, respectively, denote components of $R_{ab} - \bar{G} \cdot R_{ab}$ and $R_{ax} - \bar{G} \cdot R_{ax}$ each of which contains at least one point of A. Inasmuch as N is locally connected and y does not belong to S, it follows that $\bar{G} \cdot \bar{R}_1 \neq 0 \neq \bar{G} \cdot \bar{R}_2$. Let K denote the set $A + \bar{R}_1 + \bar{R}_2$, and let U be a component of $G_{ab} - K \cdot G_{ab}$ which contains at least one point of B. At least one point f of K is a limit point of U, because N is locally connected. Let Z denote one of the sets R_1 and R_2 such that $\bar{Z}$ does not contain f and let T denote the other one of these sets. Let g denote some point of $\bar{G} \cdot \bar{Z}$. Since f cannot belong to $\bar{G}$, there exist strongly separated regions U_f and U_g containing f and g, respectively, and such that $\bar{U}_f \cdot (\bar{G} + \bar{Z}) = 0$ and $\bar{U}_g \cdot (\bar{T} + \bar{U}) = 0$. Clearly $U + U_f + T$ contains a region V joining A and B and $Z + U_g + g$ is a region V_{ay} joining A and y, and the regions V and V_{ay} have no point in common. But V contains a region V_{ab} joining A and B and such that $\bar{V}_{ab} \subset V$; for it is only necessary to cover V with regions $[V_p]$ each lying together with its boundary wholly in V; and then take V_{ab} equal to the sum of the elements of a finite connected chain of these regions $[V_p]$ joining some point a of A and some point b of B. This is impossible, because the regions V_{ab} and V_{ay} are strongly separated and join A and B and A and y, respectively, contrary to the fact that y does not belong to S. Therefore $S = M$. Accordingly S contains a point x of B, and thus there exist two strongly separated regions R_{ab} and R_{ax} joining A and B.

(3.3) THEOREM. *Every hereditarily locally connected continuum is a rational curve.*

Proof. Let H be any hereditarily locally connected continuum. Then taking H as our space and letting $M = H$, let us obtain the countable set D as given by (3.1). Then in order to prove that H is a rational curve it suffices in view of (2.8); (b), to prove that each component of $H - D$ reduces to a single point. Suppose, on the contrary, that $H - D$ has a nondegenerate component Q. By (2.5), Q is locally connected.

Now Q can have no local separating point, (i.e., no point p which is a cut point of some region relative to Q). For suppose there exists a point p of Q

and a neighborhood R of p such that p is a cut point of the component of $R \cdot Q$ containing p. Then there exists a k such that $p \subset R_k \subset R$ and thus p separates in C some two points a and b of the component C of $R_k \cdot Q$ containing p. Thus we have a separation $C - p = C_a + C_b$ and there exists an open set G such that $G \supset C_a$, $\bar{G} \cdot C_b = 0$ and $F(G) \cdot C = p$. Then since C is a component also of $R_k \cdot (H - D)$, by applying (2.3) we obtain an open set V such that $G \cdot C = C_a \subset V$ and $F(V) \cdot R \cdot (H - D) \subset F(G) \cdot C = p$. But this gives $F(V) \cdot R_k \subset D + p$. Thus $F(V) \cdot R_k$ is countable and since it separates a and b in R_k, there must exist a subset of D which likewise separates a and b in R_k. This is impossible since $D \cdot C = 0$. Thus Q has no local separating point.

Now let A and B be two mutually exclusive open subsets of Q such that $\bar{A} \cdot \bar{B} = 0$ and hence such that $\rho(A, B) > 0$. Since Q has no cut point, it follows by (3.2) that there exist two strongly separated regions R_1 and S_1 in the space Q each of which joins A and B. Since Q has no local separating point, S_1 can have no cut point. Thus by (3.2) there exist two strongly separated regions R_2 and S_2 in S_1, (S_1 considered as a space), each of which joins A and B, because $A \cdot S_1$ and $B \cdot S_1$ are nondegenerate sets. Likewise S_2 can have no cut point, and hence by the same reasoning it follows that there exist two strongly separated regions R_3 and S_3 in the space S_2 each of which joins A and B, and so on. Continuing this process indefinitely we obtain an infinite sequence of sets $R_1, R_2, R_3, \cdots$ each of which is a region in the space Q which joins A and B and such any two of these regions are strongly separated. But then for each n, R_n is a component of $\sum_1^\infty R_i$, which is impossible in view of (2.6) because for each n, R_n joins A and B and hence $\delta(R_n) \geqq \rho(A, B)$.

Thus the supposition that $H - D$ has a nondegenerate component leads to a contradiction and, accordingly, H is a rational curve.

Note. The theorem just proved holds for generalized hereditarily locally connected continua H, because obviously any locally connected generalized continuum which is not rational contains a compact subcontinuum which is not rational.

(3.4) *In order that a continuum H should be hereditarily locally connected it is necessary and sufficient that any pair of points of any subcontinum Q of H be separated in Q by a countable set which is contained in the sum of a finite number of arbitrarily small connected subsets of Q.*

To prove the sufficiency we have only to note that by the condition any two points of a subcontinuum Q of H can be separated in Q by the sum of a finite number of arbitrarily small subcontinua of Q, and hence it follows with the aid of [I, (12.1)] that Q is locally connected.

The necessity of the condition results from the fact (3.3) that any hereditarily locally connected continuum H and therefore any subcontinuum Q of H is rational. For if a and b are any two points of Q there exists a compact and countable set K which separates a and b in Q. And since Q is locally connected it follows at once by the Borel Theorem that K is contained in the sum of a finite number of arbitrarily small regions (open or closed) in Q.

4. **Regular, rational, and n-dimensional continua.** It will be recalled that a continuum or in general a set of points M is (1) regular, (2) rational, (3) at most n-dimensional (written $\dim_p M \leqq n$) at a point $p \in M$ provided that p is contained in arbitrarily small neighborhoods whose boundaries (rel·M) are (1) finite, (2) countable, (3) of dimension less than or equal to $n - 1$, where the null set and only the null set is of dimension -1. Also M is said to be *at least n-dimensional* at a point p (written $\dim_p M \geqq n$) if M is not of dimension less than or equal to $n - 1$ at p; and M is *exactly n-dimensional* at p ($\dim_p M = n$) provided we have both $\dim_p M \leqq n$ and $\dim_p M \geqq n$. The continuum M itself is regular, rational, or of dimension less than or equal to n ($\dim M \leqq n$) provided M is regular, rational, of dimension less than or equal to n, respectively, at each of its points.

(4.1) THEOREM. *If H denotes the set of all points of a compact set K where the dimension of K is greater than or equal to n ($n = 0, 1, 2, \cdots$), then for each $p \in H$, $\dim_p \bar{H} \geqq n$; and the sum of any two compact and at most $(n-1)$-dimensional sets is itself of dimension less than or equal to $n - 1$.*

Proof. Both parts of the theorem are immediate when $n = 0$. Let us assume that both parts hold for $n - 1$ and prove from this assumption that they likewise hold for n. In this manner our proof will be completed by induction. To this end we first establish the following

LEMMA. *If two points $p, q \in H$ can be separated in $\bar{H}$ by a set of dimension less than or equal to $n - 2$, they can also be separated in K by some set of dimension less than or equal to $n - 2$.*

For by hypothesis there exists a neighborhood R of p not containing q and such that $\dim \bar{H} \cdot F(R) \leqq n - 2$. For each i ($i = 1, 2, 3, \cdots$) let E_i denote the set $F(R) \cdot [\overline{V_{1/i-1}(\bar{H})} - V_{1/i}(\bar{H})]$, where $V_{1/0}(\bar{H}) = K$ by definition. Then for each i, E_i is a compact subset of $F(R) \cdot (K - \bar{H})$. With the aid of the Borel Theorem let us cover E_i by a finite number of $1/i$-neighborhoods $G_1^i, G_2^i, \cdots, G_{k_i}^i$ such that $\bar{G}_j^i \cdot \bar{H} = 0$ and $\dim F(G_j^i) \leqq n-2$ and let $G^i = \sum_{j=1}^{k_i} G_j^i$. Then since our theorem holds for $n - 1$, we have $F(G^i) \subset \sum_{j=1}^{k_i} F(G_j^i)$, so that $\dim F(G^i) \leqq n - 2$. Set $G = \sum_1^\infty G^i$. Then $F(G) \subset \sum F(G^i) + \bar{H} \cdot F(R)$. Therefore $\dim F(G) \leqq n - 2$. For if not, then since $\dim \bar{H} \cdot F(R) \leqq n - 2$, the $(n-1)$-dimensional part of $F(G)$ cannot be contained in $\bar{H} \cdot F(R)$ and thus it contains, for some k, a point x of $F(G^k)$. But since x non $\in \bar{H}$, x belongs to at most a finite number of the sets $F(G^k)$ and is not a limit point of $F(G)$ minus all those sets $F(G^k)$ which contain x. Thus $\dim_x F(G) \leqq n - 2$, contrary to the fact that x belongs to the $(n-1)$-dimensional part of $F(G)$. Therefore $\dim F(G) \leqq n - 2$. Now if $D = R - R \cdot \bar{G}$, we have $p \subset D$ and $F(D) \subset F(G) + F(R) \cdot \bar{H}$ so that $\dim F(D) \leqq n - 2$, and since $F(D)$ separates p and q in K our lemma is proved.

Returning to the proof of our theorem, let p be any point of H. Then there exist points $q \in H$ which cannot be separated from p in K by any set of dimension less than or equal to $n - 2$. For if not, then for any $\epsilon > 0$, we could cover $F[V_\epsilon(p)]$ by a finite number of open sets $G_1, G_2, \cdots, G_k$ such that for each $i \leqq k$, $\bar{G}_i \cdot p = 0$ and $\dim F(G_i) \leqq n - 2$. Thus, our theorem holding for $n - 1$, $\dim F(\sum_1^k G_i) \leqq n - 2$, and if $D = V_\epsilon(p) - V_\epsilon(p) \cdot \sum_1^k \bar{G}_i$, we have $F(D) \subset F(\sum G_i)$ so that $\dim F(D) \leqq n - 2$. And since $\delta(D) < 2\epsilon$, we have $\dim_p K \leqq n - 2$, contrary to $p \in H$. Thus there exist such points q. By the lemma, then, p and q cannot be separated in $\bar{H}$ by any set of dimension less than or equal to $n - 2$, and it follows at once that $\dim_p \bar{H} \geqq n$.

Now for the second part of the theorem, let K_1 and K_2 be two compact sets of dimension less than or equal to $n - 1$ and suppose, contrary to what we wish to show, that $\dim (K_1 + K_2) \geqq n$. Then by the first part of our theorem, now established for n, if H denotes the part of $K_1 + K_2$ where this set is of dimension greater than or equal to n, then $\dim_p \bar{H} \geqq n$ for each $p \in H$. But then H cannot be contained in $K_1 \cdot K_2$, and hence for some $p \in H$ we have say $p \in (K_1 - K_1 \cdot K_2)$, which clearly gives $\dim_p (K_1 + K_2) = \dim_p K_1 \leqq n - 1$, contrary to $p \in H$.

(4.11) COROLLARY. *The system of all compact and at most n-dimensional sets is additive, i.e., the sum of any two such sets is likewise at most n-dimensional.*

DEFINITION. *A system G of closed sets is said to be hereditary provided that any closed subset of a set of G is itself a set of G; similarly G is said to be additive provided that the sum of any two elements of G is likewise an element of G.*

(4.2) *Let G be any hereditary and additive system of closed subsets of a continuum M. Then in order that each point of M be contained in arbitrarily small neighborhoods whose boundaries belong to G it is necessary and sufficient that any two points of M should be separated in M by some set of G.*

The necessity of the condition is obvious. To prove the sufficiency, let p be any point of M and ϵ any positive number. By hypothesis for each point x of $F[V_\epsilon(p)]$ there exists an element X of G which separates p and x in M, and thus there exists a neighborhood R_x of x such that $\bar{R}_x$ does not contain p and whose boundary is a subset of X and hence is itself a set of G (since G is hereditary). By the Borel Theorem there exists a finite number $R_1, R_2, \cdots, R_k$ of the sets R_x whose sum R covers $F[V_\epsilon(p)]$. Then since $F(R) \subset \sum_1^k F(R_i)$ and G is additive and hereditary, it follows that $F(R)$ is an element of the system G. Now if $D = V_\epsilon(p) - \bar{R} \cdot V_\epsilon(p)$, we have $F(D) \subset F(R)$, so that $F(D)$ likewise belongs to G; and since D is a neighborhood of p of diameter less than 2ϵ, our theorem is established.

(4.3) *In order that a continuum M be* (1) *regular,* (2) *rational,* (3) *of dimension less than or equal to n it is necessary and sufficient that any two points of M should be separated in M by a set which is* (1) *finite,* (2) *countable,* (3) *of dimension less than or equal to* $n - 1$, *respectively.*

This is an immediate consequence of (4.2) and of the fact that the systems (1) of all finite sets, (2) of all compact and countable sets, (3) of all compact sets of dimension less than or equal to $n - 1$ are hereditary and additive.

(4.4) *Let G be any hereditary and additive system of closed subsets of a continuum M. Then for any $p \in M$ the set Q_p of all points $x \in M$ such that p and x are not separated in M by any set of G is closed and connected.*

If a set $X \in G$ separates p and x, it also separates p from every point in some neighborhood of x, and from this it follows that Q_p is closed. To prove that it is also connected, let us suppose on the contrary that for some $x \in Q_p$ we have a separation $Q_p = S_p + S_x$. Then there exists an open set R such that $R \supset S_p$, $\bar{R} \cdot S_x = 0$, and $F(R) \cdot Q_p = 0$. Now for each point y of $F(R)$, there exists some set $X \in G$ which separates p and y in M, and hence there exists a neighborhood D_y of y with $\bar{D}_y \cdot p = 0$ and $F(D_y) \in G$. By the Borel Theorem there exists a finite number $D_1, D_2, \cdots, D_k$ of these sets D_y whose sum D covers $F(R)$. Since $F(D) \subset \sum_1^k F(D_i)$, by the hereditary and additive properties of G, we have $F(D) \in G$. Then if $V = R - R \cdot \bar{D}$, we have $p \subset V \subset R$ and $F(V) \subset F(D)$, so that $F(V) \in G$. But since $F(V)$ separates p and x, this is contrary to $x \in Q_p$. Therefore Q_p is connected.

(4.5) *Each point p of a continuum M where M is not* $\left\{\begin{array}{l}\textit{regular}\\ \textit{rational}\\ \textit{of dimension less than}\\ \quad \textit{or equal to } n\end{array}\right\}$ *is contained in a nondegenerate continuum of such points.*

To prove this we have only to take the hereditary additive systems (1) of all finite sets, (2) of all closed and countable sets, (3) of all closed sets of dimension less than or equal to $n - 1$ in M and, applying (4.4), obtain the corresponding sets Q_p. Then in each case Q_p is nondegenerate, as is seen at once by the argument given to prove (4.2).

We have shown above that the sum of any two n-dimensional continua is itself n-dimensional ($n = 1, 2, \cdots$). The sum of two regular curves may or may not be regular, as is seen with the aid of simple examples. However, the sum of two rational curves is always rational. To prove this we have only to follow the same line of argument given under (4.1), letting H denote the set of all points where M is not rational; in this way we show that $\bar{H}$ is not rational at any point of H and then using the argument in the last paragraph of that proof, it is shown that the sum of two rational curves is rational. Thus we have

(4.6) *The sum of any two rational continua is rational.*

5. Classification of curves. We shall call any one-dimensional continuum a *curve*. In the preceding sections of this chapter we have characterized each of the various types of continua considered, namely, dendrites, hereditarily locally connected continua, regular curves, rational curves, and continua of dimension less than or equal to n, by means of a separation property. In view of the fact that any subcontinuum of a continuum of any of these types is itself of the same type, so that a regular curve is "hereditarily regular," a rational curve is "hereditarily rational," and so on, then by (1.1), (3.4), and (4.3) we have the following general statement:

In order that a continuum M be
- (1) *a dendrite,*
- (2) *a regular curve,*
- (3) *hereditarily locally connected,*
- (4) *a rational curve,*
- (5) *of dimension less than or equal to n,* $(n = 1, 2, \cdots)$,

it is necessary and sufficient that any two points of any subcontinuum N of M should be separated in N by a set which is
- (a) *a single point*
- (b) *finite*
- (c) *countable and is contained in a finite number of arbitrarily small subcontinua of N,*
- (d) *countable,*
- (e) *of dimension less than or equal to* $n - 1$.

Now obviously each of the properties (a)–(e) implies the following one, i.e., any set which is (a) is (b), any set which is (b) is (c), and so on. Thus if (I) denotes the class of all dendrites, (II) the class of all regular curves, (III) the class of all hereditarily locally connected continua, (IV) the class of all rational curves and, for each $n > 0$, (V_n) denotes the class of all continua of dimension less than or equal to n, we have

$$(\mathrm{I}) \subset (\mathrm{II}) \subset (\mathrm{III}) \subset (\mathrm{IV}) \subset (\mathrm{V}_1) \subset (\mathrm{V}_2) \subset \cdots \subset (\mathrm{V}_n) \subset \cdots.$$

In other words, every dendrite is a regular curve, every regular curve is hereditarily locally connected, every hereditarily locally connected continuum is a rational curve, every rational curve is 1-dimensional, and so on. Now none of the inclusions in this relation can be reversed, i.e., there exist continua in each of the classes which belong to none of the preceding classes.

References

§1. Moore [7], Menger [1, 3], Wilder [1].

§2. (2.1) Zarankiewicz [1].
(2.2)–(2.4), (2.6), (2.8) Whyburn [17].
(2.5), (2.7) Wilder [3]. In this connection see also Kline [1] and Schweigert [1].

§3. Whyburn [17]. (The lemma (3.1) is due to Nöbeling, see Menger [1].)

§4. Menger [1, 4, 5]. Urysohn [2, 3], Hurewicz [1], Alexandroff [3]. See also Rechovsky [1], Knaster [1].

§5. Whyburn [17].

CHAPTER VI

PLANE CONTINUA

In this chapter a number of results concerning continua lying in a plane or on a sphere will be developed by elementary methods. In many cases (e.g., the Jordan Curve Theorem and the Phragmén-Brouwer Theorem) the results could be formulated and proved in much more general form which would have meaning and validity in R^n. However, this would involve considerable sacrifice of simplicity; and for our purposes the two-dimensional cases will suffice.

1. **Jordan Curve Theorem.** We begin with some geometric preliminaries, the first group of which lead to the following geometric case of this celebrated theorem.

(1.1) *Every simple closed plane polygon* separates the plane into exactly two regions and is the boundary of each of these regions.*

Proof. In general we denote our simple closed plane polygon by P and the plane containing P by π. The proof will be given in the form of a sequence of statements. We assume, of course, the usual axioms of Euclidean plane geometry (see, for example, Veblen [1], R. L. Moore [10], or Forder [1]).

(i) *Every line separates the plane into exactly two "sides."*

Proof. Let a be any line and x any point of the plane not on a. Let A be a point on a and let y be any point on the line xA such that A lies between x and y on the line xy.

Suppose that a does not separate the points x and y; then there exists a broken line $xx_1x_2 \cdots x_i \cdots y$ joining x and y, which does not intersect a. Now by the axiom "if a line intersects neither AB nor AC, then it cannot intersect BC," we have at once that the line a does not intersect xx_2; and, repeating the argument a finite number of times, it follows that the line a does not intersect xy, which contradicts the fact that this line contains the point A of xy (which lies between x and y by construction). Hence a separates the plane. That there are just two "sides" of a (i.e., components of the complement of a) follows because if z is any point not on a, not both of the segments xz and yz can intersect a.

(ii) *Every angle separates the plane into exactly two "sides."*

Proof. Let the rays AB, AC construct an angle, and let p be any point on the ray AB, and r any point on the ray AC. Draw the line pr, and let x be

* A "simple closed polygon" is a finite polygon which is at the same time a simple closed curve. A *broken line* or *path* consists of a finite number of distinct points (vertices) x_1, x_2, $\cdots$, x_n together with the open segments x_1x_2, x_2x_3, $\cdots$, $x_{n-1}x_n$ (edges) which are non-overlapping.

any point between p and r on this line, and y a point on this line such that p lies between x and y on the line xy; i.e., x and y lie on opposite sides of the line determined by AB. Draw any broken line from x to y which does not intersect the ray AB, and let q be the first intersection of this broken line with the *line* AB (such a point must exist since x and y lie on opposite sides of the line AB). Since the broken line xy did not intersect the ray AB, it follows that A must separate the points p and q in the line AB, that is, p and q must lie on opposite sides of the *line* AC. Consequently, the broken line xq must intersect the ray AC since xq lies on the same side of the line AB as does the ray AC, and no portion of the line AC other than this ray is contained on this side of AB. Thus the angle A must separate the plane, as was to be proved. Again it is immediate that A can have but two "sides."

(iii) *Every line, not through a vertex of a polygon P, meets P in an even number of points.*

Proof. Let $P \equiv A_1A_2 + A_2A_3 + \cdots + A_rA_1$, where A_i $(1 \leqq i \leqq r)$ are the vertices of the polygon P, and $A_{i-1}A_i$ are its edges. Then if a line L meets $A_{i-1}A_i$, it follows that A_{i-1} and A_i lie on opposite sides of L. Consequently, by induction, if L met P an odd number of times then the A_1 of A_1A_2 would lie on the opposite side of L from the A_1 of A_rA_1, which is impossible since they are the same point.

(iv) *If A is an angle whose vertex is not on the polygon P, and whose rays do not pass through a vertex of P, then A meets P in an even number of points.*

Proof. If A meets an edge $A_{i-1}A_i$ in one point, then A_{i-1} and A_i lie on opposite "sides" of the angle A; but if there are zero or two points of intersection, then these points lie on the same side of A. The result will now follow as in (iii).

(v) COROLLARY TO (iv). *If x is a point not on a polygon P, and if some ray from x which does not pass through a vertex of P meets P in an odd number of points, then so also does every other ray from x which contains no vertices of P.*

(vi) *Every polygon P separates the plane.*

Proof. Let AB be any line intersecting P but containing no vertices of P. From any point on AB proceed in either direction on AB until the last point of intersection of AB and P is encountered, and call this point Q. We may suppose A on the open ray from Q which does not intersect P and B on the other one; then the ray QA does not intersect P except in Q. Then by (iv), the ray QB meets P in an odd number of points distinct from Q. Let x be any point between Q and the next intersection of QB with P, and let y be any point of BA such that Q lies between x and y on BA.

Suppose that there exists a broken line $xx_1x_2 \cdots y$ from x to y which does not intersect P. Then we may adjust the ray xx_1 so that it does not pass through a vertex of P. Then, by (v), ray xx_1 must intersect P in an odd number of points, since xB does. By induction we arrive at the conclusion that ray yB must intersect P in an odd number of points, which contradicts (iii) and proves our assertion.

(vii) *No broken line separates the plane.*

For let S be the set of all points x on the broken line AB such that Ax does not separate p and q, where p and q are any two points of $\pi - AB$. Then S is open and also closed. For let y be any limit point of S. There exists a circle C with center y not enclosing either p or q and meeting Ay only in the edge which contains y and therefore only in one point x. Now $\pi - Ax$ contains an arc pq, and pq contains arcs pw and qz meeting C only in w and z, respectively. Then $pw + qz +$ (arc wz of C not containing x) is an arc pq in $\pi - Ay$.

(viii) *P is the boundary of every component of $\pi - P$.* [Corollary to (vii).]

(ix) *There are only two components of $\pi - P$.*

For let x be a point of P, not a vertex, and let C be a circle about x enclosing no vertex of P and meeting P only in the edge containing x, and therefore in only two points a and b. Then every component of $\pi - P$ must meet $C - (a + b)$ and hence contain a component of $C - (a + b)$. But there are only two components of $C - (a + b)$, and hence of $\pi - P$.

DEFINITION (Alexander). *An even chain is a closed set C which is made up of a finite number of non-intersecting edges, which are open segments, and vertices such that each vertex is on an even number of edges and such that each end point of an edge of C is a vertex of C.*

(1.2) LEMMA (Alexander). *If the vertices x and y of an even chain C are joined in C by a broken line L, then if we omit from C all (open) edges lying in L, x and y can still be joined by a broken line in the rest of C.*

Proof. Let K be the set of all points which can be so joined to x by a broken line lying in $r(C)$, the rest of C. Then every vertex of C in K except x and y is incident to an even number of edges in $r(C)$, and each of these edges is in K. The vertex x, however, is incident to an odd number of such edges. Now the total number of incidences of edges on vertices in K must be even (equal to twice the number of edges in K). Therefore K must contain some other vertex of C which is incident to an odd number of edges in $r(C)$. But y is the only such vertex different from x. Therefore $y \subset K$ and the lemma is proved.

Notation. If xy is an arc, $\widehat{xy}$ will denote the open arc $xy - (x + y)$.

(1.3) AUXILIARY THEOREM. *No simple arc separates the plane.*

Proof. Let ab be any arc in the plane, and let p and q be any two points in the plane not on ab. Since a does not separate p and q in the plane, there exist points x on ab such that ax does not separate p and q. Clearly the set S of all such points x is open in ab. It is also closed. For, suppose, on the contrary, that some point y of $ab - S$ is a limit point of S. Since p and q lie in circles not meeting ab, clearly we may suppose that y is not on the line L through p and q, for if this were the case we could replace p by a point p' near p such that the is not the case. Let w be the point of ay such that $wy \cdot L = w$. Let Q be a circle with center y and radius r which does not meet $L + aw$. Choosing y as origin, let $T(\theta)$ be the triangle with vertices (r, θ), $(r, \theta + 2\pi/3)$, and $(r, \theta - 2\pi/3)$. Let J be a circle with center y and radius small enough so that J lies entirely within $T(\theta)$ for every θ. Now since y is a limit point of S, there exists a point x of S such that the arc xy of ab lies within J. Since ax does not separate p and q, there exists a broken line A from p to q not meeting ax. This line A contains broken lines $A(p, u)$ and $A(q, v)$, respectively, such that $A(p, u) \cdot xy = u$, $A(q, v) \cdot xy = v$. Connect u and v by a broken line α lying so close to uv that it does not meet $ax + L$. Now if p' and q' are the last points on $A(p, u)$ and $A(q, v)$, respectively, which belong to pq of L, then $A(p, u) + A(q, v) + \alpha$ contains a broken line B from p' to q' such that $B \cdot pq = p' + q'$, and therefore $B + p'q'$ (of L) is a simple closed polygon P. Now by a suitable choice of θ (indeed, for all save a finite number of θ's) we have that $T(\theta) = T$ is a triangle T such that no vertex of P is on T and no vertex of T is on P. Let R denote the region obtained by adding to the interior of T that complementary region D of P which contains x, and let C be its boundary.

Then C is an even chain. For clearly C is made up of non-intersecting edges and vertices. Let z be any vertex of C. If z is a vertex of either P or T it is not on the other, and hence is incident to exactly two edges of C. If z is not a vertex of either P or T, it is on an edge π of P and an edge t of T. Thus we have exactly four segments of $T + P$ meeting in z, and of these one is within T and one within D, so that exactly two are in C. Thus in any case z is on exactly two edges of C and therefore C is a chain.

Now since $C \supset p'q'$ (of L) it follows by the lemma that $C - \widehat{p'q'}$ contains a broken line K joining p' and q'. Since $P - \widehat{p'q'} = B$, we have $K \subset B + T$. Since $xy \subset R$ and $B \cdot ax = 0$, we have $ay \cdot K \cdot B = 0$; and since $\widehat{wx} + xy \subset R$ and $aw \cdot T = 0$, we have $ay \cdot K \cdot T \subset ay \cdot C \cdot T = 0$. These together give $ay \cdot K = 0$. Therefore $A(p, u) + K + A(q, v)$ is connected and joins p and q but does not meet ay. Thus ay cannot separate p and q and hence y belongs to S, contrary to supposition. Therefore S is both open and closed in ab. But then $S = ab$, and hence ab does not separate p and q.

DEFINITION. *Any simple closed curve containing a straight line segment will be called a semi-polygon.*

(1.4) AUXILIARY THEOREM. *Any semi-polygon separates the plane π into exactly two regions.*

Proof. Let S be any semi-polygon containing a straight segment aob of a line L and an arc axb with $aob \cdot axb = a + b$. Let C be a circle with center o which neither contains nor encloses any point of axb, and which therefore meets S in only two points u and v of aob. Let C_1 and C_2 be the two arc segments of $C - (a + b)$. Then no broken line in $\pi - S$ joins points of both C_1 and C_2. For suppose such a line K joins a point p_1 of C_1 and a point p_2 of C_2 and lies otherwise without C. Then $K + p_1p_2$ is a polygon P meeting S in only one point, and that point is within C and hence on aob. But P separates a and b and hence must meet axb. Therefore $\pi - S$ is not connected. Now by (1.3), S

must be the boundary of each component of $\pi - S$. Thus each component of $\pi - S$ must meet $C - (a + b)$ and hence must contain either C_1 or C_2. Therefore there are just two such components.

(1.5) JORDAN CURVE THEOREM. *Every simple closed curve J divides the plane π into just two regions and is the boundary of each of these regions.*

Proof. There exists a straight line segment ab meeting J in exactly the two points a and b. Let ax_1b and ax_2b be the two arcs of J from a to b, and denote by S_1 and S_2, respectively, the semi-polygons $ab + ax_1b$ and $ab + ax_2b$. By (1.4), $\pi - S_1$ and $\pi - S_2$ are composed of two regions each, say $R_1 + D_1$ and $R_2 + D_2$, respectively, where $D_1 \supset ax_2b$ and $D_2 \supset ax_1b$. Now since $\pi - (S_1 + S_2) = R_1 + R_2 + D_1 \cdot D_2$, we have $\pi - J = (R_1 + R_2 + \widehat{ab}) + D_1 \cdot D_2$; and we thus have a division of $\pi - J$ into two nonvacuous sets. ($D_1 \cdot D_2 \neq 0$, because $D_1 \supset x_2$ and x_2 is a boundary point of D_2.) These two sets are also mutually separated. For $D_1 \cdot D_2$ is open, as is also $R_1 + R_2$; and no point of $\widehat{ab}$ is a limit point of $D_1 \cdot D_2$, because if so then there exists a straight segment po where $o \subset \widehat{ab}$ and $po - o \subset D_1 \cdot D_2$; and if C is a circle with center o and not enclosing any point of $p + J$, then C meets ab in two points u and v and each of the three sets R_1, R_2, and $D_1 \cdot D_2$ must contain components of $C - (u + v)$; but there are only two such components, and hence we have $(R_1 + R_2) \cdot (D_1 \cdot D_2) \neq 0$, which is impossible. Therefore $D_1 \cdot D_2$ and $R_1 + R_2 + \widehat{ab}$ are mutually separated, and thus $\pi - J$ is not connected.

By (1.3) it follows that every component of $\pi - J$ has J for its boundary. Let G_1 and G_2 be two components of $\pi - J$ and let a, b, x, and y be points on J such that arcs aob and xqy of J exist which have no common points. Then $G_1 + J$ and $G_2 + J$ contain broken lines A_1X_1 and A_2X_2, respectively, where $A_1 + A_2 \subset aob$, $X_1 + X_2 \subset xqy$, and $A_1X_1 \cdot J = A_1 + X_1$, $A_2X_2 \cdot J = A_2 + X_2$. Then A_1X_1 + (arc A_1A_2 of aob) + A_2X_2 + (arc X_1X_2 of xqy) is a semi-polygon S; and, by (1.4), $\pi - S = E_1 + E_2$. But both E_1 and E_2 must contain points of J, since each contains points both of G_1 and of G_2. Thus there is no third component of $\pi - J$, for if there were it would have to meet S which is impossible, since $S \subset G_1 + G_2 + J$.

DEFINITION. *By a θ-curve will be meant a continuum which is the sum of three simple arcs axb, ayb, azb intersecting by pairs in just their end points.*

As a direct consequence of (1.5) we get

(1.6) *A θ-curve separates the plane π into just three regions.*

Proof. Let $\theta = axb + ayb + azb$, $axb + ayb = J_1$, $axb + azb = J_2$, and $ayb + azb = J_3$. For each i ($i = 1, 2, 3$) let R_i be the complementary region of J_i which does not contain the segment $\theta - J_i$. Then $\pi - \theta = R_1 + R_2 + R_3$. For if, on the contrary, $D_1 \cdot D_2 \cdot D_3 \neq 0$, where $D_i = \pi - J_i - R_i$, then any component D of $D_1 \cdot D_2 \cdot D_3$ has all of θ as its boundary. Now $D + x' + y'$ contains an arc $x'y'$, and $R_1 + x'' + y''$ contains an arc $x''y''$ where $x', x'' \in \widehat{axb}$, $y', y'' \in \widehat{ayb}$. Set $J = x'y' + x'x'' + x''y'' + y''y'$

and $\pi - J = U + V$ where $U \supset a$. Then V must contain a point p of J_1 since it contains points of both R_1 and $\pi - R_1$. Say $p \subset axb$. But then since $p + a \subset axb \subset J_2 = F(R_2)$, J must contain a point of R_2, contrary to the fact that $J \subset D + \theta + R_1$.

Note. If axb, ayb, azb are the *edges* and R_1, R_2, R_3 the *regions* of a θ-curve θ, and if we understand that "an edge is *on* a region" means it is on the boundary of that region and "a region is *on* an edge" means it has the edge as a part of its boundary we get at once the following dual conclusions:

(1.7) I. *Each edge of θ is on exactly two regions.*

I′. *Each region of θ is on exactly two edges.*

II. *Each pair of edges of θ is on exactly one region.*

II′. *Each pair of regions of θ is on exactly one edge.*

2. **Phragmén-Brouwer Theorem. Torhorst Theorem.** It is understood throughout this section that all sets considered lie in a plane which is usually denoted by π. If K is a closed set, by a *complementary domain* of K is meant a component of the complement of K, i.e., of $\pi - K$.

(2.1) PHRAGMÉN-BROUWER THEOREM. *The boundary of every bounded complementary domain R of a closed generalized continuum N is itself a continuum.*

Proof. For $\epsilon_1 < 1$ take a hexagonal division D_1 of the plane of norm less than ϵ_1. Let H_1 be the sum of all the subdivisions in D_1 lying wholly in R. Let K_1 be any component of H_1. Assuming that D_{n-1} has been constructed, let $\epsilon_n = \rho(K_{n-1}, F(R))$, and construct a hexagonal division D_n of the plane of norm less than min $(\epsilon_n, 1/n)$. Let H_n be the sum of all subdivisions of D_n which lie wholly in R, and let K_n be the component of H_n containing K_{n-1}. Make this construction for all n, and let $F_n = F(K_n)$.

Then for each integer n, F_n is a continuum. For, firstly, the complementary system to K_n is connected. To show this let h' and h'' be any two hexagons in the complementary system to K_n, and join h' to h'' by a broken line $L = pxq$, where $p \in h'$, $q \in h''$. If L does not intersect K_n, our assertion is proved. Otherwise, L must intersect some hexagon containing boundary points of R before it meets K_n; for, if the contrary were the case, then h' would lie in K_n. Thus h', and similarly h'', lies in a connected subset of the complementary system to K_n with N, so that this complementary system is connected. Secondly, F_n is the common boundary of two regions G_1 and G_2. To prove this it will suffice to show that for any two points p, q in K, p and q can be joined in K_n by a broken line not intersecting F_n; since precisely the same argument will apply if p and q are in the complementary system to K_n. Since $p, q \in K_n$, they may be joined in K_n by a chain composed of the hexagonal divisions comprising K_n, and every consecutive pair of which have a side in common. It is then a simple matter to construct the required broken line from p to q by going from hexagon to hexagon in the chain through their common side, which, evidently, cannot lie in F_n. Thirdly, F_n is an even chain in the sense of Alexander. For, consider any hexagon of D_n which has a vertex p on F_n. Then p lies on three edges of three hexagons h, h', h'' of D_n. Consequently, since F_n is the common boundary of two regions G_1 and G_2, one of these regions, say G_2, must contain two of these three hexagons, say h' and h''. Consequently, since $p \in F_n$, G_1 contains h; therefore, the two edges of h which are incident to p lie in F_n. Since h' and h'' lie in G_2, their common side must also lie in G_2; whence p is incident to only two edges of F_n, namely the two in h which meet at p. Since the number of possible edges of F_n is finite by the nature of the subdivision, it follows that F_n is an even chain. Finally, F_n is a polygon and hence a continuum. For, let e be any edge of F_n and a, b, its vertices. Then by (1.2) there exists a broken line L in $F_n - e$ from a to b, so that $e + L$ is a simple closed polygon P in F_n; hence $\pi - P = R_1 + R_2$. Consequently, we must have either $G_1 \subset R_1$ and $G_2 \subset R_2$ or $G_1 \subset R_2$ and $G_2 \subset R_1$. In either case $F(G_1) \cdot F(G_2) = F_n \subset P$, and thus $F_n = P$, a polygon, as we asserted.

But, by construction, for a sufficiently large n we have $H_k \subset K_n$. Thus $F(R) = \lim F_n$, so that $F(R)$ is a continuum, as was to be proved.

(2.11) COROLLARY. *Any compact set which is the common boundary of two domains is a continuum.*

Since one of the domains D_1 and D_2, say D_2, is bounded, we have only to note that D_2 is a complementary domain of $\bar{D}_1$.

(2.12) *If the boundary F of a complementary domain of a generalized continuum is bounded, F is a continuum.*

(2.2) TORHORST THEOREM. *The boundary F of every complementary domain R of a locally connected continuum M is itself a locally connected continuum.*

Proof. For if F is not locally connected, then it contains a continuum of convergence $K = \lim (K_n)$, where $K_m \cdot K_n = 0$, $m \neq n$, and for some $\epsilon > 0$ we have $\delta(K_n) > 8\epsilon$ for every n. Now since M is locally connected and thus has property S, we have $M = \sum_1^n M_i$, where each M_i is a locally connected continuum of diameter less than ϵ. Now each K_n must meet each set of some pair A, B of sets (M_i) which have no common points $(A \cdot B = 0)$. Since there are only a finite number of disjoint pairs A, B in (M_i), it follows that for some such pair A, B of sets M_i there exist three of the sets (K_n), say K_1, K_2, K_3, each of which meets both A and B. Now take $\sigma > 0$ such that $8\sigma < \min [\rho(K_1, K_2), \rho(K_2, K_3), \rho(K_3, K_1)]$ and consider the regions $R_\sigma(K_i) = \sum R_\sigma(x)$, $x \in K_i$ in M $(i = 1, 2, 3)$ [$R_\sigma(x)$ is the component of $V_\sigma(x_i) \cdot M$ containing x]. These clearly are disjoint and contain arcs a_ib_i, respectively, such that $a_ib_i \cdot A = a_i$, $a_ib_i \cdot B = b_i$ $(i = 1, 2, 3)$. A contains an arc a_1a_2 and an arc a_3a, where $a_3a \cdot a_1a_2 = a$ (a may be a_3) and similarly B contains an arc b_1b_2 and an arc b_3b where $b_3b \cdot b_1b_2 = b$. Then

$$\begin{aligned} a_1b_1 + a_2b_2 + a_3b_3 + a_1a_2 + a_3a + b_1b_2 + b_3a \\ = ax_1b + ax_2b + ax_3b \\ = \theta, \text{ a } \theta\text{-curve,} \end{aligned}$$

where x_i is on a_ib_i and $\rho(x_i, A) > \sigma$ and $\rho(x_i, B) > \sigma$ $(i = 1, 2, 3)$. But since x_i lies in $R_\sigma(K_i)$ it can therefore be joined to some point of K_i by a region of diameter less than σ and hence not meeting $\theta - ux_ib$. But then if $\pi - \theta = R_1 + R_2 + R_3$, where $R_3 \supset R$, we have that $F(R_3)$ meets all three edges of θ, which is impossible.

(2.21) *If the boundary B of a complementary domain of a locally connected generalized continuum is bounded, B is a locally connected continuum.*

(2.3) *Every true cyclic element of the boundary B of a complementary domain of a locally connected continuum is a simple closed curve.*

Proof. By the Torhorst Theorem, B is locally connected. If C is a cyclic element of B, then since B can contain no θ-curve it follows that C is a simple closed curve.

(2.31) COROLLARY. *B is a regular curve.*

For, by (2.3), each of its cyclic elements is regular.

(2.4) LEMMA. *No acyclic curve separates the plane.*

Proof. For if so, then some acyclic curve A is the common boundary of two domains D_1 and D_2. But since A separates the plane it cannot be an arc, so that it must contain a triod $ao + bo + co$. Now construct a θ-curve $\theta = a'ob' + a'xb' + a'yb'$, where $a'ob' \subset ao + bo$, $a'xb' \subset D_1 + ao + bo$, $a'yb' \subset D_2 + ao + bo$. By (1.6), $\pi - \theta = R_1 + R_2 + R_3$. One of these, say R_1, contains $co - o$. Then if $F(R_1) = a'ob' + a'xb'$, then $D_2 \cdot R_1 \neq 0$; but also $D_2 \cdot (\pi - R_1) \neq 0$, since $D_2 \supset a'yb'$. This is impossible, since $D_2 \cdot F(R_1) = 0$.

(2.5) *If M is a cyclicly connected locally connected generalized continuum, the boundary B of any complementary domain R of M is a simple closed curve provided B is bounded.*

Proof. Since, if $B \neq M$, B separates the plane and M is cyclicly connected, it follows that in any case B contains a simple closed curve C [which, by (2.3), is a cyclic element of B]. Then $B = C$. For, if $B - C$ contains a point p, let px be an arc in B with $(px) \cdot C = x$ and let py be an arc in $M - x$ with $(py) \cdot C = y$. Then $px + py$ contains an arc xqy, where q lies in B. Let xay and xby be the arcs of C from x to y and set $\theta = xqy + xay + xby$, $\pi - \theta = R_1 + R_2 + R_3$, where $R_1 \supset R$. But since $F(R) \supset q + a + b$, whereas $F(R_1) =$ two of the arcs xqy, xay, xby, we have a contradiction; and hence B is a simple closed curve.

A connected set having no cut point will be said to be *cyclic.*

(2.51) *If N is a locally connected continuum in a plane π and a and b are points of π lying in different complementary domains R_a and R_b, respectively, of N, there exists a simple closed curve J in N separating a and b in π.*

For either R_a or R_b, say R_b, is bounded. Then if $M = \bar{R}_b$ it results at once that M is a cyclic locally connected continuum. Thus the boundary J of the complementary domain of M containing a is a simple closed curve and clearly J separates a and b.

(2.6) THEOREM. *If $J_1, J_2, \cdots, J_n$ are simple closed curves in a plane or sphere S and for each i, I_i and E_i are the regions on S complementary to J_i and $p \in \prod_1^n I_i \neq 0 \neq \prod_1^n E_i$, the boundary B of the component R of $S - \sum_1^n J_i$ containing p is a simple closed curve.*

Proof. Let $J = \sum J_i$, $E = \sum E_i$, $C = J + E = \sum (J_i + E_i)$. Then C is a locally connected generalized continuum. Furthermore, C is cyclic, since for any $x \in J$, each of the sets $(J_i + E_i + x) - x$ is connected and contains $\prod E_i$ and their sum is $C - x$. Hence the boundary B of the component R_0 of $S - C$ containing p is a simple closed curve (since $B \subset J$). Obviously $R_0 = R$.

(2.61) COROLLARY. *If $J_1, J_2, \cdots, J_n$ are simple closed curves in a plane π, $J = \sum J_i$, and for each i, I_i and E_i are the interior and exterior, respectively, of J_i and $p \in \prod_1^n I_i$, J contains a simple closed curve C which encloses p but encloses no point of J (thus the interior of $C \subset \prod E_i$) and also a simple closed curve D which encloses all points of $\sum_1^n I_i$ (thus the exterior of $D \subset \prod E_i$).*

3. Plane Separation Theorem. Applications.

(3.1) SEPARATION THEOREM. *If A is compact, B is closed, $A \cdot B = T$ is totally disconnected and a, b are points of $A - A \cdot B$ and $B - A \cdot B$, respectively, and ϵ is any positive number, then there exists a simple closed curve J which separates a and b and is such that $J \cdot (A + B) \subset A \cdot B$, and every point of J is at a distance less than ϵ from some point of A.*

Proof. For each point x of $A - A \cdot B$ let C_x be the circle with center x and radius less than min $[\epsilon/2, 1/2\, \rho(x, B)]$, and let I_x be the interior of this circle. Let H_1 be the set of all points x of A with $\rho(x, B) \geqq 1$, and for each $n > 1$, let H_n be the set of all points x of A with $1/n \leqq \rho(x, B) \leqq 1/(n - 1)$. Since for each n, H_n is compact, it follows that for every n there exists a finite number of the sets (I_x) with centers in H_n whose sum $K_n = \sum_{i=1}^{m_n} I_{x_i}^n$ covers H_n. Set $K = \sum_1^\infty K_n$, and let Q be the component of K which contains a. Then Q has no cut point, for every point of Q is an inner point of a circle lying in Q. Thus $\bar{Q}$ also has no cut point. Also $F(Q) \cdot (A + B) \subset A \cdot B$ and $Q \cdot B = 0$. Furthermore $\bar{Q}$ is locally connected, since $A \cdot B$ is totally disconnected and any point p of $\bar{Q}$ not in $A \cdot B$ is a point or a limit point of only a finite number of sets (I_n) lying in Q. Thus $\bar{Q}$ is a cyclicly connected locally connected continuum containing a in its interior and such that $\bar{Q} \cdot B \subset A \cdot B$. Let J be the boundary of the complementary domain of $\bar{Q}$ containing b. Then, by (2.5), J is a simple closed curve. Clearly J separates a and b, and since $J \subset F(Q)$ we have

$$J \cdot (A + B) \subset F(Q) \cdot (A + B) \subset A \cdot B.$$

Finally, since every point of $F(Q)$ is at a distance less than $\epsilon/2$ from some point of A we have the same property for J.

(3.11) Corollary (Zoretti Theorem). *If K is a component of a compact set M and ϵ is any positive number, then there exists a simple closed curve J which encloses K and is such that $J \cdot M = 0$, and every point of J is at a distance less than ϵ from some point of K.*

For, by [I, (9.3)], there exists a separation of M into two mutually separated sets A and B_0, where $A \supset K$, and every point of A is at a distance less than $\epsilon/2$ from some point of K. Let r be a ray emerging from some point b such that $\rho(r, K) > 2\epsilon$. Set $B = B_0 + b$, and apply (3.1) obtaining the curve J every point of which is at a distance less than $\epsilon/2$ from A. Then J cannot enclose b, since $J \cdot r = 0$ and every point of J is at a distance less than $\epsilon/2 + \epsilon/2 = \epsilon$ from some point of K.

(3.12) Corollary. *If p is a regular point of a compact set K, then for each $\epsilon > 0$ there exists a simple closed curve J enclosing p with $\delta(J) < \epsilon$ and such that $J \cdot K$ is finite and of power less than or equal to the order of p in K.*

(3.2) *If the boundary F of a bounded domain R is a locally connected continuum, then R has property S.*

Proof. If, on the contrary, R does not have property S, then there exists a positive number d and an infinite sequence $P \equiv (p_1, p_2, \cdots)$ of points of R converging to a point p of F and such that no two of these points lie together in any connected subset of R of diameter less than d. Now, by (2.31), F is a regular curve. Hence, taking $\epsilon = d/16$, there exists in B an ϵ-neighborhood U of p whose boundary K is finite. Let r be a ray emerging from b where $\rho(r, F) > d$. Set $A = \bar{U} = U + K$ and $B = F - U + b$ so that $A \cdot B = K$; take $a = p$, and apply (3.1), obtaining a simple closed curve J separating p and b and such that $J \cdot F \subset K$ and every point of J is at a distance less than ϵ from some point of U. Then, since $J \cdot r = 0$, J must enclose p, and as $\delta(U) < \epsilon$, we have $\delta(J) < 2\epsilon$. Now for each point x of P lying within J there exists an arc xy in R such that $(xy) \cdot J = y$. Then since $J - J \cdot F = s_1 + s_2 + \cdots + s_k$ where each s_i is an open arc segment, and since J encloses infinitely many points (x) of P, it follows that for some i, $(1 \leqq i \leqq k)$ there exist two points x_1 and x_2 of P within J such that the corresponding points y_1 and y_2 both belong to s_i. But then $x_1y_1 + s_i + x_2y_2$ is a connected subset of R containing two points of P and since $\delta(x_iy_i) \leqq \delta(J) < 2\epsilon = d/8$, and $\delta(s_i) \leqq \delta(J) < d/8$, we have $\delta(x_1y_1 + s_i + x_2y_2) < d$, contrary to the definition of P. Therefore, R has property S.

(3.3) *If the boundary of a region R is a simple closed curve C, then R is uniformly locally connected.*

Proof. For if not, then there exists a $d > 0$ and two sequences $X \equiv x_1, x_2, \cdots$ and $Y \equiv y_1, y_2, \cdots$ of points in R each converging to one and the same point p of C and such that for no n do x_n and y_n lie together in any connected subset of R of diameter less than d. Take $\epsilon = d/16$. Then since p is a point of order 2 of C, by (3.12), there exists a simple closed curve J enclosing p, with $\delta(J) < \epsilon$ and $J \cdot C = a + b$, two points. Now $J - (a + b) =$

$\widehat{axb} + \widehat{ayb}$, where $\widehat{axb}$ and $\widehat{ayb}$ are arc segments one of which, say $\widehat{axb}$, lies within C and the other without C. Now for some n, say $n = k$, x_k and y_k both lie within J. But then if $x_k u$ and $y_k v$ are arcs in R with $(x_k u) \cdot J = u$ and $(y_k v) \cdot J = v$, we have $u + v \subset \widehat{axb}$ and hence $x_k u + \widehat{axb} + y_k v$ is a connected subset of R of diameter less than d containing x_k and y_k, contrary to the definitions of X and Y.

(3.31) COROLLARY. *If p is any point on a simple closed curve C, then for each $\epsilon > 0$ there exists an arc apb of C and an arc axb such that axb is within (without) C and $\delta(apb + axb) < \epsilon$.*

(3.32) COROLLARY. *If o is any inner point of an arc aob and ϵ is any positive number, then there exists an ϵ-simple closed curve J enclosing o and meeting aob in exactly two points u and v, where we have $u \subset ao$, $v \subset ob$.*

(3.4) *No arc or semi-open arc separates any plane region.*

Proof. Let R be any plane region and let uv be an arc such that $uv - v \subset R$. Clearly it suffices to show that $R - \alpha$ is connected where $\alpha = uv \cdot R$. Let p and q be any two points of $R - \alpha$ and let pq be an arc in $\pi - \alpha$ from p to q. Now choosing $A = uv$, $B = F(R) + pq$, $a = u$, $b = p$ and applying (3.1) (which we can do since $A \cdot B \subset v$) we obtain a simple closed curve J which separates p and u and hence separates $p + q$ and α, since pq and α are connected, and such that $J \cdot (A + B) \subset v$. Thus we have $(J - J \cdot v) \subset R$, since $J - J \cdot v$ is connected and must meet R but does not meet $F(R)$. Now there exist arcs px and qy in R with $px \cdot J = x$ and $qy \cdot J = y$. Thus we have $[px + qy] \cdot \alpha = 0$ and hence $px + qy + J - J \cdot v$ is a connected subset of $R - \alpha$ containing $p + q$. Accordingly $R - \alpha$ is connected.

(3.41) COROLLARY. *If X is the sum of a finite number of disjoint arcs each lying in R except possibly for one end point, then $R - R \cdot X$ is connected.*

(3.5) *If R is a plane region with connected boundary $F(R)$ and ab is a simple arc such that $ab - (a + b) = \alpha \subset R$ while $a + b \subset F(R)$, then $R - \alpha$ is the sum of exactly two disjoint regions.*

Proof. Let o be any point of α. By (3.12) there exists a simple closed curve J lying wholly in R and inclosing o and meeting aob in exactly two points u and v, where $u \subset ao$, $v \subset ob$. Let J_1 and J_2 be the two arc segments of $J - (u + v)$ from u to v. Then J_1 and J_2 are in different components of $R - \alpha$. For if not, then there exists an arc xwy in $R - \alpha$ such that $xwy \cdot J = x + y$ and $x \subset J_1$, $y \subset J_2$. Then $xwy + J = xwy + xuy + xvy = \theta$, where θ is a θ-curve; and $\pi - \theta = R_1 + R_2 + R_3$, where R_1 is the interior of J and therefore contains o. Since $F(R)$ is connected and $F(R) \cdot \theta = 0$ and $F(R) \cdot R_1 = 0$ by the definition of J, we have $a + b \subset F(R) \subset R_2$ or $a + b \subset F(R) \subset R_3$, say the former. But this is impossible, because $au - u + bv - v \subset R_2$ and thus $F(R_2) = J$ which is incompatible with $F(R_1) = J$. Therefore $R - \alpha$ has at least two components (one containing C_1 and one containing C_2). Now it follows by (3.41) that the boundary of each component of $R - \alpha$ contains α. Therefore each component of $R - \alpha$ must meet and hence contain a component of $J - (u + v)$. But there are only two such components, namely C_1 and C_2. Hence $R - \alpha$ has just two components.

4. **Accessibility. Regions and their boundaries.** A point p is said to be *accessible* from a set R provided that if $x \epsilon R$ there exists a simple arc xp lying in $R + p$. A point p is *regularly accessible* from R provided that for any $\epsilon > 0$ a $\delta > 0$ exists such that if x is any point of R with $\rho(x, p) < \delta$, there exists a simple arc xp in $R + p$ with $\delta(xp) < \epsilon$. If R is a plane region with connected boundary B, a point p of B is said to be *accessible from all sides from* R (Schoenflies) provided that if ab is any cross cut in R (i.e., an arc in $R + a + b$ where $a + b \subset B$), p is accessible from each of the [at most two, by (3.5)] components into which ab cuts R which has p on its boundary.

(4.1) THEOREM. *If R is a region in a generalized locally connected continuum, in order that a point p of $F(R)$ be regularly accessible from R it is necessary and sufficient that $R + p$ be locally connected.*

The necessity is obvious; and the sufficiency results at once from the fact that $R + p$ is a G_δ-set and thus by [II, (2.4), (5.5)] any region in $R + p$ is arc-wise connected.

DEFINITION. *An unbounded plane region with bounded boundary $F(R)$ will be said to have property S provided that if I is the interior of a circle enclosing $F(R)$, then $I \cdot R$ has property S.*

We next establish a cycle of propositions. From (4.1) and [I, (15.2)] we have

(a) *If a plane region R has property S, every point of $F(R)$ is regularly accessible from R.*

(b) *If a point p of the boundary of a plane region R is regularly accessible from R it is accessible from all sides from R.*

For let ab be any cross cut in R. Clearly we may suppose $a \neq p \neq b$, since a simple application of (3.1) shows both a and b to be accessible from each region of $R - ab$. Take $\epsilon < \rho(p, ab)$. By the regular accessibility of p, a δ exists such that any point x of R with $\rho(x, p) < \delta$ lies in an arc xp with $\delta(xp) < \epsilon$ and $xp - p \subset R$. Thus $xp \cdot ab = 0$ and hence $xp - p$ lies wholly in the component of $R - ab$ which contains x. Thus p is accessible from every such component of which it is a limit point.

(c) *If every point of the connected boundary $F(R)$ of a plane region R is accessible from all sides from R, then $F(R)$ is locally connected.*

Proof. For if not, then $F(R)$ contains a continuum of convergence $K = \lim [K_n]$ where $K_n \cdot K \neq 0$ for every n. But now if a and b are any two points of K, then $F(R) - (a + b)$ is not connected. For by virtue of the accessibility hypothesis, there exists an arc ab with $ab - (a + b) = \alpha \subset R$ and by (3.5), $R - \alpha = R_1 + R_2$ where R_1 and R_2 are regions. Let o be a point of α and applying (3.12), we obtain a simple closed curve J enclosing o but neither containing nor enclosing any point of $F(R)$ and meeting α in exactly two points u and v, where $u \subset ao$, $v \subset ob$. Let C_1 and C_2 be the two arc segments of $J - (u + v)$ where we choose the indices so that $C_1 \subset R_1$ and $C_2 \subset R_2$. Now if we set $A = F(R_1)\cdot F(R) - (a + b)$ and $B = F(R_2)\cdot F(R) - (a + b)$. Then we must have $A\cdot B = 0$. For if $A\cdot B$ contains a point p, then p is accessible from both R_1 and R_2 and hence there exist arcs xp and yp such that $xp - p \subset R_1$, $yp - p \subset R_2$, $xp\cdot J = x$ and $yp\cdot J = y$. Hence if we set $xp + yp = xyp$, then $xpy + J = xpy + xvy + xvy = \theta$, a θ-curve; and we have $\pi - \theta = G_1 + G_2 + G_3$ where G_1 is the interior of J. Now since a and b are without J we may suppose $a \subset G_2$. Then $G_3 \supset b$, because neither G_1 nor G_2 can have both u and v on its boundary and $(au - u + bv - v)\cdot\theta = 0$. But then since $a + b \subset K$, there exists an n such that $K_n\cdot G_2 \neq 0 \neq K_n\cdot G_3$, and this is impossible because $F(R)\cdot\theta = p \subset K$ and hence $\theta\cdot K_n = 0$. Thus we have $A\cdot B = 0$. And since $\bar{A}\cdot\bar{B} = a + b$ (whence $\bar{A}\cdot B = \bar{B}\cdot A = 0$) and $A + B = F(R) - (a + b)$, we have a separation of $F(R) - (a + b)$ into two nonvacuous mutually separated sets A and B. Thus we have shown that $F(R) - (a + b)$ is not connected. Hence every pair of points of K contains a local separating point of $F(R)$ and thus K contains uncountably many local separating points of $F(R)$. This is impossible by [III, (3.1)]. Thus the supposition that $F(R)$ is not locally connected leads to a contradiction and (c) is established.

(d) *If the bounded locally connected continuum F is the boundary of a region R, then R has property S.*

This is the same as (3.2) if R is bounded. If R is unbounded and C is a circle enclosing F and I is its interior and a is an arc with $ab\cdot C = a$, $ab\cdot F = b$, and B is an arbitrarily small arc of C with one end point $= a$, then since, by (3.4), $R - (ab + C - B)$ is connected it follows at once that $I\cdot R - ab$ is connected and since $F + C + ab$ is locally connected, therefore, by (3.2), $I\cdot R - ab$ (a complementary domain of $F + C + ab$) has property S. Therefore $I\cdot R - ab + ab = I\cdot R$ has property S.

Summing up, the chain of propositions (a)–(d) gives

(4.2) THEOREM. *If R is any plane region whose boundary $F(R)$ is a bounded continuum, then the following properties are equivalent:*

(i) *that R have property S,*

(ii) *that every point of $F(R)$ be regularly accessible from R,*

(iii) *that every point of $F(R)$ be accessible from all sides from R,*

(iv) *that $F(R)$ be locally connected.*

DEFINITION. *A plane continuum M will be called an E-continuum provided that for any $\epsilon > 0$ there are at most a finite number of complementary domains of M of diameter greater than ϵ.*

(4.3) *If p is any point of an E-continuum E which is not on the boundary of any complementary domain of E, then for every $\epsilon > 0$ there exists an ϵ-subcontinuum N of E such that p lies in a bounded complementary D domain of N.*

Proof. Let C be a circle with center p and radius less than $\epsilon/4$ such that any complementary domain of E intersecting C is of diameter less than $\epsilon/4$. Let N_0 be the continuum $C + \sum F(R_i)$ where $R_1, R_2, \cdots$ are the complementary domain of E intersecting C and let

$$N = N_0 - \sum C \cdot R_i = C \cdot E + \sum F(R_i).$$

Then $N \subset E$ and $\delta(N) < \epsilon$. Further, N is a continuum because if we had a separation $N = A + B$, we would have the separation

$$N_0 = A_0 + B_0,$$

where $A_0 = A +$ all sets $C \cdot R_i$ such that $F(R_i) \subset A$ and $B_0 = B +$ all sets $C \cdot R_i$ such that $F(R_i) \subset B$ and this contradicts the connectedness of N_0. Finally, the complementary domain D of N containing p lies within C and thus is of diameter less than ϵ.

(4.31) COROLLARY. *If H denotes the sum of the boundaries of the complementary domain of an E-continuum E, then E is locally connected at every point of $E - H$.*

For if $p \subset E - H$, then choosing N so that $N = F(D) = F(\pi - \bar{D})$ and setting $A = \bar{D} \cdot E$, $B = (\pi - D) \cdot E$ we have $A \cdot B = N$, $A + B = E$, A, B closed. Therefore A is connected and since $\delta(A) \leqq \delta(\bar{D}) < \epsilon$ and p is an inner point of A rel. E, the corollary results.

(4.4) THEOREM. *In order that a plane continuum M be locally connected it is necessary and sufficient that* (1) *M be an E-continuum and* (2) *that if R is any complementary domain of M and $F(R)$ is its boundary, then one of the conditions* (i)–(iv) *in* (4.2) *be satisfied.*

Proof. The condition is sufficient. For let p be any point of M. If p is on the boundary of no complementary region of M then by (4.31), M is locally connected at p. If p is on the boundary of at least one complementary domain of M, then let $R_1, R_2, \cdots$ be those domains whose boundaries $F_1, F_2, \cdots$ contain p and let ϵ be any positive number. Clearly we may suppose each R_i bounded, since a slight modification of M would produce this situation without altering M in the vicinity of p. Since by (4.2), for each n, F_n is locally connected and since $\lim_{n \to \infty} \delta(F_n) = 0$ it follows that $F = \sum_1^\infty F_n$ is locally connected. [For if n is sufficiently large $\sum_n^\infty F_i$ is an ϵ-continuum and $\sum_1^n F_i$ is locally connected.] Thus there exists an $\epsilon/2$-region K in F containing p. Now since p is not on the boundary of any complementary domain of the E-continuum $E = M + \sum_1^\infty R_n$, by (4.31), E is locally connected at p and hence there exists an $\epsilon/2$-subcontinuum H of E containing p as an inner point rel. E and such that $H \cdot F \subset K$. Then $(H + \bar{K}) \cdot M$ is of diameter less than ϵ and contains p in its interior rel. M. But also $(H + \bar{K}) \cdot M = H \cdot M + \bar{K} = \bar{K} + H - H \cdot \sum_1^\infty R_n$ is connected; for if $\bar{K} + H - H \cdot \sum R_n = A + B$, we have $\bar{K} \subset A$ and hence $B \subset H - H \cdot \sum R_n - H \cdot \bar{K}$. Thus B is open in H, since $H \cdot F(\sum R_n) =$

$H \cdot F \subset K \subset A$. Thus B is both open and closed in H, contrary to the fact that H is connected.

The condition is also necessary. For by the Torhorst Theorem we have that if M is locally connected, then the boundary of each of its complementary domains is locally connected and hence (2) is satisfied. Now let us suppose, contrary to (1), that there exists an $\epsilon > 0$ and an infinite collection $[G]$ of distinct complementary domains each of diameter greater than ϵ. Taking $\sigma = \epsilon/8$ and recalling that M has property S, we can write $M = \sum_1^n M_i$ where each M_i is a locally connected continuum of diameter less than σ. Now the boundary F or each domain G of $[G]$ contains three points a, b, c such that the distance between any two of these points is greater than σ. Thus no M_i contains two of these points and hence each F intersects each of three sets A, B, C of $[M_i]$ which have no common points $[A \cdot B = A \cdot C = B \cdot C = 0]$. Now since there are only a finite number of such disjoint triples A, B, C in $[M_i]$, it follows that for at least one such triple A, B, C there exist three (in fact infinitely many) domains G_1, G_2, G_3, of $[G]$ such that the boundary of each of these domains meets all three of the sets A, B, C. Let a_i, b_i, c_i be points on $F(G_i) \cdot A$, $F(G_i) \cdot B$, $F(G_i) \cdot C$ ($i = 1, 2, 3$). Now by virtue of (2), there exist arcs a_1b_1 and a_2b_2 with $\widehat{a_1b_1} \subset G_1$, $\widehat{a_2b_2} \subset G_2$ and arcs xc_1 and yc_2 with $xc_1 - c_1 \subset G_1$, $yc_2 - c_2 \subset G_2$, $xc_1 \cdot a_1b_1 = x$, $yc_2 \cdot a_2b_2 = y$. There exist arcs a_1a_2, b_1b_2, c_1c_2 in A, B, C, respectively. Then $a_1b_1 + a_2b_2 + xc_1 + yc_2 + a_1a_2 + b_1b_2 + c_1c_2 = xa_1y + xb_1y + xc_1y = \theta$, a θ-curve and $\pi - \theta = R_1 + R_2 + R_3$ where $R_1 \supset G_3$. But then since $F(G_3) \supset a_3 + b_3 + c_3$ and A, B, and C each meet only one edge of θ it follows that $F(R_3)$ meets all three edges of θ, which is impossible. Therefore M is an E-continuum and the proof is complete.

From (4.3), (4.4) and (2.51) we have at once

(4.5) *If the point p of a locally connected plane continuum E is not on the boundary of any complementary domain of E, then for any $\epsilon > 0$, E contains a simple closed curve of diameter less than ϵ enclosing p.*

5. Characterization of the sphere.

(5.1) THEOREM. *If S is a locally connected continuum such that*

(a) *S contains at least one simple closed curve,*

(b) *every simple closed curve of S separates S,*

(c) *no simple arc of a simple closed curve of S separates S; then S is homeomorphic with the surface of a sphere.*

Proof. We first establish a series of lemmas:

(i) *Condition* (c) *is equivalent to the condition* (d) *that every simple closed curve J of S be the boundary of each component of $S - J$.*

It follows immediately that (d) implies (c). To prove the reverse implication, suppose that the boundary $F(R)$ of some component R of the complement of a simple closed curve J of S is a proper subset of J. Then $F(R)$ is contained in a simple arc t of J; and clearly t separates S, contrary to (c).

(ii) *S is cyclicly connected.*

By (a), S contains a simple closed curve J. Let C be the cyclic element of S containing J, and assume that there exists a point p in $S - C$. Let R be the component of $S - C$ containing p. Then $F(R)$ is a single point, and hence a cut point of S. But this contradicts (c), since $F(R)$ lies on some simple closed curve, as it is in a true cyclic element C.

(iii) *If J is a simple closed curve in S, then S cannot contain three simple arcs m, m', and n spanning* J and having the following properties: $\langle m \rangle$† and $\langle m' \rangle$ are in different components of $S - J$; n does not meet either m or m' and its end points are separated on J by the end points of both m and m'.*

Assuming the existence of three such arcs let a and b be the end points of n, and let A and B be the components of $J - (m + m') \cdot J$ which contain a and b, respectively. Clearly A and B are open arcs which are joined by n. Moreover, $J + m + m' - A - B$ is a simple closed curve which we shall call J', and which obviously contains both of the simple arcs m and m'. Since J' separates S and $A + B + n$ is connected and does not intersect J', there exists a component D of $S - J'$ not intersecting this set and hence not intersecting J. Thus D is contained in a single component E of $S - J$. But since $D + \langle m \rangle + \langle m' \rangle$ is connected and does not intersect J, this gives $\langle m \rangle + \langle m' \rangle \subset E$ contrary to hypothesis. This contradiction proves (iii).

(iv) *Every simple closed curve J of S separates S into exactly two components.*

Let $\langle a \rangle$ and $\langle b \rangle$ be any two open arcs of J, and A any component of $S - J$. Then every point of $\langle a \rangle + \langle b \rangle$ is a limit point of A so that the set $K = A + \langle a \rangle + \langle b \rangle$ is connected. From the local connectivity of S it follows that A is open in $S - J$. Hence K is open in $\bar{A}$. Otherwise K would contain a limit point p of $\bar{A} - K$. This point cannot lie in A, since A is open in $S - J$ so that it must lie in either $\langle a \rangle$ or $\langle b \rangle$. But this is impossible since both $\langle a \rangle$ and $\langle b \rangle$ are open in J and $\bar{A} - K$ is a subset of J. Consequently, K is open in $\bar{A}$.

It follows easily that $\bar{A}$ is locally connected, since S is, and therefore that K is arcwise connected.

Now if we assume that $S - J$ has at least three components R, H, and T, and make use of the fact that any set K constructed as above is arcwise connected, we may easily construct three simple arcs m, m', and n which satisfy the conditions of (iii) and yield the desired contradiction. Therefore, $S - J$ cannot contain more than two components. Hence it must contain exactly two, since J separates S by virtue of (b).

(v) *If a, b, c are three simple arcs with common end points but otherwise disjoint, then $S - (a + b + c)$ is the sum of three components having $(a + b)$, $(b + c)$, and $(c + a)$ as their boundaries.*

By (iv), $(a + b)$ separates S into exactly two components, one of which

* An arc t spans a point set M if and only if t has its end points but nothing else in common with M.

† If t be a closed arc, then $\langle t \rangle$ will denote the corresponding open arc.

contains $\langle c\rangle$, and both of which have $(a + b)$ as their boundary. Consequently in $S - (a + b + c)$, one component, A_c, is separated from $\langle c\rangle$ by $(a + b)$ and has this simple closed curve as its boundary. Similarly, components A_a and A_b are separated from $\langle a\rangle$ and $\langle b\rangle$, respectively, by $(b + c)$ and $(c + a)$, respectively, and have these two simple closed curves as their respective boundaries. We must show that

$$A_a + A_b + A_c = S - (a + b + c).$$

In other words if B_a, B_b, B_c are the regions $S - \bar{A}_a$, $S - \bar{A}_b$, $S - \bar{A}_c$, respectively, then since

$$S - (a + b + c) = A_a + A_b + A_c + B_a \cdot B_b \cdot B_c,$$

we have to show that $B_a \cdot B_b \cdot B_c = 0$. If this is not so, let $p \in B_a \cdot B_b \cdot B_c$. Since $p + \langle a\rangle \subset B_a$, there exists an arc px in B_a, where $x \in \langle a\rangle$, such that $px - x \subset B_a \cdot B_b \cdot B_c$. Similarly, since $p + \langle c\rangle \subset B_c$ there exists an arc py in B_c, $y \in \langle c\rangle$, with $py - y \subset B_a \cdot B_b \cdot B_c$. Thus $px + py$ contains an arc xy such that

$$xy - (x + y) \subset B_a \cdot B_b \cdot B_c.$$

Clearly there exists an arc $x'y'$ with $x'y' - (x' + y') \subset A_b$ and $x' \in \langle a\rangle$, $y' \in \langle c\rangle$. But then since the end points of b separate on $(a + c)$ both x and y and x' and y', the three arcs xy, $x'y'$ and b satisfy the conditions in (iii) with $J = (a + c)$. Thus we have a contradiction and hence (v) is proved.

Now by (iv) any simple closed curve J divides S into exactly two components. We shall prove that the closure C of either of these components is homeomorphic with a 2-cell, that is, with a plane triangle plus its interior, in such a way that J corresponds to the boundary of the 2-cell. The theorem is then an immediate consequence.

A subdivision of C by means of a graph*

$$G = J + \sum_{i \leq n} a_i$$

will be called a *C-complex* (C), if the graph G can be constructed by taking the following step n times: add to $J + \sum_{i<m} a_i$, a spanning arc a_m of this set in C. If we call a single point a 0-cell, the homeomorph of an interval a 1-cell, and the homeomorph of a plane triangle and its interior a 2-cell, it will be natural to call the 0-, 1-, and "2-cells" of (C) the 0- and 1-cells of G, and the closures of the components of $C - G$, respectively. By the boundary of a 2-cell we shall mean the collection of all 1-cells in the interiors of which the 2-cell has limit points.

* By graph we shall mean linear graph, that is, any point set which consists of the sum of a finite number of simple arcs, degenerate or nondegenerate, disjoint except for their end points.

We now prove:

(vi) *If a C-complex (C) is given, we can find a subdivision of the 2-simplex* C′ by means of polygons such that the resulting complex (C′) is isomorphic* with (C).*

We prove this lemma by induction on the number of spanning arcs used in constructing the graph G defining (C).

The lemma follows at once for the case when G consists of the simple closed curve J itself; since J may be sent into the triangle by means of a homeomorphism, the interior of J into the interior of the triangle so that the isomorphism has been established.

Secondly, assuming that the lemma has been proved for G_1, a graph defining $(C)_1$, we shall prove it for $G_2 = G_1 + a$, where a is an arc in C spanning G_1. This will establish the induction and complete the proof. Since a spans G_1 in C, it follows that $\langle a \rangle$ lies in a single component or open "2-cell" A of $(C)_1$. This "2-cell" A has a simple closed curve for its boundary, divided by the end points of a into two simple arcs b' and b''. Corresponding to these, since the theorem holds for $(C)_1$, we shall have the arcs B' and B'', and a 2-cell A' in $(C')_1$. In A' we can construct a line segment a' joining the end points of B' (or B''). This will divide A' into two 2-cells E and F with boundaries $B' + a'$ and $B'' + a'$. Consequently, by (v), A is divided into two components with boundaries $a + b'$ and $a + b''$, so that the isomorphism has been extended to $(C)_2$, and the lemma follows.

(vii) *The closure of each component of $S - G$ is cyclic.*

The boundary of each component of $S - G$ is a simple closed curve, so that A contains a simple closed curve and hence a true cyclic element E. The proof that $E = A$ now follows exactly as in (ii).

(viii) *Any pair of points p and q of C can be separated in C by a spanning arc of J.*

If p lies outside of J we may construct a spanning arc H of J in C which passes through p. Assume that H passes through q also. Then H together with either of the simple arcs into which its end points divide J will form a simple closed curve J', which, by (v), will divide C into exactly two components one of which will have J' as its boundary. In the closure of this component we may construct a spanning arc of J' having its end points separated on J' by $p + q$. Using this arc and the proper subarc or subarcs of H we may easily find a simple arc spanning J, and passing through p but not containing q.

Thus, in any event, we may always find a simple arc H spanning J and passing through p but disjoint with q. By (v) C will be divided by H into exactly two components. Let K be the one of these components which contains q. In $\bar{K}$ construct a spanning arc of $\bar{K} - K$ passing through q and having its end points on $\bar{K} - K - \bar{H}$. This will divide K into exactly two components, one

* By a 2-simplex we mean a plane triangle and its interior. Two complexes are said to be isomorphic if we can set up a one-to-one correspondence between their cells such that the relation "on the boundary of" remains invariant under that correspondence.

of which will have a simple closed curve J'' containing p and q as its boundary. Let Z be the closure of this component. Now J'' will contain two subarcs of J which separate p and q on J''. Let r be a spanning arc of J'' in Z having its end points on these two subarcs of J. Then, by (v), it follows that p and q will lie in different components of $C - r$, which proves the lemma if neither p nor q lies in J.

If either p or q lies in J, or if both of them do, the proof is only simplified.

To prove that C is homeomorphic with a 2-cell we first construct a finite number of simple arcs, not necessarily forming a graph, and dividing C into arbitrarily small pieces.

For a given $\delta > 0$ we cover C by a finite number of closed sets V_i ($i = 1, \cdots, n$) of diameter less than $\delta/2$. Because V_i is compact we can find for any point p of $C - V_i$ a finite set of spanning arcs of J such that p is separated from any point of V_i by at least one of those arcs. The set of points p separated from any point of V_i by one of those arcs contains a neighborhood of p, so that we can find a finite set of such finite sets of arcs, such that if $V_i \cdot V_j = 0$, any point p of V_i is separated from any point q of V_j by one of those arcs. Doing this for all pairs of sets V_i having no points in common, we find a finite set of spanning arcs of J, $\alpha_1, \cdots, \alpha_p$, such that any pair of points, taken out of any pair of sets V_i having no point in common, is separated by one of those arcs.

We construct a sequence of C-complexes $(C)_i$ defined by graphs G_i contained in $J + \sum \alpha_i$ and approximating $J + \sum \alpha_i$. They will depend on a sequence of numbers ϵ_i, $\epsilon_i > \epsilon_{i+1} > 0$, $\epsilon_i \to 0$. G_{-1} is equal to J and equal to G_0^0. If the graph G_i^j has been defined and $j < p$, we take for G_i^{j+1} the sum of G_i^j and all subarcs of α_{j+1}, spanning G_i^j, such that the end points of each subarc cannot be joined in G_i^j by an arc of diameter less than ϵ_i (for instance, $G_0^1 = J + \alpha_1$ if ϵ_0 is not too big). The number of such arcs is finite because G_i^j is locally connected, so that G_i^{j+1} is really a graph. If G_i^j has been defined and $j = p$, we put $G_i^p = G_i = G_{i+1}^0$.

The graphs G_i have the property that for a certain m, G_m contains a spanning arc of J separating any pair of points taken from any pair of sets V_i having no point in common. If this were not true, we could find a sequence of such pairs of points p_i, q_i such that G_i contains no arc separating p_i and q_i. For some sequence of integers v_i, p_{v_i} and q_{v_i} will both converge. We call the limits p and q. They will be in some sets V_a and V_b having no point in common, so that there will be an arc α_l separating them, with end points x and y. No G_i will contain an arc separating them. For any such arc would separate p_{v_j} and q_{v_j} for some $v_j > i$ and be contained in G_{v_j}, contrary to our assumption.

We join p and q to J by two arcs γ_1 and γ_2 not meeting α_l and determine a number k such that ϵ_k is less than the distance from $\gamma_1 + \gamma_2$ to α_l. If U is an ϵ_k-neighborhood of α_l in C, then x and y must be in one connected set of $U' = U \cdot G_k$. For if $U_x + U_y$ is a separation of U' between x and y, $U_x \cdot G_k^{l-1}$ will contain a last point r of α_l and $U_y \cdot G_k^{l-1}$ will contain a first point s after r of α_l. Then G_k must contain an arc of diameter less than ϵ_k between r and s and this arc must be in U', which is absurd Then U' must contain an arc $\beta = xy$ and β must contain a spanning arc of J separating the end points on J of γ_1 and γ_2, and accordingly separating p and q.

The C-complex $(C)_m$ defined by G_m is of mesh less than δ, for any two points on the closure of the same 2-cell of $(C)_m$ must be on the same or adjoining sets V_i, so that their distance must be less than δ.

We construct a polygonal subdivision $(C')_m$ of C' similar to $(C)_m$ and a polygonal subdivision $(C')_\delta$ of $(C')_m$ (taking care that any new end point is in the interior of some old 1-cell) of mesh less than δ, finally a subdivision $(C)_\delta$ of $(C)_m$ similar to $(C')_\delta$.

We have seen how to construct for C and C' similar subdivisions $(C)_\delta$ and $(C')_\delta$ of mesh less than δ. But then we can construct a sequence of such pairs of similar subdivisons $(C)_{\delta_i}$ and $(C')_{\delta_i}$, where $\delta_i \to 0$, for we can repeat the process applied to C, for the closure of any 2-cell of $(C)_{\delta_i}$.

The sets of 0-cells of all $(C)_{\delta_i}$'s and $(C')_{\delta_i}$'s form two one-to-one corresponding everywhere dense point sets in C and C'. This correspondence is uniformly continuous both ways. Let $\epsilon > 0$ be given; find n such that $2\delta_n < \epsilon$, now find $\delta > 0$ such that two points of distance less than δ in C or in C' are in or on the boundary of the same or adjoining 2-cells of $(C)_{\delta_n}$ or $(C')_{\delta_n}$. If such a δ did not exist, we could find two sequences of points p_i and q_i convergent to the same point p of C (or C') and not in adjoining 2-cells. We can assume that all p_i and all q_i are on the closure of the same 2-cell of $(C)_{\delta_n}$ [or $(C')_{\delta_n}$] because there is only a finite number of such 2-cells to choose from. But the point p belongs to the closure of both 2-cells so that they were adjoining after all.

If two vertices of any subdivision of C (or C') have a distance less than δ, their corresponding vertices in C' (or C) are in or on the boundary of the same or adjacent 2-cells of $(C')_{\delta_n}$ (or $(C)_{\delta_n}$), so that their distance is less than ϵ. This makes the correspondence uniformly continuous.

Now the correspondence of the two sets of vertices can be extended to a topological transformation of their closures. But these closures are identical with C and C', so that C and C' are homeomorphic. This can of course be extended immediately to prove that S is homeomorphic with the surface of a sphere.

(5.2) THEOREM. *In order that a locally connected continuum C be a closed 2-cell it is necessary and sufficient that $C = R + J$ where J is a nondegenerate simple closed curve, R is a nonvacuous region bounded by all of J, $(R \cdot J = 0)$, and if γ is any spanning arc of J in C, $\langle\gamma\rangle$ cuts R into exactly two regions R_a and R_b bounded by $\gamma + \alpha$ and $\gamma + \beta$, respectively, where α and β are the two arcs into which the end points of γ separate J.*

Proof. We first prove that the sets $\bar{R}_a$ and $\bar{R}_b$ have the same properties as are assumed for C in the theorem with respect to the simple closed curves $\alpha + \gamma$ and $\beta + \gamma$, respectively. As soon as this is accomplished, our result follows from the proof of (5.1), starting at the end of the proof of (iv), because from this point on everything needed is available and the conclusion there established is that C is homeomorphic with the closed 2-simplex.

To this end let η be a spanning arc of $\alpha + \gamma$ with $\langle\eta\rangle$ in R_a.

Case I. η has one end point x on $\langle\gamma\rangle$ and the other y on $\langle\alpha\rangle$, let α_1, α_2 and γ_1, γ_2 be the arcs into which α and γ, respectively, are divided by the end points of η where α_1 and γ_1 have a common end point as do also α_2 and γ_2. Of the two regions into which $\langle\eta + \gamma_1\rangle$ divides R let R_1 be the one bounded by $\alpha_1 + \eta + \gamma_1$ and let D_1 be the other. Similarly let R_2 and D_2 be the regions of $R - \langle\eta + \gamma_2\rangle$ so chosen that R_2 is bounded by $\alpha_2 + \eta + \gamma_2$. Clearly we have only to show that $R_1 + R_2 + \langle\eta\rangle \supset R_a$.

If this is not so, let D be a component of $R_a - (R_1 + R_2 + \langle\eta\rangle)$. Since D is open in R_a and hence in C, $F(D)\cdot R$ cannot lie wholly either in $\eta + \gamma_1$ or in $\eta + \gamma_2$ because this would give D a third component of $R - \langle\eta + \gamma_1\rangle$ or of $R - \langle\eta + \gamma_2\rangle$, respectively. Thus $F(D)$ has points in both $\gamma_1 - x$ and $\gamma_2 - x$. This gives an arc t with end points u and v in $\gamma_1 - x$ and $\gamma_2 - x$, respectively, and $\langle t\rangle \subset D$. But $t + \gamma - \langle uxv\rangle$ (of γ) is a spanning arc s of J which does not separate $\langle\alpha\rangle$ and $\langle\beta\rangle$ in C because $\alpha + \eta + R_b + \beta$ is connected and does not intersect s.

Case II. η has one end point x in common with one end point of γ and its other end point y on $\langle\gamma\rangle$.

Let R_1 be the component of $R - \langle w\rangle$ bounded by $\alpha + w$ where w is the spanning arc $\eta + \gamma - \langle xy\rangle$ of γ, and let $R_2 = R_a - (R_1 + \langle\eta\rangle)$. Since obviously $R_1 \subset R_a$ and $F(R_2) = \eta + xy$ of γ, we have only to show that R_2 is connected.

If this is not so, let D_1 and D_2 be distinct components of R_2. Now since any arc obtained from $\eta + xy$ (of γ) by omitting an open segment with one end point at x can be made a part of a spanning arc of J it follows just as in Case I that both $F(D_1)$ and $F(D_2)$ must contain nondegenerate arcs ux and vx of η and xy (of γ), respectively. Let $ux + vx = uxv$. But then we could construct arcs a_1b_1 and a_2b_2 in $D_1 + a_1 + b_1$ and $D_2 + a_2 + b_2$, respectively, such that a_1, b_1, a_2, b_2 are on the arc uxv in the order $u, a_1, a_2, x, b_1, b_2, v$. This is impossible, because $\gamma - \langle b_2x\rangle$ (of γ) $+ b_2a_2 + a_2x$ (of η) is a spanning arc of J with same end points as γ which does not separate $\langle\alpha\rangle$ and $\langle\beta\rangle$ in C since $\alpha + R_1 + a_1b_1 + R_b + \beta$ is connected.

The case where both end points of η are on $\langle\gamma\rangle$ is very similar to, but considerably simpler than, Case II. The proof is essentially the same except that now *any arc* in $\eta + xy$ (of γ) is a part of a spanning arc of J so that $F(D_1) = F(D_2) = \eta + xy$ (of γ).

The remaining cases, namely, where both end points of η are on $\langle\alpha\rangle$, where both coincide with end points of γ and where one is an end point of γ and the other is on $\langle\alpha\rangle$ reduce at once, respectively, to two instances of Case I, two instances of Case II and one each of Cases I and II by drawing an auxiliary arc.

Thus, since $\bar{R}_a$ obviously is a locally connected continuum, $\bar{R}_a$ (and similarly $\bar{R}_b$) has all properties assumed for C in the theorem and the proof is now completed, as shown in the first paragraph.

References

§1. Jordan [1], Veblen [2], Alexander [1]. See also Menger [1].

§2. (2.1) Brouwer [2]. See also Alexandroff [4].
(2.2) Torhorst [1]. See also Kuratowski [4], Whyburn [1], Wilder [4].
(2.5)–(2.6) Moore [6]. See also Cleveland [1].

§3. (3.1) Zoretti [1], Moore [11], Lubben [2].
(3.2) Moore [5].
(3.3) Moore [12].
(3.4) Moore [6].
(3.5) Schoenflies [1], Carathéodory [1].

§4. (4.1) Whyburn [18]. In this connection see also Aitchison [1].
(4.2) (a), (b): Whyburn [19], (c): Schoenflies [1], (d): Moore [5].
(4.3) Whyburn [20].
(4.4) Schoenflies [1]. See also references under (4.2).
(4.5) Whyburn [21, 20].

§5. Moore [6], Zippin [1]. For complete history and references as well as essentially the proof given here, see van Kampen [1].

CHAPTER VII

SEMI-CONTINUOUS DECOMPOSITIONS AND CONTINUOUS TRANSFORMATIONS

1. **Upper semi-continuous collections.** A collection of sets G is said to be *upper semi-continuous* provided that if $g \in G$ and U is any neighborhood of g (i.e., open set containing g) then there exists a neighborhood V of g such that if h is any set of G intersecting V then $h \subset U$. It results at once that the elements of any such collection are disjoint from one another.

For example, it is readily verified that the elements of any null sequence of disjoint closed sets form an upper semi-continuous collection as do also the components of any compact set.

(1.1) THEOREM. *In a compact metric space M a necessary and sufficient condition that a collection G of closed sets be upper semi-continuous is that for any sequence $g_1, g_2, \cdots$ of elements of G with* $\liminf (g_n) \cdot g \neq 0$, *where $g \in G$, we have* $\limsup (g_n) \subset g$.

To prove the necessity of the condition we suppose G upper semi-continuous but suppose contrary to our condition that for some sequence $g_1, g_2, \cdots$ in G and some $g \in G$ we have $\liminf (g_n) \cdot g \neq 0$ but $L - g \cdot L \neq 0$ where $L = \limsup (g_n)$. Let $x \in L - g \cdot L$ and let U be a neighborhood of g such that $\overline{U} \cdot x = 0$. Let V be the neighborhood of g associated with U by the upper semi-continuity of G. Then for some integer N we have $g_n \subset U$ for $n > N$ since $g \cdot \liminf (g_n) \neq 0$. But clearly this gives x non $\in L$ since $\overline{U} \cdot x = 0$. Thus the condition is necessary.

The condition is also sufficient. For suppose the condition satisfied but suppose G is not upper semi-continuous. Then for some $g \in G$ and some neighborhood U of g there exists no corresponding neighborhood V of g such that $h \cdot V \neq 0$ always implies $h \subset U$ where $h \in G$. Thus if we take a sequence of positive numbers δ_n converging monotonically to 0 and such that $V_{\delta_1}(g) \subset U$, then for each n there exists a $g_n \in G$ such that

(i) $$g_n \cdot V_{\delta_n}(g) \neq 0 \neq g_n \cdot (M - U).$$

Let (g_{n_i}) be a convergent subsequence of (g_n) and let $L = \lim (g_{n_i})$. Then since M is compact it follows by (i) that we have

$$L \cdot g \neq 0 \neq L \cdot (M - U);$$

and this contradicts our condition.

(1.11) *If, in a compact space, the collection G of disjoint closed sets is the sum of three subcollections G_1, G_2 and G_3 where G_1 is the collection of components of a* *closed set, G_2 is a null sequence and G_3 is a collection of single points, then G is upper semi-continuous.*

(1.2) THEOREM. *The same result as in* (1.1) *is obtained if M, instead of being compact, is locally compact provided all the sets of G are continua.*

The necessity proof is the same as for (1.1) since the compactness of M was not used there. To prove the sufficiency we proceed exactly as in (1.1) until we obtain the convergent subsequence (g_{n_i}) with limit L. Now since M is locally compact and g is compact a simple application of the Borel Theorem gives us a neighborhood U_0 of g such that $\overline{U}_0 \subset U$ and $\overline{U}_0$ is compact. We then have, for each i,

(i)
$$g_{n_i} \cdot V_{\delta_{n_i}}(g) \neq 0 \neq g_{n_i} \cdot (M - U_0).$$

Since each g_{n_i} is a continuum this gives

(ii)
$$g_{n_i} \cdot F(U_0) \neq 0 \qquad \text{for each } i.$$

And since $F(U_0)$ is compact we have at once

$$L \cdot F(U_0) \neq 0 \neq L \cdot g$$

contrary to our condition, since $g \subset U$.

2. **Upper semi-continuous decompositions.** A collection G of subsets of a metric or metrizable space M is said to give an *upper semi-continuous decomposition of M* provided that (1) the sets of G are compact, (2) every point of M belongs to a set of G and (3) G is upper semi-continuous. We suppose we have such a decomposition G of M.

We wish to define an associated space M' whose points will be the sets of G and which we shall call the *decomposition space* or the *hyperspace of the deposition.* In order to define neighborhoods in such a space, however, we need the following lemma.

(2.1) *If U is any open set in M containing at least one element g of G, then the set U_0 of all points lying on elements h of G, such that $h \subset U$, is open.*

Let $x \in U_0$. Then $x \in h \in G$ where $h \subset U$. Since G is upper semi-continuous, there exists an open set V such that $h \subset V \subset U$ and such that if k is any element of G intersecting V we have $k \subset U$. But this gives $k \subset U_0$ for every such k. Whence, $x \in h \subset V \subset U_0$ so that U_0 is open.

The decomposition space M' may now be defined as follows: the points g' of M' are the sets of g of G and if g' is any "point" of M' then by a *neighborhood* of g' in M' is meant any set of "points" U' in M' containing g' and such that the set U in M of all points on sets $g \in G$ such that $g' \in U'$ is open in M. We proceed to prove

(2.2) THEOREM. *If M is a separable metric space, so also is M'.*

To do this we shall show that M' is a regular perfectly separable topological space [see I, (5.3)]. Taking the four axioms for a topological space in order, we see that Axioms (1) and (2) [I, §2] are automatically satisfied by the definition of neighborhood in M'. To prove Axiom (3), let U' and V' be neighborhoods, and let $g' \epsilon U' \cdot V'$. The corresponding sets U and V in M are open in M and hence so also is $U \cdot V$. Furthermore $U \cdot V$ corresponds to the set $U' \cdot V'$, i.e., $(U \cdot V)' = U' \cdot V'$. Thus $U' \cdot V'$ is a neighborhood of g' in M' and Axiom (3) is satisfied. To show that Axiom (4) holds, let $g', h' \epsilon M'$, $g' \neq h'$. Since the sets g and h in M are disjoint and closed, there exists an open set U such that $g \subset U \subset M - h$. (Note: We could take $U = M - h$.) Let U_0 be the set given by (2.1). Then U_0 is open and $g \subset U_0 \subset M - h$. Accordingly U_0' is a neighborhood of g' in M' not containing h' and Axiom (4) is established. Thus M' is a topological space.

We prove next that M' is regular. To this end let $g' \epsilon M'$ and let U' be any neighborhood of g'.

Since U is open in M and contains g there exists [I, (5.1)] an open set V in M such that

(i) $$g \subset V \subset \bar{V} \subset U.$$

Let V_0 be the set associated with V by the lemma (2.1). Then V_0 is open in M and

(ii) $$g \subset V_0 \subset \bar{V}_0 \subset U.$$

Thus V_0' is a neighborhood of g' in M'. It remains to show that $\overline{V_0'} \subset U'$. To this end let $h' \epsilon M' - U'$. Then $h \subset M - U$. Thus by (ii) there exists an open set W in M such that $h \subset W \subset M - V_0$. Let W_0 be the set associated with W by (2.1). Then W_0 is open in M and $h \subset W_0 \subset M - V_0$. Accordingly, W_0' is a neighborhood of h' in M' and $W_0' \subset M' - V_0'$. Thus no $h' \epsilon M' - U'$ can be a limit point of V_0' so that $\overline{V_0'} \subset U'$.

Finally, to prove M' perfectly separable, let $R_1, R_2, \cdots$ be a fundamental sequence of neighborhoods in M. For each finite set of positive integers $(n_1, n_2, \cdots, n_k)$ let

$$R(n_1, n_2, \cdots, n_k) = \sum_1^k R_{n_i},$$

and let $R_0(n_1, \cdots, n_k)$ be the set associated with $R(n_1, \cdots, n_k)$ in (2.1). Then the sets $[R_0'(n_1, \cdots, n_k)]$ are neighborhoods in M' and since they are countable they may be arranged into a sequence $[{}_iR_0'(n_1, \cdots, n_k)]$. We shall show that this is a fundamental sequence of neighborhoods in M'. To this end let $g' \epsilon M'$ and let U' be any neighborhood of g' in M'. Then since the set g is compact and the set U is open in M it follows by the Borel Theorem that g can be covered by a finite collection $R_{n_1}, R_{n_2}, \cdots, R_{n_k}$ of the sets R_i such that

$$R(n_1, \cdots, n_k) = \sum_1^k R_{n_i} \subset U.$$

This gives $R_0(n_1, \cdots, n_k) \subset U$ and thus $R'_0(n_1, \cdots, n_k) \subset U'$; and since for some i we have $R'_0(n_1, \cdots, n_k) = {}_iR'_0(n_1, \cdots, n_k)$, therefore for this i

$$g' \subset {}_iR'_0(n_1, \cdots, n_k) \subset U$$

and the theorem is proved.

3. Relations between upper semi-continuous decompositions and continuous transformations.

(3.1) *If M' is the hyperspace of an upper semi-continuous decomposition G of a separable metric space M, there exists a continuous transformation $T(M) = M'$ such that for each $g' \in M'$, $T^{-1}(g') = g \in G$.*

Proof. For each $x \in M$, let g be the element of G containing x and let $T(x) = g'$ where g' is the point of M' given by the element g of G. Then T is single-valued and $T(M) = M'$. To prove T continuous, let $x \in M$ and let U' be any neighborhood of $T(x) = g'$ in M'. By the definition of neighborhood in M' it follows that the set U of points in M on sets g of G such that $g' \in U'$ is open in M. Hence if $y \in U$ we have $y \in h \in G$ and $h \subset U$ so that $T(y) = h' \in U'$. Hence T is continuous.

(3.11) Corollary. *If M is $\begin{Bmatrix} \text{compact} \\ \text{connected} \end{Bmatrix}$, so also is M'.*

This is immediate since the properties involved are invariant under continuous transformations.

Note. Henceforth we shall speak of the transformation T defined under (3.1), as the transformation *associated with* the upper semi-continuous decomposition G of M.

(3.2) *If M is locally connected or locally compact, so also is M'.*

Proof. Let M be locally connected. To prove M' locally connected, let $g' \in M'$ and let U' be any neighborhood of g' in M'. Then U is a neighborhood of the set g in M. Since M is locally connected and g is compact it follows by an application of the Borel Theorem that g is contained in the sum W of a finite number of connected open sets $V_1, V_2, \cdots, V_n$ in M each of which intersects g and is a subset of U. Let V be the sum of all sets $h \in G$ such that $h \subset W$. Then V is open in M and we have $g \subset V \subset W \subset U$. The set V' is a neighborhood of g' in M'. Let $h' \in V'$. Then $h \subset V$. Hence for some $i \leqq n$ we have $h \cdot V_i \neq 0$. Since V_i is connected and the transformation T associated with G is continuous, $T(V_i)$ is connected; and since $h \cdot V_i \neq 0 \neq g \cdot V_i$ we have

$$h' + g' \subset T(V_i) \subset U.$$

Accordingly, M' is locally connected.

Now suppose M is locally compact. Let $g' \in M'$ and let $g = T^{-1}(g')$. Then $g \in G$ and since g is compact it follows at once that there exists a conditionally compact neighborhood U of g in M. Let V be the sum of all elements h of G such that $h \subset U$. Then V is a conditionally compact neighborhood of g and

$T(V) = V'$ is a neighborhood of g'. Now since $\bar{V}$ is compact, and T is continuous, therefore $T(\bar{V})$ is compact. But obviously $T(V) \subset T(\bar{V})$. Hence $T(V)$ is a conditionally compact neighborhood of g' and thus M' is locally compact.

(3.3) *Let M and N be separable metric spaces and suppose $T(M) = N$ is continuous and either* (a) *M is compact or* (b) *M is locally compact and each $T^{-1}(x)$, for $x \in N$, is a continuum. Then the collection $[T^{-1}(x)]$, $x \in N$, gives an upper semi-continuous decomposition of M; and if M' is the hyperspace of this decomposition, there exists a* (1-1) *continuous transformation $T_1(M') = N$.*

Proof. The sets $T^{-1}(x)$ are compact and disjoint and they fill up M since T is single-valued and continuous. To prove the first statement it remains only to show that the collection $[T^{-1}(x)]$ is upper semi-continuous. Since in any case the conditions of either (1.1) or (1.2) are satisfied, it suffices to show that if $g_1, g_2, \cdots$ is any sequence of this collection, then $\liminf (g_n) \cdot g \neq 0$ implies $\limsup (g_n) \subset g$, where g is a set of the collection $[T^{-1}(x)]$.

To this end let p be any point of $\limsup (g_n)$ and let $a \in \liminf (g_n) \cdot g$. We must prove that $p \in g$. There exists a sequence of integers (n_i) and two sequences of points (p_{n_i}) and (a_{n_i}) such that

$$a_{n_i} + p_{n_i} \subset g_{n_i}, \qquad p_{n_i} \to p, \; a_{n_i} \to a.$$

Since T is continuous, this gives

$$T(p_{n_i}) \to T(p), \qquad T(a_{n_i}) \to T(a).$$

And since $T(p_{n_i}) = T(a_{n_i})$, we have $T(p) = T(a) = T(g)$. Accordingly, $p \in g$, since $g = T^{-1}T(g)$.

To establish the latter part of the theorem we define $T_1(g') = T(g)$, for each $g' \in M'$, where g is the element of the collection $[T^{-1}(x)]$, $x \in N$, giving the point g' of M'. Since $g = T^{-1}T(g)$, clearly T_1 as thus defined is (1-1). To prove it continuous, let $g' \in M'$ and let $g_i' \to g'$ in M'. By the above proof we have at once

$$\limsup (g_i) \subset g,$$

where g_i and g are the elements of $[T^{-1}(x)]$ giving g_i' and g', respectively. Since T is continuous and $T(g_i)$ and $T(g)$ are single points of N, we have at once

$$T_1(g_i') = T(g_i), \qquad T(g) = T_1(g').$$

Accordingly, T_1 is continuous.

(3.4) THEOREM. *If M is a compact metric space, any upper semi-continuous decomposition of M is equivalent to a continuous transformation defined on M, and conversely—in the sense that, given any upper semi-continuous decomposition G of M with hyperspace M', there exists a continuous transformation $T(M) = M'$ whose inverse sets $T^{-1}(x)$, $x \in M'$, are exactly the elements of G and, on the other hand, given any continuous transformation $T(M) = N$, the inverse sets $[T^{-1}(x)]$ give an upper semi-continuous decomposition of M whose hyperspace is homeo-*

*morphic with N and whose associated transformation is topologically equivalent to T.**

This is an immediate consequence of (3.1) and (3.3). For in case M is compact, both M' and N in (3.3) are compact; and hence, by [II, (1.4)], T_1 is a homeomorphism. Further, if $f(M) = M'$ is the transformation associated with the decomposition $[T^{-1}(x)]_{x \in N}$ of M, then by the definition of T_1 it follows that

$$T_1 f(x) = T(x)$$

for each $x \in M$. Accordingly f and T are topologically equivalent.

4. **Particular kinds of transformations.** In this section several types of transformations to be treated more fully later will be defined and their relation to the associated decomposition clarified. A and B will denote metric spaces and all transformations mentioned will be assumed single-valued and continuous.

(4.1) *Monotone transformations.* A transformation $f(A) = B$ is said to be *monotone* provided that for each point $y \in B$, $f^{-1}(y)$ is a continuum. It will be noted that a monotone function of a single real variable on an interval is characterized by this property. If $f(A) = B$ is monotone, the decomposition of A associated with f, i.e., into the sets $[f^{-1}(x)]_{x \in B}$, is an upper semicontinuous decomposition into continua. Conversely, if G is an upper semi-continuous decomposition of A into continua, the associated transformation $f(A) = A'$ is monotone (A' is the hyperspace of G). Thus as a consequence of (3.4) we have the

THEOREM. *Any monotone transformation $f(A) = B$ on a compact space A is equivalent to an upper semi-continuous decomposition of A into continua. Conversely, any upper semi-continuous decomposition of A into continua with hyperspace A' is equivalent to a monotone transformation $f(A) = A'$.*

(4.2) *Non-alternating transformations.* A transformation $f(A) = B$ is said to be *non-alternating* provided that for no two points x, y of B does there exist a separation

$$(*) \qquad A - f^{-1}(x) = A_1 + A_2$$

such that $f(A_1) \cdot f(A_2) \supset y$. In other words, for any $x \in B$ and any separation (*), we have

$$f(A_1) \cdot f(A_2) = 0,$$

or

$$A_i = f^{-1}(f(A_i)), \qquad i = 1, 2.$$

It will be noted that this definition says exactly that *the transformation is non-alternating provided the associated collection $[f^{-1}(x)]_{x \in B}$ is non-separated.* Thus

* Two transformations $f(A) = B$ and $g(A') = B'$ are topologically equivalent provided there exist homeomorphisms $h_1(A) = A'$ and $h_2(B') = B$ such that $f(x) \equiv h_2 g h_1(x)$, $x \in A$.

a non-alternating transformation is equivalent to an upper semi-continuous non-separated decomposition of A in the same sense as the term equivalence was used above in (4.1).

We recall that a non-separated collection $[G]$ of sets in a space M is said to be *saturated* provided that if $G_1 \epsilon [G]$ and $p \epsilon M - G_1$, there exists an element G_2 of $[G]$ which separates p and G_1 in M.

(4.21) THEOREM. *If M is compact, the elements of any non-separated saturated collection $[G]$ of sets in M are compact and disjoint and the collection $[G]$ is upper semi-continuous.*

Proof. Let $G_1 \epsilon [G]$ and $p \epsilon M - G_1$. There exists an element G_2 of M and a separation

$$M - G_2 = M_1 + M_2, \qquad \text{where } M_1 \supset G_1, M_2 \supset p.$$

Accordingly, p is not a limit point of G_1 and thus G_1 is closed. Also p could not belong to any element G_3 of $[G]$ which intersects G_1, since any such G_3 is necessarily $\subset M_1 + G_2$. Hence the elements of $[G]$ are disjoint.

To show the upper semi-continuity, let $G \epsilon [G]$ and let $G_1, G_2, \cdots$ be a sequence of elements of $[G]$ such that $G \cdot \liminf (G_i) \supset$ a point y. Let $x \epsilon \limsup (G_i)$. If x non ϵ G, there exists an element G' of $[G]$ and a separation

$$M - G' = M_x + M_y, \qquad \text{where } M_x \supset x, M_y \supset G \supset y.$$

But then M_y contains almost all the sets (G_i), contrary to $x \epsilon \limsup (G_i)$. Hence $x \epsilon G$, so that $[G]$ is upper semi-continuous.

(4.22) THEOREM. *If M is compact and $[G]$ is a non-separated saturated collection in M, there exists a non-separated saturated collection $[Q]$ in M containing $[G]$ and filling up M.*

Proof. For each point $x \epsilon M$ let Q_x be x together with all points y of M which are not separated in M from x by any single element of $[G]$. Let $[Q]$ be the collection of all sets Q_x. Then clearly if $x \epsilon G \epsilon [G]$ we have $Q_x \equiv G$ so that every element of $[G]$ is an element of $[Q]$. Obviously $[Q]$ fills up M. Also, $[Q]$ is non-separated. For let $x, y \epsilon Q \epsilon [Q]$, and suppose we had a separation

$$M - Q_p = M_x + M_y, \qquad \text{where } Q_p \epsilon [Q], x \epsilon M_x, y \epsilon M_y.$$

Since x non ϵ Q_p, there exists an element G of $[G]$ and a separation

$$M - G = N_p + N_x \qquad \text{where } N_p \supset Q_p, N_x \supset Q_x = Q.$$

But this gives the separation

$$M - G = M_x \cdot N_x + [M_y \cdot N_x + N_p] \qquad \text{between } x \text{ and } y,$$

contrary to $x + y \subset Q$.

Finally, $[Q]$ is saturated. For let $Q_x \epsilon [Q]$, $p \epsilon M - Q_x$. Then by definition of Q_x, there exists an element G of $[G]$ which separates p and x in M. Since $G \epsilon [Q]$, this shows that $[Q]$ is saturated.

(4.23) THEOREM. *If $[G]$ is any non-separated saturated collection of sets filling up a continuum M, therefore giving an upper semi-continuous decomposition of M, and $f(M) = N$ is the associated continuous transformation, f is non-alternating and N is a dendrite.*

Proof. That f is non-alternating results at once from the definition of non-separated. To show that N is a dendrite, let $a, b \epsilon N$ and set

$$A = f^{-1}(a), \qquad B = f^{-1}(b).$$

Since $A, B \epsilon [G]$ and $[G]$ is saturated, there exists an element X of $[G]$ so that we have a separation $M - X = M_1 + M_2$, $M_1 \supset A$, $M_2 \supset B$. Thus $N - x = f(M_1) + f(M_2)$, where $x = f(X)$ and $f(M_1) \cdot f(M_2) = 0$, since f is non-alternating. Now since M is compact and f is continuous, it follows that $f(M_1)$ and $f(M_2)$ are separated and obviously $f(M_1) \supset a$, $f(M_2) \supset b$. Hence any two points a, b of N are separated in N by a third point x of N; and accordingly N is a dendrite.

(4.24) THEOREM. *Conversely, if M is a continuum and $f(M) = N$ is a non-alternating transformation of M onto a dendrite N, the collection $[f^{-1}(y)]$, $y \epsilon N$, is non-separated and saturated.*

For the non-separatedness of this collection follows from the non-alternating property of f and the fact that the collection is saturated follows from the continuity of f together with the fact that N is a dendrite.

Note. That, in case M is locally connected, there actually exist non-separated collections in M, and hence there actually exist non-alternating transformations of M onto dendrites, is a consequence of III, (2.31).

(4.3) *Interior transformations.* If every set open in A maps into a set open in B, the transformation $f(A) = B$ is said to be *interior*. This property holds for transformations generated by analytic functions of a complex variable; and as will be borne out by results in later chapters, interior transformations behave in many respects similarly to the analytic transformations.

A collection $[G]$ of sets in a metric space A is said to be *lower semi-continuous* provided that if G is any element of $[G]$ and $\epsilon > 0$, there exists an open set V containing G and such that if G' is any element of $[G]$ intersecting V, we have $V_\epsilon(G') \supset G$. If the elements of $[G]$ are compact, clearly this is equivalent to saying that for any $G \epsilon [G]$ and any $\epsilon > 0$ a $\delta > 0$ exists such that $G' \cdot V_\delta(G) \neq 0$ implies $V_\epsilon(G') \supset G$ for $G' \epsilon [G]$. Likewise, in this case, upper semi-continuity of $[G]$ means that for any $G \epsilon [G]$ and any $\epsilon > 0$, a $\delta > 0$ exists such that $G' \cdot V_\delta(G) \neq 0$ implies $G' \subset V_\epsilon(G)$.

Now a collection $[G]$ which is both upper semi-continuous and lower semi-continuous will be said to be *continuous*. Thus if the elements of $[G]$ are compact sets, $[G]$ is continuous if and only if for each $G \epsilon [G]$ and each $\epsilon > 0$, a $\delta > 0$ exists such that $G' \cdot V_\delta(G) \neq 0$ implies both

$$G' \subset V_\epsilon(G) \text{ and } G \subset V_\epsilon(G'), \qquad G' \epsilon [G].$$

A collection $[G]$ of sets in A will be said to be *continuous in the limit sense* provided that if $G \epsilon [G]$ and (G_n) is a sequence of sets of $[G]$ such that $G \cdot \lim \inf (G_n) \neq 0$, then (G_n) converges to G. If A is compact, this is equivalent to ordinary continuity of the collection as defined above.

(4.31) THEOREM. *If $f(A) = B$ is interior, the collection $[f^{-1}(x)]_{x \epsilon B}$ is continuous in the limit sense. If A is compact and $[f^{-1}(x)]_{x \epsilon B}$ is continuous, $f(A) = B$ is interior.*

Proof. Let $X, X_1, X_2, \cdots$ be elements of $[f^{-1}(x)]_{x \epsilon B}$ such that $X \cdot \lim \inf (X_n) \neq 0$. Let $f(X) = x$, $f(X_n) = x_n$ and let $y \epsilon X$. If y non $\epsilon \lim \inf (X_n)$, there exists an open set U containing y and such that $U \cdot X_n = 0$ for infinitely many integers n. But since $f(U)$ is open and contains x and $x_n \to x$, $U \cdot X_n \neq 0$ for almost all n. Therefore $y \epsilon \lim \inf (X_n)$ so that $X \subset \lim \inf (X_n)$. Since $\lim \sup (X_n) \subset X$ by continuity of f, this gives $X = \lim (X_n)$.

On the other hand let A be compact, let $[f^{-1}(x)]_{x \epsilon B}$ be continuous and let U be any open set in A. If $f(U)$ is not open, there exists a sequence $x_1, x_2, \cdots$ in $B - f(U)$ converging to a point x of $f(U)$. Now the compactness of A and the continuity of our collection give $f^{-1}(x) \cdot \lim \inf (f^{-1}(x_n)) \neq 0$. Hence $U \cdot f^{-1}(x_n) \neq 0$ for almost all n. But this gives $f(U) \supset x_n$ for almost all n contrary to $x_n \epsilon B - f(U)$. Thus f is interior.

COROLLARY. *If A is compact, in order that the transformation $f(A) = B$ be interior it is necessary and sufficient that the collection $[f^{-1}(x)]_{x \epsilon B}$ be continuous.*

As a further consequence of (4.31) and its proof we have the following theorem of Eilenberg.

(4.32) THEOREM. *In order that the transformation $f(A) = B$ be interior it is necessary and sufficient that for each sequence (y_n) in B converging to a point $y \epsilon B$,* $\lim f^{-1}(y_n) = f^{-1}(y)$.

As an immediate consequence we have the following interesting result:

(4.33) THEOREM. *If A and B are metric and $f(A) = B$ is (1-1), continuous and interior, f is a homeomorphism.*

(4.4) *Light transformations.* If for each $x \epsilon B$, $f^{-1}(x)$ is totally disconnected, the transformation $f(A) = B$ is said to be *light*. Since, for compact sets A, this condition is equivalent to the condition

$$\dim f^{-1}(x) = 0, \qquad x \epsilon B,$$

a light transformation is sometimes called a 0-dimensional one. Clearly a light transformation is equivalent in the sense used above in this section to an upper semi-continuous decomposition of A into closed totally disconnected sets.

(4.41) *If A is compact, in order that a continuous transformation $f(A) = B$ be light it is necessary and sufficient that for any $\epsilon > 0$ a $\delta > 0$ exist such that if X is any continuum in B of diameter less than δ, any component of $f^{-1}(X)$ is of diameter less than ϵ.*

The sufficiency is immediate. For if we suppose the condition satisfied and that there is a nondegenerate component H of $f^{-1}(p)$ for some $p \epsilon B$, taking $\epsilon < \delta(H)$ we can find a continuum X in B with $p \epsilon X$ and $\delta(X) < \delta$, whereas the component of $f^{-1}(X)$ containing X is necessarily of diameter greater than ϵ.

To prove the necessity, let us suppose, on the contrary, that there exists a sequence $X_1, X_2, \cdots$ of continua in B with $\delta(X_i) \to 0$ and a sequence $H_1, H_2, \cdots$ of components of $f^{-1}(X_1), f^{-1}(X_2), \cdots$, respectively, with $\delta(H_i) \geqq \epsilon > 0$ for each i. Clearly we may suppose the sequences $[X_i]$ and $[H_i]$ convergent. But this gives $\lim X_i = p \epsilon B$, $\lim H_i = H \subset f^{-1}(p)$, and since $\delta(H) \geqq \epsilon$ this is contrary to the hypotheses that f is light.

(4.5) *A-set reversing transformations.* It will be recalled that a continuous transformation $f(A) = B$ is said to be *A-set reversing* provided that for each $x \epsilon B$, $f^{-1}(x)$ is an A-set in A. If A is a semi-locally-connected continuum, $[f^{-1}(x)]_{x \epsilon B}$ is an upper semi-continuous collection of A-sets filling up A and any transformation associated with such a decomposition of A is A-set reversing. Now since for no $\epsilon > 0$ can there be infinitely many disjoint A-sets in A all of diameter greater than ϵ, it follows that *if $f(A) = B$ is A-set reversing, $[f^{-1}(x)]_{x \epsilon B}$ is a null collection.*

5. **Semi-closed sets and collections.** *A* set K in a metric space S will be said to be *semi-closed* provided each component of K is closed and any convergent sequence of components of K whose limit set intersects $S - K$ converges to a single point of $S - K$.

Similarly, a collection of disjoint sets is said to be semi-closed if each set of G is closed and any convergent sequence of sets of G whose limit set intersects $S - G^*$ converges to a single point of $S - G^*$, where G^* denotes the point set which is the sum of all the sets of the collection G.

For example, any closed set is semi-closed, as is also any totally disconnected set or the sum of any closed set and any set of dimension zero. Any null collection of disjoint closed sets (i.e., a collection having only a finite number of elements of diameter greater than any preassigned $\epsilon > 0$) is semi-closed. The collection of components of any closed set K is semi-closed, as is also this collection together with an arbitrary null collection of disjoint closed sets, no one of which intersects K.

We begin with some results giving fundamental relations between these sets and collections and upper semi-continuous collections.

(5.1) Theorem. *If a collection G of disjoint closed sets in a compact space S is upper semi-continuous, then in order that G be semi-closed it is necessary and sufficient that the decomposition of S into the sets of G and the individual points of $S - G^*$ be upper semi-continuous.*

This theorem follows immediately from the definitions of semi-closed and of upper semi-continuous collections.

(5.11) COROLLARY. *If G is any upper semi-continuous collection of disjoint closed sets filling up S, and if G_0 is the set of all nondegenerate elements of the collection G, then any subcollection G_1 of G such that $G_0 \subset G_1 \subset G$ is semi-closed.*

(5.2) *The collection G of all components of any semi-closed set K in a compact space S is upper semi-continuous.*

For if this were not so, there would exist a convergent sequence $g_1, g_2, \cdots$ of sets of G such that if $L = \lim (g_i)$, then for some $g \in G$ we have

$$L \cdot g \neq 0 \neq L \cdot (S - g).$$

Since L is connected and g is a component of K, it follows that $L \cdot (S - K) \neq 0$. But then (g_i) would necessarily converge to a single point of $S - K$, contrary to $L \cdot g \neq 0$.

(5.3) THEOREM. *If K is a semi-closed subset of a compact space S, there exists a monotone transformation $T(S) = S'$ of S into a compact space S' such that for each $y \in S'$, $T^{-1}(y)$ is either a component of K or a single point of $S - K$. Thus T is a homeomorphism on $S - K$.*

For by (5.2) and (5.1) the decomposition of S into the components of K and the individual points of $S - K$ is upper semi-continuous. Hence by (4.1) this decomposition is equivalent to a monotone transformation $T(S) = S'$ satisfying all the requirements of (5.3).

(5.4) THEOREM. *In order that K be semi-closed it is sufficient, and in case the space is compact it is also necessary, that for each $\epsilon > 0$ the sum of all the components of K of diameter greater than or equal to ϵ be a closed set.*

Proof. To prove the sufficiency, let $g_1, g_2, \cdots$ be any convergent sequence of components of K whose limit set L intersects $S - K$. Then this sequence must be a null sequence. For if it contains an infinite subsequence (g_{n_i}), each element of which is of diameter greater than some preassigned $\epsilon > 0$, it would follow from our hypothesis that $\sum g_{n_i}$ is contained in a closed set which in turn is contained in K, and this is impossible since (g_{n_i}) converges to L and $L \cdot (S - K) \neq 0$. Thus (g_i) is a null sequence and hence L must reduce to a single point of $S - K$.

To prove the necessity, let us suppose S is compact, let $\epsilon > 0$ be given, and let K_ϵ be the sum of all components of K of diameter greater than or equal to ϵ. Then if K_ϵ is not closed, it readily follows that there exists in K_ϵ a convergent sequence of components $g_1, g_2, \cdots$ of K whose limit set L is not contained in K_ϵ. But this is impossible because $\delta(g_i) \geqq \epsilon$ $(i = 1, 2, \cdots)$ gives $\delta(L) \geqq \epsilon$. Also L is connected, and since K is semi-closed, we must have $L \subset K$. Thus L is contained in some component g of K which in turn belongs to K_ϵ.

(5.41) COROLLARY. *If S is compact and $K \subset S$ is semi-closed, the sum of all nondegenerate components of K is an F_σ.*

We note also the following results involving additivity properties of semi-closed collections.

(5.5) *If X and Y are disjoint subsets of a compact space S and $[G_x]$ and $[G_y]$ are upper semi-continuous semi-closed collections in S filling up X and Y, respectively, $[G_x] + [G_y]$ is an upper semi-continuous semi-closed collection in S filling up $X + Y$.*

This follows at once from the definitions.

(5.6) *If $[G]$ is any null family of disjoint closed sets in a metric space S and if then each set G of $[G]$ be decomposed upper semi-continuously, the resulting total collection is semi-closed and upper semi-continuous.*

For let $[g]$ be the resulting collection and let $A = G^* = g^*$ be the set obtained by adding together all elements of $[G]$ or $[g]$. If $p \in S - A$ and $\lim g_n \supset p$ where $g_n \in [g]$, we have $\liminf G_n \supset p$ where $G_n \in [G]$ and $G_n \supset g_n$. Since (G_n) is a null sequence, this gives $\lim G_n = \lim g_n = p$. Thus $[g]$ is semi-closed.

Now let $g \in [g]$ and let U be any open set containing g. Then if G is the element of $[G]$ containing g, there exists an open set V_1 with $g \subset V_1 \subset U$ and such that any element of $[g]$ which lies in G and intersects V_1 must lie wholly in U because the decomposition of G is upper semi-continuous. Now since $[G]$ is a null family and g is compact, there exists an open set V with $g \subset V \subset V_1$ and such that any element of $[g]$ which is not in G and which intersects V must lie wholly in U. Thus *any* element of $[g]$ which intersects V lies in U and accordingly $[g]$ is upper semi-continuous.

(5.7) THEOREM. *Let $T(A) = B$ be continuous, where A and B are compact, and let B_1 be a subset of B such that $[T^{-1}(x)]$, $x \in B_1$, is a semi-closed collection. Then T can be factored into the form $T(A) = T_2T_1(A) = T_2(A') = B$, where T_1 is (1-1) on $A - T^{-1}(B_1)$ and T_2 is (1-1) on $T_2^{-1}(B_1)$.*

Proof. Let $A_1 = T^{-1}(B_1)$. Let A' be the hyperspace of the upper semi-continuous decomposition of A into the sets $[T^{-1}(x)]$, $x \in B_1$, and the individual points of $A - A_1$ and let $T_1(A) = A'$ be the associated transformation. For each $x \in A'$, let $T_2(x) = TT_1^{-1}(x)$. Then since T is constant on any set $T_1^{-1}(x)$, $x \in A'$, it follows that T_2 is single-valued and continuous and for each $p \in A$, $T(p) = T_2T_1(p)$. Also, T_1 is (1-1) on $A - A_1$ by definition; and T_2 is (1-1) on $T_1(A_1)$, because if $x, y \in T_1(A_1)$, $x \neq y$, we have $T^{-1}T_2(x) = T_1^{-1}(x) \neq T_1^{-1}(y) = T^{-1}T_2(y)$ so that $T_2(x) \neq T_2(y)$.

(5.8) Now suppose A is compact and $f(A) = B$ is continuous. For each $y \in B$ let us decompose $f^{-1}(y)$ upper semi-continuously into a collection $[g]_y$ so that if $y_1, y_2, \cdots$ is a sequence of distinct points in B and $g_1, g_2, \cdots$ is a convergent sequence of sets where $g_n \in [g]_{y_n}$ and $L = \lim [g_n] \subset f^{-1}(y_0)$, then $L \subset$ some $g_0 \in [g]_{y_0}$. It results as in (5.6) that the decomposition of A into the sets $[g]_y$ for all $y \in B$ is upper semi-continuous. Let $f_1(A) = A'$ be the associated transformation and for each $x' \in A'$ let us define

$$f_2(x') = ff_1^{-1}(x').$$

It is readily seen that this transformation is continuous and for each $x \in A$ we have

$$f_2 f_1(x) = f(x).$$

Thus we have a *factorization* of f corresponding to the decompositions of the sets $f^{-1}(y)$, $y \in B_0$.

(5.81) Now let B_1 be a subset of B such that the collection $[G_1] = [f^{-1}(y)]_{y \in B_1}$ is a null family and so that $[G_2] = [f^{-1}(y)]_{y \in B - B_1}$ is semi-closed. Then if each element of $[G_1]$ be decomposed upper semi-continuously in an arbitrary way, the resulting collection will be semi-closed (by 5.6) and upper semi-continuous. Thus if this collection is augmented by $\left\{\begin{matrix}\text{(1) the individual points of } A - f^{-1}(B_1) \\ \text{(2) the sets of } [G_2]\end{matrix}\right\}$ by (5.5) we obtain an upper semi-continuous decomposition of A. If $f_1(A) = A'$ is the associated transformation and f_2 is defined as above so that

$$f(x) = f_2 f_1(x), \qquad x \in A,$$

then $\left\{\begin{matrix}\text{(1) } f_1 \text{ is (1-1) on } A - f^{-1}(B_1) \\ \text{(2) } f_2 \text{ is (1-1) on } f_1 f^{-1}(B - B_1)\end{matrix}\right\}$.

An interesting case where the conditions of (5.81) [hence also (5.8)] are satisfied is the following

(5.82) *If A is a semi-locally-connected continuum, $f(A) = B$ is monotone and B_1 is a subset of B on which f is A-set reversing, the conditions of* (5.81) *are satisfied.* (Thus the elements of $[f^{-1}(y)]_{y \in B_1}$ may be decomposed (upper semi-continuously) at will to yield factorizations of f.)

Proof. Since the elements of the collection $[f^{-1}(y)]$, $y \in B_1$, are disjoint A-sets they must form a null collection. To prove that $[f^{-1}(y)]$, $y \in B - B_1$, is semi-closed, let $X_1, X_2, \cdots$ be a convergent sequence of elements of this collection whose limit set L intersects $f^{-1}(B_1)$. Let y be a point of B_1 such that $L \cdot f^{-1}(y) \neq 0$. Since $f^{-1}(y)$ is an A-set and the sets X_i are disjoint continua not intersecting $f^{-1}(y)$ and since by continuity of f, $L \subset f^{-1}(y)$, it follows from [IV, (4.1)] that L must reduce to a single point. This proves (5.82).

6. **Null collections.** We recall that a collection $[G]$ is a *null collection* or a *null family* provided that for any $\epsilon > 0$ at most a finite number of elements of $[G]$ are of diameter greater than ϵ. The results here given are largely a recapitulation of what has already been proved in the preceding sections as applied in particular to null collections.

We note first

(6.1) *Any null collection of disjoint closed sets is upper semi-continuous and semi-closed.*

From §5 we have

(6.2) *If $f(A) = B$ is continuous and $[f^{-1}(y)]_{y \in B}$ is a null collection of compact sets, then for any subset B_1 of B, f can be factored in the form $f = f_2 f_1$ so that*

$f_1(A) = A'$ is equivalent to f on $f^{-1}(B_1)$ and is (1-1) on $A - f^{-1}(B_1)$ and f_2 is (1-1) on $f_1[f^{-1}(B_1)]$.

(6.3) *Let $f(A) = B$ be continuous, suppose $[f^{-1}(y)]_{y \in B}$ is a null collection of compact sets and let B_1 be a subset of B. For each $y \in B_1$ let $f^{-1}(y)$ be decomposed upper semi-continuously in an arbitrary way. Then f can be factored into the form $f = f_3 f_2 f_1$ where f_1 is equivalent to f on $A - f^{-1}(B_1)$, f_2 is equivalent to the separate decompositions of $f^{-1}(y)$ for each $y \in B_1$ and (1-1) on $f_1(A) - f_1 f^{-1}(B_1)$ and f_3 is equivalent to f on $f_2 f_1 f^{-1}(B_1)$ and (1-1) elsewhere.*

To achieve this factorization we first apply (6.2) to factor f into the form $f = f_2' f_1$. Then use the decomposition of $f_1(A)$ effected by decomposing the sets $f_1 f^{-1}(y)$, $y \in B_1$, as given to set up the transformation $f_2 f_1(A) = A''$. Finally

$$f_3(x'') = f f_1^{-1} f_2^{-1}(x'').$$

As interesting cases where such a factorization result can be applied we note

(6.4) *If A is a semi-locally-connected continuum and $f(A) = B$ is A-set reversing, the collection $[f^{-1}(y)]_{y \in B}$ is a null collection.*

(6.5) *If A is a hereditarily locally connected continuum and $f(A) = B$ is monotone, $[f^{-1}(y)]_{y \in B}$ is a null collection.*

7. **Homeomorphism of the original and the image set.** If $f(A) = B$ is a continuous transformation, the question of under what conditions on f will B be homeomorphic with A is of fundamental significance. If f is a topological transformation this is obviously the case. However, in many cases of importance a much weaker condition on the transformation will insure that B and A are necessarily homeomorphic even though f itself is not a homeomorphism.

In this section only a weak form of the general problem will be handled, namely that in which all but one of the sets $f^{-1}(y)$, $y \in B$, reduce to single points. This case, however, suffices as a basis for far reaching conclusions which will be developed in a later chapter.

(7.1) THEOREM. *Suppose A is compact, $f(A) = B$ is continuous, $f^{-1}(y)$ reduces to a single point except for one point $y = b \in B$ and suppose that for each $y \in B$ there exists a point $x \in A$ such that $A - f^{-1}(y)$ is homeomorphic with $A - x$. Then B is homeomorphic with A.*

Proof. Let a be a point so that we have a homeomorphism.

$$h(A - a) = A - f^{-1}(b).$$

Then since $f[A - f^{-1}(b)] = B - b$ is a homeomorphism, so also is

$$g(A - a) = f[h(A - a)] = f[A - f^{-1}(b)] = B - b.$$

Let us define

$$g(a) = b.$$

Let (x_n) be a sequence in A converging to a. The set $[g(x_n)]$ must have a limit point p in B; and since if $p \neq b$, $g^{-1}(p) \subset A - a$ whereas g maps $A - a$ topologically onto $B - b$, it follows that $p = b$. Thus g is continuous and since it is (1-1) it therefore maps A onto B topologically.

This argument shows essentially that h^{-1} has a continuous extension $h_1(A) = A$ to A which generates the same decomposition of A as does f. Then when $g(A) = B$ is defined by $g(x) = fh_1^{-1}(x)$, $x \in A$, g is 1-1 and continuous and thus is topological. From this viewpoint we readily obtain the following generalization to a vastly less restrictive situation.

We define a transformation $f(A) = B$ to be *closed* provided the image of every closed set in A is closed in B. Then we have

(7.2) THEOREM. *Suppose $f(A) = B$ is closed and continuous where A and B are metric spaces. If there exists a closed continuous transformation $h(A) = A$ of A onto A which generates the same decomposition of A as does f, then B is homeomorphic with A.*

For in this case the transformation $g(A) = B$ defined by $g(x) = fh^{-1}(x)$, $x \in A$, is readily seen to be (1-1); and both g and g^{-1} are closed since h and f are both closed and continuous. Hence g is a homeomorphism. (See Note 9, p. 280.)

REFERENCES

§§1–3. Moore [13, 1], Alexandroff [5], Kuratowski [5], Kerékjártó [1], Vietoris [1]. See also Roberts [3].

§4. (4.1) Moore [13], Morrey [1], Vietoris [2].

(4.2) Whyburn [22, 23].

(4.3) Stoïlow [1], Eilenberg [4].

(4.4) Whyburn [24].

§5. Whyburn [25]. In this connection see also Wilson [1]. In connection with §7, see Wardwell [1].

CHAPTER VIII

GENERAL PROPERTIES. FACTORIZATION

1. **Preliminaries.** We begin with some results concerning single-valued and continuous transformations which are of lemma character.

(1.1) If $f_1(A) = A'$, $f_2(A') = B$ are single-valued transformations and $f = f_2f_1$, i.e., $f(x) = f_2f_1(x)$ for each $x \in A$, we have

(i) for any $X \subset A'$, $f_2(X) = f_2f_1f_1^{-1}(X) = ff_1^{-1}(X)$;

(ii) for any $Y \subset B$, $f_2^{-1}(Y) = f_1f_1^{-1}f_2^{-1}(Y) = f_1f^{-1}(Y)$;

(iii) for any $Y \subset B$, $f^{-1}(Y) = f_1^{-1}f_2^{-1}(Y)$.

The first equality in (i) results from $f_1f_1^{-1}(X) = X$ and the second from $f_2f_1 = f$. The first equality in (ii) is merely $f_1f_1^{-1}[f_2^{-1}(Y)] = f_2^{-1}(Y)$ and the second results from (iii) which in turn is a direct consequence of $f = f_2f_1$.

(1.2) *If $f(A) = B$ is continuous and Q is a quasi-component of a subset A_0 of A, $f(Q)$ is contained in a single quasi-component of $f(A_0)$.*

For if there were a separation

$$f(A_0) = B_1 + B_2$$

between $x_1 \in B_1 \cdot f(A)$ and $x_2 \in B_2 \cdot f(Q)$,

$$A_0 = A_0 \cdot f^{-1}f(A_0) = A_0 \cdot f^{-1}(B_1) + A_0 \cdot f^{-1}(B_2)$$

would be a separation of A_0 between a point of $A_0 \cdot f^{-1}(x_1)$ and a point of $A_0 \cdot f^{-1}(x_2)$, both of which belong to Q.

(1.3) *If A is compact, $f(A) = B$ is continuous and E is any subset of B, then for any separation $f^{-1}(E) = F_1 + F_2$, where both $f(F_1)$ and $f(F_2)$ intersect a single quasi-component Q of E, there exists a point $p \in E$ such that $f^{-1}(p)$ intersects both F_1 and F_2.*

For if there were no such point p, $f(F_1)$ and $f(F_2)$ would be disjoint. Then since $E = f(F_1) + f(F_2)$ cannot be a separation [because $Q \cdot f(F_1) \neq 0 \neq Q \cdot f(F_2)$], there exists a point p in $f(F_1)$, say, which is a limit point of $f(F_2)$. But then $f^{-1}(p)$ would contain a limit point of $f^{-1}f(F_2) = F_2$ (by the compactness of A and the continuity of f) and this is impossible since $f^{-1}(p) \subset F_1$.

2. **Characterizations of monotone and non-alternating transformations.** If $f(A) = B$ is a single-valued transformation, a subset X of A which is the inverse of its transform under f will be called an *inverse set.* In other words a set X satisfying

$$X = f^{-1}f(X)$$

is an inverse set. Clearly for any subset Y of B, $f^{-1}(Y)$ is an inverse set.

A continuous transformation $f(A) = B$ has been defined to be non-alternating provided that for any $x \epsilon B$ and any separation

$$A - f^{-1}(x) = A_1 + A_2,$$

the sets A_1 and A_2 are inverse sets [see VII, (4.2)]. Clearly this is equivalent to saying that for any $y \epsilon B$, $f^{-1}(y)$ is contained in a single quasi-component of any set of the form $A - f^{-1}(x)$, $x \epsilon B - y$, and this in turn is the same as saying that the quasi-components of sets of the form $A - f^{-1}(x)$, $x \epsilon B$, are inverse sets. Thus we have

(2.1) *In order that a continuous transformation $f(A) = B$ be non-alternating it is necessary and sufficient that the quasi-components of any set of the form $A - f^{-1}(x)$ be inverse sets.*

We now apply the lemmas in the preceding section to obtain further useful characterizations of monotone and non-alternating transformations.

(2.2) *If A is compact and $f(A) = B$ is continuous, in order that f be monotone it is necessary and sufficient that the inverse of every connected set in B be connected.*

The sufficiency of the condition is immediate, since any single point in B is a connected set.

To prove the necessity, let E be a connected subset of B. By (1.3) if there were a separation

$$f^{-1}(E) = F_1 + F_2$$

there would be a point $p \epsilon E$ whose inverse would intersect both F_1 and F_2. This is impossible because $f^{-1}(p)$ is connected. Hence $f^{-1}(E)$ is connected.

(2.21) *Unicoherence is invariant under monotone transformations on continua.*

For let $f(A) = B$ be monotone and suppose $B = H + K$ where H and K are continua. Then $A = f^{-1}(H) + f^{-1}(K)$. Since $f^{-1}(H)$ and $f^{-1}(K)$ are continua, $f^{-1}(H) \cdot f^{-1}(K) = f^{-1}(H \cdot K)$ is a continuum. Accordingly $H \cdot K$ is a continuum.

(2.22) *The properties of being* (i) *an hereditarily locally connected continuum* (ii) *a regular curve* (iii) *a rational curve are invariant under monotone transformations.*

For let $f(A) = B$ be monotone where A is compact. Then if B has a continuum of convergence K which is the limit of a sequence of disjoint continua $[K_i]$ all of diameter greater than ϵ and none intersecting K, and if $\delta > 0$ is chosen so that any set of diameter less than δ in A maps into a set of diameter less than ϵ, $[f^{-1}(K_i)]$ is a sequence of disjoint continua in A all of diameter greater than δ and hence a suitably chosen subsequence will converge to a non-degenerate subcontinuum of $f^{-1}(K)$. Thus if B has a continuum of convergence, so has A, and (i) is established. To prove (ii) and (iii), let $a, b \epsilon B$. Then if A is a regular (rational) curve, $f^{-1}(a)$ and $f^{-1}(b)$ can be separated in A by a finite (countable closed) set E; and it follows at once from (1.3) that $f(E)$ separates a and b in B, so that B is regular (rational).

(2.3) THEOREM. *If A is compact, in order that a continuous transformation*

$f(A) = B$ be non-alternating it is necessary and sufficient that for any $x \in B$ and any quasi-component Q of $B - x$, $f^{-1}(Q)$ be a quasi-component of $A - f^{-1}(x)$.

Proof. To establish the sufficiency, let $x, y \in B$ and let Q be the quasi-component of $B - x$ containing y. Then since $f^{-1}(Q) \supset f^{-1}(y)$, $f^{-1}(x)$ cannot separate $f^{-1}(y)$ in A.

To prove the necessity, let us choose $E = B - x$ and apply (1.3). Then if there existed a separation of $f^{-1}(E)$ between two points of $f^{-1}(Q)$, there would be a point $p \in E$ such that $f^{-1}(x)$ separates $f^{-1}(p)$ in A, contrary to the non-alternating property of f. Thus

(i) $$f^{-1}(Q) \subset K,$$

where K is some quasi-component of $A - f^{-1}(x)$. But by (1.2) $f(K)$ is contained in some single quasi-component of $B - x$; and since $f(K) \supset Q$, this must be Q. Thus

(ii) $$f(K) \subset Q.$$

Hence by (i)

$$f(K) = Q.$$

Whence,

$$f^{-1}(Q) \supset K.$$

Thus by (i)

$$f^{-1}(Q) = K.$$

(2.31) Corollary. *If A is a locally connected continuum, in order that $f(A) = B$ be non-alternating it is necessary and sufficient that for each $x \in B$ and each component Q of $B - x$, $f^{-1}(Q)$ be connected.*

(2.4) Theorem. *Let $f(A) = B$ be continuous where A is a semi-locally-connected continuum. In order that f be non-alternating it is necessary and sufficient that:*

(i) *for each $y \in B$ and each $a, b \in f^{-1}(y)$, $E(a, b) \subset f^{-1}(y)$;*

(ii) *f be non-alternating on each cyclic element of A.*

Proof. To prove the necessity of (i), suppose there were a point $p \in E(a, b) - E(a, b) \cdot f^{-1}(y)$. Let $f(p) = x$. Then since $f^{-1}(x)$ contains p but contains neither a nor b, $f^{-1}(x)$ separates the two points a and b of $f^{-1}(y)$ in A, contrary to the non-alternating property of f.

To show condition (ii) necessary, let E be a true cyclic element of A, let $f(E) = E'$ and $x, y \in E'$. If there existed a separation of $E - E \cdot f^{-1}(x)$ between two points y_1, y_2 of $f^{-1}(y)$, by [IV, (3.22)] there would exist a separation of $A - E \cdot f^{-1}(x)$ between y_1 and y_2 contrary to hypothesis. Thus f is non-alternating on E.

To establish the sufficiency of the conditions, we suppose them satisfied and let $x, y \in B$. Suppose there does exist a separation

$$A - f^{-1}(x) = A_1 + A_2$$

where $a \subset A_1$, $b \subset A_2$ and $a, b \in f^{-1}(y)$. Let

$$K = E(a, b) + a + b,$$

$$K_1 = K \cdot A_1, \qquad K_2 = K \cdot A_2 .$$

Then the sets K_1 and K_2 are compact and disjoint. Since K is naturally ordered, there exists a last point a_k in K_1; and since all the points of K following a_k are in K_2 and hence cannot have a_k as a limit point, there exists a first point b_k of K following a_k and $b_k \subset K_2$. Then a_k and b_k are conjugate, because there is no point of K between them. Hence they belong to some cyclic element E of A; but then $E \cdot f^{-1}(x)$ separates the two points a_k and b_k of $E \cdot f^{-1}(y)$ in E contrary to (ii). This completes the proof.

(2.41) *Any non-alternating transformation on a dendrite is necessarily monotone.*

For if A is a dendrite, $f(A) = B$ is non-alternating, $y \in B$ and $a, b \in f^{-1}(y)$, then since $E(a, b) + a + b$ is the simple arc ab in A it results that $f^{-1}(y)$ is connected.

3. Composite transformations. Product theorems. We consider a continuous transformation $f(A) = B$ on a compact set A which is obtained by applying successively two continuous "factor" transformations

$$f_1(A) = A',$$

$$f_2(A') = B$$

so that $f_2 f_1(x) = f(x)$ for each $x \in A$. We proceed to establish the following product theorems.

(3.1) *If f is interior, so also is f_2.*

This results from (1.1), (i), because if X is an open subset of A', $f_1^{-1}(X)$ is open and hence $ff_1^{-1}(X) = f_2(X)$ is open by virtue of the interiority of f.

(3.2) *If f is monotone, so also is f_2.*

This follows by (1.1) (ii), because if Y is connected, $Y \subset B$, $f^{-1}(Y)$ is connected by the monotoneity of f and thus $f_1 f^{-1}(Y) = f_2^{-1}(Y)$ is connected. Hence, by (2.2), f_2 is monotone.

(3.3) *If f is non-alternating, so also is f_2.*

For, taking $Y = B - y$, where $x, y \in B$, we have $f^{-1}(x)$ contained in a single quasi-component Q of $f^{-1}(Y)$. Thus, by (1.2), $f_1 f^{-1}(x)$ is contained in a single quasi-component of $f_1 f^{-1}(Y)$. By (1.1), (ii),

$$f_1 f^{-1}(x) = f_2^{-1}(x),$$

$$f_1 f^{-1}(Y) = f_2^{-1}(Y) = A - f_2^{-1}(y).$$

Accordingly, f_2 is non-alternating.

(3.4) *If f_1 and f_2 are monotone, so also is f.*

For if Y is any connected set in B, $f_2^{-1}(Y)$ and $f_1^{-1}f_2^{-1}(Y)$ are connected since f_1 and f_2 are monotone. Thus $f^{-1}(Y)$ is connected and hence f is monotone.

(3.5) *If f_1 is monotone and f_2 is non-alternating, f is non-alternating.*

For if $Y = B - y$, where $x, y \in B$, $f_2^{-1}(x)$ lies in a single quasi-component Q of $f_2^{-1}(Y)$ because f_2 is non-alternating. Then, taking $E = f_2^{-1}(Y)$ and applying (1.3), since for each $p \in E, f_1^{-1}(p)$ is connected it follows that $f_1^{-1}(Q)$ lies in a single quasi-component of $f_1^{-1}(E) = f_1^{-1}f_2^{-1}(Y) = f^{-1}(Y)$. Since $f_1^{-1}(Q) \supset f_1^{-1}f_2^{-1}(x) = f^{-1}(x)$, it follows that f is non-alternating.

(3.6) *If A' is semi-locally-connected, then if f_2 is A-set reversing and f_1 is non-alternating, f is non-alternating.*

For let $x \in B$ and let Q be a quasi-component of $B - x$. Then $f_2^{-1}(Q)$ is a quasi-component of $A' - f_2^{-1}(x)$ and hence is a component of $A' - x'$ for some $x' \in f_2^{-1}(x)$ because $f_2^{-1}(x)$ is an A-set in A'. Hence by (2.3), $f_1^{-1}f_2^{-1}(Q)$ is a quasi-component of $A - f_1^{-1}(x')$. Since

$$f_1^{-1}f_2^{-1}(Q) = f^{-1}(Q)$$

and

$$f^{-1}(Q) \subset A - f^{-1}(x) \subset A - f_1^{-1}(x'),$$

using the fact that $f^{-1}(Q)$ is open, it follows that $f^{-1}(Q)$ is a quasi-component of $A - f^{-1}(x)$. Accordingly, f is non-alternating.

(3.7) *If f_1 and f_2 are interior, so also is f.*

4. **Factorization.** If $f(A) = B$ is continuous, any representation of f in the form

$$f(x) = f_2f_1(x), \qquad x \in A,$$

where the transformations

$$f_1(A) = A', \qquad f_2(A') = B$$

are continuous will be called a *factorization* of f. Numerous results concerning the relation between the composite transformations f and its factors f_1 and f_2 were established in the preceding section. Also, in Chapter VII, certain sufficient conditions were developed. By far the most definite and useful existence theorem for factorization is the following

(4.1) Factor theorem. *If A is compact and $f(A) = B$ is continuous there exists a unique factorization*

$$f(x) = f_2f_1(x), \qquad x \in A$$

such that

$$f_1(A) = A'$$

is monotone and

$$f_2(A') = B$$

is light.

Proof. Let us decompose A into the components of the sets $f^{-1}(y)$, $y \in B$. In other words we form the collection $[g]$ of sets such that for each $x \in A$, the set g of $[g]$ containing x is the component of $f^{-1}f(x)$ which contains x. Since for each fixed $y \in B$, $f^{-1}(y)$ is thus decomposed upper semi-continuously, and since if $y_1, y_2, \cdots$ is a sequence in B converging to $y \in B$ and $g_{y_1}, g_{y_2}, \cdots$ are components of $f^{-1}(y_1), f^{-1}(y_2), \cdots$ converging to a limit set L then L is connected and contained in $f^{-1}(y)$ and hence is in some single element g of $[g]$, it follows that this decomposition is upper semi-continuous [compare with VII, (5.8)]. Let $f_1(A) = A'$ be the transformation associated with this decomposition. Since the sets $g \in [g]$ are connected, f_1 is monotone.

Now for each $x' \in A'$ let us define

$$f_2(x') = ff_1^{-1}(x').$$

Obviously we have

$$f(x) = f_2f_1(x), \qquad x \in A.$$

Now f_2 is continuous. For if $x_1', x_2', \cdots$ is a sequence in A' converging to $x' \in A'$, we have

$$\limsup f_1^{-1}(x_n') \subset f_1^{-1}(x').$$

Thus almost all the sets $f_1^{-1}(x_n')$ lie in an arbitrarily chosen neighborhood of $f_1^{-1}(x')$ and hence the sequence $[ff_1^{-1}(x_n')] = [f_2(x_n')]$ must converge to $ff_1^{-1}(x') = f_2(x')$ by virtue of the continuity of f.

Also f_2 is light. For if X' is any component of $f_2^{-1}(y)$, $y \in B$, then since f_1 is monotone, $f_1^{-1}(X')$ is connected and hence is contained in a single component X of $f^{-1}(y) = f_1^{-1}f_2^{-1}(y)$. But $f_1(X)$ is a single point $x' \in A'$ by the definition of f_1. Hence $X' = x'$ is a single point.

Now to prove that this monotone-light factorization is unique, let

$$f(x) \equiv g_1g_2(x)$$

be any such factorization, i.e.,

$$g_1(A) = A''$$

is monotone and

$$g_2(A'') = B$$

is light. Let $x' \in A''$. Since g_1 is monotone, $g_1^{-1}(x')$ is connected; and since if $y = g_2(x')$,

$$g_1^{-1}(x') \subset g_1^{-1}g_2^{-1}(y) = f^{-1}(y),$$

$g_1^{-1}(x')$ is contained in some single component X of $f^{-1}(y)$. But since $g_1(X)$ is connected, $g_1(X) \subset g_2^{-1}(y)$, and $g_2^{-1}(y)$ is totally disconnected, $g_1(X)$ must reduce to the single point x'. Hence $g_1^{-1}(x') = X$ and therefore g_1 is equivalent to f_1. That g_2 must be equivalent to f_2 results now from

$$g_2(x') = fg_1^{-1}(x'), \qquad x' \in A'',$$

and

$$f_2(x') = ff_1^{-1}(x'), \qquad x' \in A'.$$

DEFINITION. *The factorization just obtained will be called the monotone-light factorization of the continuous transformation* $f(A) = B$.

As a direct consequence of (4.1) and (3.1) we have the following factor theorem for interior transformations:

(4.2) THEOREM. *Any interior transformation* $f(A) = B$ *on a compact set* A *can be factored uniquely into the form*

$$f(x) = f_2f_1(x), \qquad x \in A,$$

so that the transformation

$$f_1(A) = A'$$

is monotone and

$$f_2(A') = B$$

is interior and light.

5. **Retractions.** We recall that a continuous transformation $f(X) = Y$ is said to be *retracting* and to *retract* X into Y provided $X \supset Y$ and $f(x) = x$ for every $x \in Y$, i.e., every point of Y is *fixed* under f. In Chapter IV A-sets in semi-locally-connected continua were characterized by means of such retractions. We propose here to sharpen the result in this connection somewhat by obtaining the following

(5.1) THEOREM. *If* A *is an* A*-set in a connected set* M, *there exists one and only one non-alternating retraction of* M *into* A, *namely the transformation*

$$r(x) = x \qquad \text{for } x \in A,$$

$$r(x) = F(G_x) \qquad \text{for } x \in M - A,$$

where G_x *is the open set in* $M - A$ *which contains* x *and is bounded by a single point of* A. *Furthermore* $r(x)$ *is* A*-set reversing and hence monotone in the extended sense* [*i.e.*, $r^{-1}(y)$ *is connected for each* $y \in A$].

Proof. In the first place, $r(x)$ as here defined is continuous, since $M - A$ is made up of a finite or null sequence of disjoint open sets $[G_i]$ each bounded by a single point of A.

Also since for each $y \in A$, $r^{-1}(y) = y +$ all sets G_i with $F(G_i) = y$, it is clear that $M - r^{-1}(y)$ is an open set bounded by y. Hence $r^{-1}(y)$ is an A-set—in fact a nodal set—and therefore $r(x)$ is A-set reversing.

Now to prove that $r(x)$ is the only possible non-alternating retraction of M into A, let $f(M) = A$ be any such retraction, suppose $y = r(p) \neq f(p)$ for some p and let G_y be the sum of all those open sets G_i in $M - A$ as mentioned above which are bounded by y. Since $[G_i]$ is a null sequence, it follows that $F(G_y) = y$. Now by definition of r we have $p \in G_y$. But if $x = f(p)$ then since $x = f(x)$, if we let $C = M - G_y$ we have the separation

$$M - f^{-1}(y) = [G_y - G_y \cdot f^{-1}(y)] + [C - C \cdot f^{-1}(y)]$$

between the two points p and x of $f^{-1}(x)$ contrary to the fact that f is non-alternating. Accordingly

$$f(x) = r(x), \qquad x \in M.$$

6. The mapping of cyclic elements and A-sets under monotone and non-alternating transformations. Let A be a continuum, let K be the set of all cut points of A and suppose $f(A) = B$ is non-alternating. Under this assumption we obtain first some conclusions concerning the mapping of cut points and simple links under f.

For reference we list a part of (2.4) as

(6.11) *If $y \in B$ and $a, b \in f^{-1}(y)$, then $E(a, b) \subset f^{-1}(y)$.*

Now let $x \in K$. If $B - f(x)$ is connected, $A - f^{-1}f(x)$ is contained in a single quasi-component of itself and hence is connected. Thus since x is a cut point of A, either $f(x)$ is a cut point of B or $f^{-1}f(x)$ contains an open set in B. Hence in particular, *if f is interior or light, any cut point of A maps into a cut point of B.* In any case we have

(6.12) *All save possibly a countable number of points of $f(K)$ are cut points of B. Thus for any simple link E_b of B, $E_b \cdot f(K)$ is countable.*

As a result of (6.11) we have

(6.13) *For any $y \in B - f(K)$, there exists a simple link or an end point E_a^y of A which contains $f^{-1}(y)$.*

Now let E be any simple link in B, let $x, y \in E_b - E_b \cdot f(K)$, let E_a^x and E_a^y be determined as in (6.13) so that $E_a^x \supset f^{-1}(x)$ and $E_a^y \supset f^{-1}(y)$. Suppose $E_a^y \neq E_a^x$. Then in any case there exists a point p separating $E_a^x - p$ and $E_a^y - p$ in A. Since $p \in K$, $x \neq f(p) \neq y$. Thus $f^{-1}f(p)$ separates $f^{-1}(x)$ and $f^{-1}(y)$ in A. Accordingly, by (2.3), $f(p)$ would separate x and y in B; and this is absurd since $x, y \in E_b$. Thus we have

(6.14) *For any simple link E_b of B there exists a simple link E_a of A such that $E_a \supset f^{-1}[E_b - E_b \cdot f(K)]$.*

For by the reasoning just given we have only to set $E_a = E_a^x$ for any $x \in E_b - E_b \cdot f(K)$ and we have $E_a = E_a^x = E_a^y \supset f^{-1}(y)$ for any $y \in E_b - E_b \cdot f(K)$.

This leads readily to

(6.2) THEOREM. *If A is a continuum and $f(A) = B$ is non-alternating, then for any simple link E_b of B there exists one and only one simple link E_a of A such that $f(E_a) \supset E_b$. For any other simple link E in A, $f(E) \cdot E_b$ contains at most one point.*

To prove this we have only to let E_a be determined as in (6.14) so that

$$f(E_a) \supset E_b - E_b \cdot f(K).$$

Then since $E_b \cdot f(K)$ is countable [see (6.12)], and $f(E_a)$ is compact, this gives

$$f(E_a) \supset E_b.$$

The uniqueness of E_a results from (6.14). To prove the last statement, suppose $f(E) \cdot E_b$ contained more than one point. Then since this set is connected it would contain a point y of $E_b - E_b \cdot f(K)$; and by (6.14) this would give

$$E = E_a.$$

(6.21) *The property of being a dendrite is invariant under non-alternating transformations.*

For if B has a true cyclic element so also has A.

(6.22) *Under the conditions of* (6.2), *if a subcontinuum A_0 of A has the property that "$E_a \cdot A_0$ nondegenerate" implies "$E_a \subset A_0$" for any simple link E_a of A, $f(A_0)$ has this same property in B.*

For if $E_b \cdot f(A_0)$ is nondegenerate (E_b a simple link of B), this set contains at least two points not belonging to $f(K)$. Hence the set E_a corresponding to E_b in (6.14) must intersect A_0 in at least two points. Thus $E_a \subset A_0$ so that $f(A_0) \supset f(E_a) \supset E_b$.

(6.23) *If A is a semi-locally-connected continuum and $f(A) = B$ is non-alternating, A-sets in A map onto A-sets in B.*

For in this case the property mentioned in (6.22) is equivalent to the property of being an A-set [see IV, (3.3)].

Without the assumption of semi-local-connectedness, however, we can obtain the latter conclusion by direct reasoning as will now be shown.

(6.3) THEOREM. *If A is a continuum and $f(A) = B$ is non-alternating, A-sets in A map onto A-sets in B.*

Proof. Let A_0 be an A-set in A, let $A - A_0 = \sum G_i$ where $[G_i]$ is a finite or null sequence of disjoint open sets, each G_i bounded by a single point x_i of A_0. For each i let

$$H_i = G_i - G_i \cdot f^{-1}f(x_i)$$

and

$$K_i = f(H_i).$$

Then

$$B - f(A_0) = \sum K_i.$$

Furthermore, since H_i is an open inverse set, K_i is open ($i = 1, 2, \cdots$). Also since $G_i + x_i$ is compact, so is $K_i + f(x_i)$ for each i. Finally, by uniform continuity, $[K_i]$ is a finite or a null sequence since $[H_i]$ is a finite or a null sequence.

Concerning end points, we have the following

(6.4) *If A is a continuum and $f(A) = B$ is non-alternating, then for any end point p of B, $f^{-1}(p)$ is connected.*

For let $U_1 \supset U_2 \supset U_3 \supset \cdots$ be a sequence of neighborhoods in B closing down on p and such that $F(U_i)$ is a single point p_i. For each i, let Q_i be the quasi-component of $B - p_i$ containing p. By (2.3), $f^{-1}(Q_i)$ is a quasi-component of $A - f^{-1}(p_i)$. Hence $f^{-1}(Q_i)$ lies in a component W_i of $\overline{f^{-1}(U_i)}$. Now clearly $f^{-1}(p) = \lim W_i$. Accordingly, $f^{-1}(p)$ is connected.

It need hardly be remarked that all the preceding results hold for monotone transformations because any monotone transformation is non-alternating. We mention this fact because some of the principal applications to be made of these theorems have to do with monotone transformations.

Perhaps the most basic of the results established up to this point is (6.2) to the effect that any simple link of the image space is covered by the image of some simple link of the original space. In case our original and image spaces are semi-locally-connected continua, this result admits considerable refinement as will be obtained in the next theorem.

(6.5) THEOREM. *Let A and B be semi-locally-connected continua, let $f(A) = B$ be $\left\{\begin{matrix}\text{non-alternating}\\ \text{monotone}\end{matrix}\right\}$, let E_b be any true cyclic element of B and let E_a be the cyclic element in A such that $f(E_a) \supset E_b$. If $r(x)$ and $s(x)$ are the monotone retractions of A onto E_a and B onto E_b, respectively, the transformations $z(E_a) = sf(E_a) = E_b$ and $q(A) = zr(A) = sfr(A) = E_b$ are $\left\{\begin{matrix}\text{monotone}\\ \text{non-alternating}\end{matrix}\right\}$.*

Proof. By (6.3), $f(E_a)$ is an A-set in B and hence is semi-locally-connected. Then since, by (4.1), $s(x)$ is A-set reversing, it follows from (3.6) that $z(x)$ is non-alternating. Now if $f(x)$ is monotone, it results from (3.4) that $z(x)$ is monotone since both f and s are monotone.

Since $r(x)$ is monotone and $z(x)$ is non-alternating (resp. monotone) it follows by (3.5), (resp. (3.4)), that $q(x)$ is non-alternating (resp. monotone).

(6.51) COROLLARY. *If A and B are semi-locally-connected continua and $f(A) = B$ is non-alternating (resp. monotone), any cyclic element of B is the image of some cyclic element of A or of all of A under a non-alternating (resp. monotone) transformation.*

7. Interior transformations.

(7.1) *If $f(R) = S$ is single-valued, in order that a subset Q of R be an inverse set, i.e., $Q = f^{-1}f(Q)$, it is necessary and sufficient that for any subset X of R,*

$$f(Q\cdot X) = f(Q)\cdot f(X).$$

Proof. To establish the sufficiency of the condition suppose there existed a point $x \in f^{-1}f(Q) - Q$. Then, taking $X = x$ we have

$$f(Q \cdot X) = 0.$$

Whereas

$$f(Q) \cdot f(X) = f(x),$$

since $f(x) \subset f(Q)$.

To prove the necessity, let Q be an inverse set. Obviously $f(Q \cdot X) \subset f(Q) \cdot f(X)$. To prove the reverse inclusion, let $x \in f(Q) \cdot f(X)$. Then $f^{-1}(x) \subset Q$, $f^{-1}(x) \cdot X \neq 0$. Thus if $y \in f^{-1}(x) \cdot X$, we have $y \in Q \cdot X \cdot f^{-1}(x)$, a relation which gives $f(y) = x \in f(Q \cdot X)$.

(7.2) *If $f(R) = S$ is an interior transformation on R, and Q is any inverse set in R, f is an interior transformation on Q.*

Proof. Let X_q be any open subset of Q. There exists an open subset X of R such that $X_q = X \cdot Q$. By hypothesis $f(X) = U$ is open in S, and by (7.1)

$$f(X_q) = f(Q \cdot X) = f(Q) \cdot f(X) = f(Q) \cdot U.$$

Thus $f(X_q)$ is open in $f(Q)$ and the lemma follows.

(7.3) *If $f(A) = B$ is continuous, then for any conditionally compact set $R \subset A$*

(i) $$\overline{f(R)} = f(\bar{R}),$$

(ii) $$\overline{f(R)} - f(R) \subset f(\bar{R} - R).$$

Thus if f is an interior transformation and R is open in A,

(iii) $$F[f(R)] \subset f[F(R)],$$

where in general $F(X)$ denotes the boundary of X.

(7.31) Corollary. *The Menger-Urysohn order of a point in a locally compact set A is never increased when A undergoes an interior transformation. The properties of being* (i) *a curve of order less than n,* (ii) *a regular curve,* (iii) *a rational curve, are invariant under interior transformations.* (See Note 1, p. 279.)

(7.4) Theorem. *Let A and B be compact and let $f(A) = B$ be interior. Then if R is any connected open set in B, every quasi-component Q of $f^{-1}(R)$ maps onto all of R under f.*

Proof. Suppose, on the contrary, that for some $p \in R$, $f^{-1}(p) \cdot Q = 0$. Then since $f^{-1}(p)$ is compact and is $\subset f^{-1}(R)$, it follows by an application of the Borel Theorem that there exists an open set U such that

(i) $$Q \subset U \subset f^{-1}(R), \qquad F(U) \cdot f^{-1}(R) = 0, \qquad U \cdot f^{-1}(p) = 0.$$

Now $f(U) = V$ is open, and we have

(ii) $$V \supset f(Q) \subset R, \qquad V \subset R - p.$$

Whence, since R is connected, $F(V) \cdot R \neq 0$. But by (7.3), $F(V) \subset f[F(U)]$ and, by (i), $f[F(U)] \cdot R = 0$. Thus our theorem is proved.

(7.41) COROLLARY. *If A is locally connected, if B_0 is any closed set in B and R is any component of $B - B_0$, then $f^{-1}(R)$ has just a finite number of components and each one of these components maps onto all of R under f.*

(7.42) COROLLARY. *If A is locally connected and B is connected and cyclic (i.e., without cut points), then for each $x \in B$, $A - f^{-1}(x)$ has just a finite number of components and each of these components maps onto $B - x$ under f.*

(7.43) *Let $f(M) = N$ be interior, where N is connected and M is arbitrary. For any inverse set M_0 in M which is both open and closed in M, $f(M_0) = N$.*

For $f(M_0)$ and $f(M - M_0)$ are disjoint open sets in N by interiority of f. Hence $f(M_0)$ must coincide with N.

(7.5) THEOREM. *Let $f(A) = B$ be an interior transformation, where A is compact, and let C be any continuum in B. Then for each component K of $f^{-1}(C) = Q$, we have $f(K) = C$.*

For by (7.2), the transformation $f(Q) = C$ is interior. Thus, since C is connected and open in C, it follows by (7.4) that each component of Q maps onto all of C under f.

(7.6) *If $f(A) = B$ is interior, where A and B are arbitrary metric spaces, any compact nodal subset of A maps onto a nodal subset of B under f.*

For let N be a compact nodal subset of A. If $N \cdot \overline{A - N} = 0$, N is open in A; and hence $f(N)$ is open in B so that $f(N) \cdot \overline{B - f(N)} = 0$. Thus we may suppose $N \cdot \overline{A - N} = x \in A$. Since $N - x$ is open in A, it follows that $f(N) - f(x)$ is open in B. Accordingly, $f(N) \cdot \overline{B - f(N)} \subset f(x)$, so that $f(N)$ is a nodal subset of B.

(7.61) *If $f(A) = B$ is interior and B is a cyclic continuum, any compact nondegenerate nodal subset of A maps onto all of B under f.*

Clearly this follows from (7.6) since the only nondegenerate nodal subset of B is B itself and the image of a nondegenerate nodal set in A cannot be degenerate.

(7.7) *Let $f(A) = B$ be interior, where A and B are semi-locally-connected continua. For any true cyclic element E_b of B there exists a true cyclic element E_a of A such that $f(E_a) \supset E_b$. (Furthermore the transformation $f[E_a \cdot f^{-1}(E_b)] = E_b$ is interior except possibly at one point.)*

Proof. Let K be the product of all A-sets in A containng $f^{-1}(E_b)$. Then K is an A-set and hence it has at least one node N. If N were an end point of K there would exist an (arbitrarily small) nodal set M in K such that $M \cdot f^{-1}(E_b)$ is nondegenerate and so that $f[M \cdot f^{-1}(E_b)] \neq E_b$; and this is contrary to (7.61) because $M \cdot f^{-1}(E_b)$ is a nodal subset of $f^{-1}(E_b)$ and the transformation $f[f^{-1}(E_b)] = E_b$ is interior by (7.2). Hence N is a true cyclic element of K and thus also of A. Accordingly, since $N \cdot f^{-1}(E_b)$ is a nondegenerate nodal subset of $f^{-1}(E_b)$, we have only to take $E_a = N$ and our result follows by (7.61).

In order to get a more specific inversion theorem for cyclic elements under interior transformations we develop next certain results concerning the relation between interiority and local separation.

DEFINITION. *A subset K of M is said to locally separate a set M at a point $x \in K$ provided that for every sufficiently small neighborhood V of x in M, $V - V \cdot K$ is disconnected. A mapping $f(A) = B$ is said to be interior at a point $x \in A$ provided that for every open set U in A containing x, $f(x)$ is an interior point of $f(U)$.*

(7.81) THEOREM. *Let $f(A) = B$ be continuous where A is locally compact and suppose E is the set of points in A where f is not interior. For any $x \in E \cdot \overline{(A - f^{-1}f(E))}$ such that x is a component of $f^{-1}f(x)$ and any neighborhood U of x, $f(E \cdot U)$ locally separates B at $f(x)$.*

Proof. Suppose this is not true. We may suppose U chosen arbitrarily small, conditionally compact and such that $0 < d = \rho\{f(x), f[F(U)]\}$. There exists a neighborhood V of $y = f(x)$ such that $\delta(V) < d/2$ and the set $X = V - f(E \cdot U) \cdot V$ is connected. Now since x is a limit point of $A - f^{-1}f(E)$, there exists a point $p \in U$ such that $f(p) \in X$. Thus $X \cdot f(U) \neq 0$. If $f(U)$ did not contain X, then since $f(\overline{U})$ is compact and $X \cdot f(\overline{U}) \subset X \cdot f(U) + X \cdot f[F(U)] = X \cdot f(U)$ there would exist a point z of $X \cdot f(U)$ which was a limit point of $X - X \cdot f(U)$; and if $q \in U \cdot f^{-1}(z)$, since f is interior at q, z would be interior to $f(U)$ which is impossible. Therefore $f(U) \supset X$. Whence $f(U) \supset V$ because $V - X \subset f(U)$. Thus $f(x)$ is interior to $f(U)$, and since U was arbitrarily small it results that f is interior at x contrary to $x \in E$. Accordingly $f(E \cdot U)$ locally separates B at $f(x)$.

Note. The condition $x \in \overline{A - f^{-1}f(E)}$ is essential. For if A consists of a circle C plus its interior I and a second circle D, of the same size tangent to C at a point p and f merely reflects D onto C by a reflection in the tangent line through p so that $f(C + I + D) = B = C + I$, we have $\overline{A - f^{-1}f(E)} = \bar{I} = B$ and for $x \in A - B$, $f(E)$ does not locally separate B at $f(x)$.

(7.82) THEOREM. *If $f(A) = B$ is interior, X is a closed subset of B and $f^{-1}(X)$ does not locally separate A at a point $x \in f^{-1}(X)$, then X does not locally separate B at $f(x)$.*

Proof. Let U be any neighborhood of $f(x)$. By hypothesis there exists a neighborhood V of x such that $f(V) \subset U$ and the set $Q = V - V \cdot f^{-1}(X)$ is connected. Then $f(V)$ is a neighborhood (arbitrarily small) of $f(x)$ and the set

$$f(Q) = f(V) - f(V) \cdot X$$

is connected.

DEFINITION. *A generalized continuum which is not separated locally at any point by a compact 0-dimensional set will be said to be locally a Cantorian manifold of dimension greater than or equal to 2.*

(7.83) THEOREM. *If A is locally a Cantorian manifold of dimension greater than or equal to 2, so also is any interior light image of A.*

Proof. Let $f(A) = B$ be interior and light. By interiority, B is a generalized continuum. Let X be any compact 0-dimensional subset of B. Since f is light and continuous, $\dim f^{-1}(X) = 0$. Hence $f^{-1}(X)$ does not locally separate A at any point. Therefore by (7.82), X does not locally separate B at any point.

It may be noted that *any local Cantorian manifold M is locally connected.* For if $p \in M$ then since the null set (0) does not locally separate M at p, there exists an arbitrarily small open set V containing p such that $V - (0) = V$ is connected.

(7.831) (Compare Stoïlow, [2, p. 122].) *If A and B are locally Cantorian manifolds of dimension greater than or equal to 2, $f(A) = B$ is continuous and light and there exists a set $E \subset A$ such that both E and $f(E)$ are punctiform and f is interior at every point of $A - E$, then f is interior on A.*

(*Note.* A set which contains no continuum is said to be *punctiform.*)

This follows at once from (7.83) and (7.81) since if $f(E)$ locally separated B at a point p there would exist a region R in B which is separated by $f(E)$ and hence by some closed subset X of $f(E) \cdot R$. Since $f(E)$ is punctiform, $\dim X = 0$. Hence there would exist a region $R' \subset R$ which was separated by a compact subset of X, contrary to the fact that B is locally a Cantorian manifold. Hence $f(E)$ cannot locally separate B at any point; but if there were a point $x \in E$ where f is not interior, x would belong to $\overline{A - f^{-1}f(E)}$ (since f is light) and by (7.81), $f(E)$ would locally separate B at $f(x)$.

Incidentally we have now established the fact that a *generalized continuum M is a local Cantorian manifold of* $\dim \geqq 2$ *if and only if M is locally connected and no region in M is separated by any punctiform set.*

We prove next an inversion theorem for A-sets under light transformations.

(7.91) THEOREM. *Let $f(M) = N$ be continuous and light where M and N are semi-locally-connected continua and suppose that for each $y \in N$, $f^{-1}(y)$ disconnects no cyclic element of M. Then for each A-set N_a in N, each component of $f^{-1}(N_a)$ is an A-set in M.*

Proof. Let K be a component of $f^{-1}(N_a)$. If K is not an A-set, there exists a component R of $M - K$ such that $\bar{R} - R = F$ contains more than one point. Since any two points of F are conjugate, there exists a cyclic element E of M containing F. Since $R \cdot E$ is nonvacuous and connected, it follows that $R \cdot E$ is not a subset of $f^{-1}(N_a)$. Let S be a component of $E - E \cdot f^{-1}(N_a)$. Then $f(S)$ is a connected subset of $N - N_a$ and hence $\bar{S} - S$ must map into the single point p which is the boundary of the component of $N - N_a$ containing $f(S)$. Therefore $\bar{S} - S \subset f^{-1}(p)$. Thus S is a component of $E - E \cdot f^{-1}(p)$, and since $K \cdot E$ is connected and nondegenerate, $S + E \cdot f^{-1}(p) \neq E$. Hence $f^{-1}(p)$ disconnects E, contrary to hypothesis. Accordingly, K is an A-set.

It may be noted that the hypothesis that $f^{-1}(y)$, $y \in N$, does not disconnect any cyclic element of M is satisfied in case M is unicoherent or in case each true cyclic element of M is a Cantorian manifold of dimension greater than or equal to 2 (i.e., a generalized continuum not disconnected by any 0-dimensional subset).

As applied to interior transformations, we get

(7.911) *If* $f(M) = N$ *is interior and light, where* M *and* N *are semi-locally-connected continua, and for no* $y \in N$ *does* $f^{-1}(y)$ *disconnect any cyclic element of* M*, then for each* A*-set* N_a *in* N*,* $f^{-1}(N_a)$ *is the sum of a finite number of disjoint* A*-sets in* M *each of which maps interiorly onto* N_a.

For by (7.5) each component of $f^{-1}(N_a)$ maps onto N_a. Thus there are only a finite number of such components since by (7.91) each of them is an A-set in A. As each component K of $f^{-1}(N_a)$ is then open in $f^{-1}(N_a)$, the mapping is interior on each K.

(7.92) THEOREM. *Let* $f(A) = B$ *be interior and light, where* A *is a locally connected continuum each true cyclic element of which is locally a Cantorian manifold of dimension greater than or equal to* 2. *For each true cyclic element* E_b *of* B*,* $f^{-1}(E_b)$ *is the sum of a finite number of cyclic elements of* A *each of which maps interiorly onto* E_b *under* f.

Proof. Let K be any component of $f^{-1}(E_b)$. By (7.911) K is an A-set in A mapping interiorly onto E_b. Since each node of K maps onto E_b, there can be only a finite number of nodes of K. Hence if E_a is any true cyclic element in K, E_a contains just a finite set F of cut points of K. Hence $f(F) = H$ is finite. Now, by (7.83), E_b is locally a Cantorian manifold of dimension greater than or equal to 2, because the mapping of any node N of K onto E_b is interior. [To see this let $p = N \cdot \overline{K - N}$. Then $f(N) = E_b$ is interior except at p. If it is not interior at p, there must exist a point $q \in N - p$ such that $f(q) = f(p)$. By interiority at q and (7.82) it results that $f(p)$ does not locally separate E_b. But then by (7.81), the mapping $f(N) = E_b$ is interior at p also.] Hence $E_b - H$ is connected. Thus $E_a - E_a \cdot f^{-1}(H) = G$ is connected, since $f^{-1}(H)$ is 0-dimensional. But G is a component of $K - K \cdot f^{-1}(H)$ since $f^{-1}(H) \supset F$. Hence G maps onto all of $E_b - H$ and thus E_a maps onto all of E_b under f. Since the mapping $f(E_a) = E_b$ is interior on $E_a - F$ and F is finite it follows from (7.831) that this mapping is interior everywhere on E_a.

Since each true cyclic element of K maps interiorly onto E_b, there can be only a finite number of such elements in K; and since clearly K can have no arc of cut points [for $f(K) = E_b$ is an interior mapping] it follows that K is the sum of a finite number of true cyclic elements each mapping interiorly onto E_b. As $f^{-1}(E_b)$ has only a finite number of components K, our theorem follows.

8. **Quasi-monotone transformations.** If A is any continuum, a continuous transformation $f(A) = B$ is said to be *quasi-monotone* provided that for any continuum K in B with a nonvacuous interior, $f^{-1}(K)$ has just a finite number of components and each of these maps onto K under f.

It results at once that *any monotone mapping on a continuum is quasi-monotone.* It will be shown below that for locally connected continua A, interior transformations on A are also quasi-monotone; and conversely any quasi-monotone transformation is the resultant of a monotone mapping followed by a light interior one. Thus on locally connected continua, quasi-monotone transformations represent exactly the common ground shared by monotone and interior transformations.

In each of the next five results it is assumed that A is a locally connected continuum.

(8.1) THEOREM. *In order that a mapping $f(A) = B$ on a locally connected continuum A be quasi-monotone it is necessary and sufficient that each component of the inverse of any region R in B map onto R under f.*

To prove the necessity of the condition let Q be any component of $f^{-1}(R)$, let $p \epsilon Q$ and let $x \epsilon R$. There exists a region R' such that $f(p) + x \subset R' \subset \bar{R}' \subset R$. Let Q' be the component of $f^{-1}(\bar{R}')$ containing p. Obviously $Q' \subset Q$. But by hypothesis $f(Q') \supset \bar{R}' \supset x$. Thus $f(Q) \supset x$. Whence $f(Q) \supset R$, since x was an arbitrary point of R.

To prove the sufficiency we prove first that if our condition is satisfied:

(*) *If K is any continuum in B and C is any component of $f^{-1}(K)$, then $f(C) = K$.*

For if this is not so, there exists a region U in A containing C such that $F(U) \cdot f^{-1}(K) = 0$ and such that if $H = f^{-1}(K) \cdot U$, then $f(H) \neq K$. Since H, and therefore $f(H)$, is compact, there is a point $p \epsilon f(H)$ which is a limit point of $K - f(H)$. Since $F(U) \cdot f^{-1}(K) = 0$ we have $f[F(U)] \cdot K = 0$. Let R be the component of $B - f[F(U)]$ containing p and let Q be the component of $f^{-1}(R)$ containing a point $q \epsilon H \cdot f^{-1}(p)$. Then by hypothesis $f(Q) = R$. But since R contains points of $K - f(H)$ and since Q contains no points of $F(U)$ and hence is contained in U, this contradicts the definition of H.

Now to complete the proof we have only to show that if K (a continuum in B) has interior points, then there are only a finite number of components $f^{-1}(K)$. If K has interior points it contains a region R in B. Let $p \epsilon R$. By hypothesis every component of $f^{-1}(R)$ intersects $f^{-1}(p)$. Since $f^{-1}(p)$ is compact it is therefore covered by a finite number of components of $f^{-1}(R)$; and since clearly no point belongs to more than one component of $f^{-1}(R)$, there are only a finite number of such components. Now by (*) each component of $f^{-1}(K)$ contains a component of $f^{-1}(R)$. Therefore there are only a finite number of components of $f^{-1}(K)$.

(8.11) COROLLARY. *Any interior transformation on a locally connected continuum is quasi-monotone.*

(8.2) THEOREM. *If A is a locally connected continuum and $f(A) = B$ is light, then in order that f be quasi-monotone it is necessary and sufficient that it be interior.*

The sufficiency being a consequence of (8.11), we proceed to prove the necessity. Let f be quasi-monotone, let G be any open set in A and let p be any point of $f(G)$. Since f is light, if $q \epsilon G \cdot f^{-1}(p)$, there exists a region Q in A such that $Q \subset G$ and $F(Q) \cdot f^{-1}(p) = 0$. Let R be the component of $B - f[F(Q)]$ containing p. Then if Q' is the component of $f^{-1}(R)$ containing q, we have $Q' \subset Q$ because $Q' \cdot F(Q) = 0$; and by (8.1), $f(Q') = R$. Whence $f(G) \supset f(Q) \supset f(Q') = R \supset p$; and since R is open in B, $f(G)$ is open in B.

(8.3) THEOREM. *Let $f(A) = B$ be continuous, where A is a locally connected continuum, and let $f(x) = f_2f_1(x)$, $x \in A$, be any factorization of f into continuous factors $f_1(A) = A'$ and $f_2(A') = B$. Then*
(i) *if f is quasi-monotone so also is f_2;*
(ii) *if f_1 and f_2 are quasi-monotone, so also is f.*

Proof of (i). Let R be a region in B and R' a component of $f_2^{-1}(R)$. If Q is a component of $f_1^{-1}(R')$, Q is a component of $f^{-1}(R)$ and hence

$$R = f(Q) = f_2f_1(Q) \subset f_2(R') \subset R.$$

Hence $f_2(R') = R$ and thus f_2 is quasi-monotone by (8.1).
Proof of (ii). Let R be any region in B and Q any component of $f^{-1}(R)$. Since f_1 is quasi-monotone and $f^{-1}(R) = f_1^{-1}[f_2^{-1}(R)]$, $f_1(Q)$ is a component of $f_2^{-1}(R)$; and thus since f_2 is quasi-monotone, $f_2f_1(Q) = f(Q) = R$. Hence, by (8.1), f is quasi-monotone.

(8.4) THEOREM. *In order that a continuous transformation $f(A) = B$ on a locally connected continuum A be quasi-monotone it is necessary and sufficient that it admit of factorization in the form $f(x) = f_2f_1(x)$ where $f_1(A) = A'$ is monotone and $f_2(A') = B$ is light and interior.*

The sufficiency results at once from (8.3), (ii), since by (8.2) and the remark at the beginning of this section both f_1 and f_2 are quasi-monotone.
To prove the necessity, let $f(A) = B$ be quasi-monotone. Applying the general factorization theorem, we get a factorization $f(x) = f_2f_1(x)$ where f_1 is monotone and f_2 is light. By (8.3), f_2 is quasi-monotone and hence by (8.2) it is interior.
Thus we have at once

(8.5) THEOREM. *A topological property of a locally connected continuum A is invariant under quasi-monotone transformations if and only if it is invariant under both monotone and light interior transformations.*

Unicoherence is an example of such a property. This will result from the next theorem to be proved.

DEFINITION. *For any continuum X and any division $X = X_1 + X_2$ where X_1 and X_2 are continua, let $r(X_1, X_2)$ be one less than the number of components of $X_1 \cdot X_2$. Then let us define the degree of multicoherence $r(X)$ of X to be the least upper bound of $r(X_1, X_2)$ for all possible such divisions.* [*See* IV, §12.]

For example, if X is unicoherent, $r(X) = 0$, and conversely. If X is a simple closed curve, a circular ring, or a Möbius band, $r(X) = 1$, if X is a θ-curve, $r(X) = 2$, and so on.

(8.6) THEOREM. *If X is any continuum and $f(X) = Y$ is quasi-monotone, then $r(Y) \leq r(X)$.*

Proof. Let $Y = H + K$ be any decomposition of Y into continua H and K.

If either H or K, say H, is Y, we have $H \cdot K = K$ so that $r(H, K) = 0 \leqq r(X)$. Thus we may suppose neither H nor K is Y and hence both H and K have interior points (rel. Y). Then by quasi-monotoneity, the components of $f^{-1}(H)$ and $f^{-1}(K)$ taken together give a decomposition of X into a finite number of continua. Hence there exists a component X_1 of $f^{-1}(H)$ or of $f^{-1}(K)$, say of $f^{-1}(H)$, such that the set $X_2 = f^{-1}(K) + [f^{-1}(H) - X_1] = f^{-1}(K) +$ all components of $f^{-1}(H)$ except X_1, is connected. For if a continuum is the sum of a finite number of continua E_i, there is some one of these sets, say E_j, such that $\sum_{i \neq j} E_i$ is connected. Hence $X = X_1 + X_2$ is a division of X into continua. Furthermore, since by quasi-monotoneity, $f(X_1) = H$ we have

$$f(X_1 \cdot X_2) = H \cdot K.$$

Hence there are as many components in $X_1 \cdot X_2$ as in $H \cdot K$, i.e.,

$$r(H, K) \leqq r(X_1, X_2).$$

Accordingly,

$$r(Y) = \text{l.u.b. } [r(H, K)] \leqq \text{l.u.b. } [r(X_1, X_2)] = r(X).$$

(8.61) *Unicoherence is invariant under quasi-monotone transformations.*

(8.62) *If X is a continuum and $f(A) = B$ is monotone, $r(Y) \leqq r(X)$.*

(8.63) *If A is a locally connected continuum and $f(A) = B$ is interior, $r(B) \leqq r(A)$.*

9. The relative distance transformation. We shall consider a connected and locally connected metric space which we denote by M. For each pair of points x and y of M let $\rho^*(x, y)$ be the greatest lower bound of the aggregate of diameters $[\delta(C)]$ of all connected sets C in M which contain both x and y. Now for each pair x, y in M, the number $\rho^*(x, y)$ is a finite non-negative real number. Obviously this is true if $\delta(M)$ is finite, which we may suppose without loss of generality. Even without assuming $\delta(M)$ finite, however, we may still show that every $\rho^*(x, y)$ is finite by using the fact that x and y can be joined in M by a finite simple chain of connected regions each of diameter less than 1 which is thus a connected set $C \supset x + y$ and with $\delta(C) \leqq k$, where k is the number of links in the chain. Also, as will be shown below, the numbers $\rho^*(x, y)$ have distance character in that they satisfy the distance axioms necessary to define a metric space. Let M^* denote the space whose points are exactly the points of M but in which "distance" is defined by means of the above definitions, i.e., the distance $\rho(x^*, y^*)$ between two points x^* and y^* of M^* is the number $\rho^*(x, y)$ determined in the above manner for the points x and y of M, where x and y are the points in M identified with the points x^* and y^*, respectively, in M^*. Thus M^* is the space obtained from M by changing all distances $\rho(x, y)$ into distances $\rho^*(x, y)$. If we regard this operation as a transformation f, then we have that $f(M) = M^*$. We denote points of M by $x, y, \cdots$ and corresponding points in M^* by x^*, y^*, $\cdots$. Distances in each space are denoted by the operator ρ. Thus we have always

(i) $$\rho^*(x, y) = \rho(x^*, y^*).$$

We shall call M^* the *relative distance space* for M and the mapping $f(M) = M^*$ the *relative distance transformation*.

(a) PROPERTIES OF THE TRANSFORMATION

(9.1) THEOREM. *If M is connected and locally connected,*

(1) $\rho^*(x, y) \geqq \rho(x, y)$;

(2) *diameters of connected sets are invariant under f, i.e., $\delta(C) = \delta(C^*)$, where C is any connected set;*

(3) *the numbers $\rho^*(x, y)$ have distance character, i.e., $\rho^*(x, y) > 0$, $x \neq y$; $\rho^*(x, x) = 0$; $\rho^*(x, y) = \rho^*(y, x)$; $\rho^*(x, z) \leqq \rho^*(x, y) + \rho^*(y, z)$; thus M^* is a metric space;*

(4) *f is a topological transformation; thus M^* is homeomorphic with M;*

(5) *property S is invariant under f;*

(6) *M^* is uniformly locally connected;*

(7) *the inverse of f is uniformly continuous;*

(8) *if M is uniformly locally connected, then f is uniformly continuous, and conversely;*

(9) *all spherical neighborhoods in M^* are connected, i.e., for each point p^* of M^* and $r > 0$, the set $V_r(p^*)$ of all points x^* in M^* with $\rho(x^*, p^*) < r$ is connected;*

(10) *if $w(\tilde{M}^*) \subset \tilde{M}$ is the extension of f^{-1} to $\tilde{M}^*$ and $\tilde{M}^* - M^* = A$, $\tilde{M} - M = B'$, we have*

$$w(A) = B \subset B';$$

(11) *w is light; indeed, for any $p \in w(\tilde{M}^*)$ and any $\epsilon > 0$, $w^{-1}(p)$ is the sum of a finite or countable number of sets (F_i) each of diameter less than ϵ and with $\rho(F_i, F_j) \geqq \epsilon/4$ for $i \neq j$ [thus each F_i is both open and closed in $w^{-1}(p)$].*

Proof. If $C \supset x + y$, then $\delta(C) \geqq \rho(x, y)$. Therefore, g.l.b. $[\delta(C)] = \rho^*(x, y) \geqq \rho(x, y)$, where "g.l.b." means "greatest lower bound," and (1) is proved.

By (1) we have $\delta(C) \leqq \delta(C^*)$. Since $x + y \subset C$ implies $\rho^*(x, y) \leqq \delta(C)$, therefore we have $\delta(C^*) =$ l.u.b. $[\rho^*(x, y)] \leqq \delta(C)$, where x and y are in C and "l.u.b." means "least upper bound." These two inequalities give $\delta(C) = \delta(C^*)$, which is (2).

The first two relations in (3) are corollaries to (1) and (2), respectively. The third results from the fact that $x + y \subset C$ is the same relation as $y + x \subset C$. The fourth follows from the fact that if, in general, C_{ab} denotes a connected set containing $a + b$, then any set $C_{xy} + C_{yz}$ is a set C_{xz}. For then since $\delta[C_{xy} + C_{yz}] \leqq \delta(C_{xy}) + \delta(C_{yz})$, we have

$$\text{g.l.b. } [\delta(C_{xz})] \leqq \text{g.l.b. } [\delta(C_{xy})] + \text{g.l.b. } [\delta(C_{yz})],$$

which is the same as $\rho^*(x, z) \leqq \rho^*(x, y) + \rho^*(y, z)$.

That the transformation f is (1-1) follows from (3). By virtue of the local connectivity of M it follows that when $\rho(x_n, x) \to 0$, also $\rho^*(x_n, x) = \rho(x_n^*, x^*) \to 0$; and by (1) we have that $\rho(x_n, x) \to 0$ when $\rho^*(x_n, x) \to 0$.

Therefore f and f^{-1} are continuous and thus f is topological, which proves (4).

If M has property S, then for any $\epsilon > 0$, $M = \sum_1^k C_i$, where each C_i is connected and $\delta(C_i) < \epsilon$. Then since by (2), $\delta(C_i^*) = \delta(C_i)$, we have $M^* = \sum_1^k C_i^*$, where $\delta(C_i^*) < \epsilon$. Thus M^* has property S, and (5) is proved.

To prove (6), let ϵ be any positive number. Take a number $\delta = \epsilon/2$. Then if x^* and y^* are any two points of M^* with $\rho(x^*, y^*) = \rho^*(x, y) < \delta$, by the definition of $\rho^*(x, y)$ there exists a connected set C in M containing $x + y$ and with $\delta(C) < 2\delta = \epsilon$. Then $C^* \supset x^* + y^*$, and, by (2), $\delta(C^*) = \delta(C) < \epsilon$. Thus M^* is uniformly locally connected.

Property (7) follows immediately from (1). For if ϵ is any positive number, then taking $\delta = \epsilon$ we have by (1) that if x^* and y^* are points in M^* with $\rho(x^*, y^*) = \rho^*(x, y) < \delta = \epsilon$, then $\rho(x, y) \leqq \rho^*(x, y) < \epsilon$.

If M is uniformly locally connected, then for any $\epsilon > 0$, a $\delta > 0$ exists such that when $\rho(x, y) < \delta$, there exists a connected set C in M with $x + y \subset C$ and $\delta(C) < \epsilon$; thus it follows that $\rho(x, y) < \delta$ implies $\rho^*(x, y) = \rho(x^*, y^*) < \epsilon$, which proves that f is uniformly continuous. If, conversely, T is uniformly continuous, then for any $\epsilon > 0$, a $\delta > 0$ exists such that $\rho(x, y) < \delta$ implies $\rho(x^*, y^*) < \epsilon$; therefore, by definition of $\rho^*(x, y)$, there exists a connected set C in M with $x + y \subset C$ and $\delta(C) < \epsilon$, which proves M uniformly locally connected. Thus (8) is established.

Finally, (9) results from the fact that $\rho(x^*, p^*) < r$ implies the existence of a connected set C in M with $x + p \subset C$ and $\delta(C) < r$; this gives $C \subset V_r(p^*)$, from which relations the connectivity of $V_r(p^*)$ is manifest.

In (10), $\tilde{M}$ and $\tilde{M}^*$, of course, stand for the complete enclosures of M and M^*, respectively. Since, by property (7), f^{-1} is uniformly continuous, it follows by [II, (2.3)] that the definition of f^{-1} can be extended to include the points of $\tilde{M}^* - M^*$ in one and only one way so that if we call the resulting transformation w, then w will be single-valued and continuous and will be equal to f^{-1} on M^*. The image under w of any point p of $\tilde{M}^* - M^*$ must be a point of $\tilde{M} - M$; for if $w(p) = x$ where $x \subset M$, then $x^* \subset M^*$ and hence $x^* \neq p$; but then M^* contains a subset N^* having p but not x as a limit point; hence, w being continuous, the subset $N = w(N^*)$ of M has x as a limit point, contrary to the fact that M and M^* are homeomorphic. Thus we have proved that

(ii) $$w(\tilde{M}^* - M^*) \subset \tilde{M} - M,$$

or

$$w(A) = B \subset B'$$

as was to be proved.

To prove (11), let $M_1, M_2, \cdots$ be the components of $M \cdot V_{\epsilon/3}(p)$ having p as a limit point. For each i, let F_i be the set of all points x of $w^{-1}(p)$ such that if $x_1, x_2, \cdots$ is any sequence in M^* converging to x, then almost all of the points $w(x_1), w(x_2), \cdots$ are in M_i and of course the sequence converges to p. Since for any $x \in w^{-1}(p)$ and any sequence $x_k \to x$, where $x_k \in M^*$ for all k, there exists an i such that almost all $w(x_k)$ lie in M_i [because for $y_i \in M_i \cdot V_{\epsilon/6}(p)$, $y_j \in M_j \cdot V_{\epsilon/6}(p)$, we have $\rho^*(y_i, y_j) \geqq \epsilon/6$ for $i \neq j$], it follows that $w^{-1}(p) = F_1 + F_2 + \cdots$. Now if $x \in F_i, y \in F_j, i \neq j$, we have $\rho^*(x, y) = \lim \rho^*(x_k, y_k) = \lim \rho^*[w(x_k), w(y_k)] \geqq \epsilon/4$, where $x_k \to x$, $y_k \to y$, and $x_k, y_k \in M^*$. Hence $\rho(F_i, F_j) \geqq \epsilon/4$. Thus since $w^{-1}(p)$ is closed, each F_i is closed and is open in $w^{-1}(p)$.

(9.2) THEOREM. *If M is connected and has property S, $\tilde{M}$ and $\tilde{M}^*$ are compact and w maps $\tilde{M}^*$ onto $\tilde{M}$ carrying M^* topologically onto M and A $(= M^* - M)$ continuously onto $B = B'$ $(= \tilde{M} - M)$.*

For by property (6) of (9.1), M^* also has property S. Then since in the spaces $\tilde{M}$ and $\tilde{M}^*$, respectively, $\bar{M} = \tilde{M}$ and $\bar{M}^* = \tilde{M}^*$ it follows that $\tilde{M}$ and $\tilde{M}^*$ have property S. Thus $\tilde{M}$ and $\tilde{M}^*$ are complete spaces having property S, and hence they must be compact. Now in this case it follows that every point of $\tilde{M} - M$ must be the image under w of some point of $\tilde{M}^* - M^*$. For let p be any point $\tilde{M} - M$. Then since, as is at once clear, $M + p$ is locally connected at p, there exists in M a sequence of connected sets $U_1 \supset U_2 \supset U_3 \supset \cdots$ such that $\delta(U_i) \to 0$ and for each n, $p \subset \bar{U}_n$. For each i, choose a point x_i in U_i. Then since, for each m, $\sum_m^\infty x_i \subset U_m$, it follows that $x_1^*, x_2^*, \cdots$ is a fundamental sequence in M^*. Hence this sequence has a limit point q in $\tilde{M}^*$. The point q cannot be in M^*, for the sequence $x_1, x_2, \cdots$ has as its limit point the point p of $\tilde{M} - M$. Thus $q \subset \tilde{M}^* - M^*$, and as w is continuous, $w(q) = p$. Thus we have shown that $w(\tilde{M}^* - M^*) \supset \tilde{M} - M$. By virtue of (ii) we have, in case M has property S, that

$$\text{(iii)} \qquad w(\tilde{M}^* - M^*) = \tilde{M} - M.$$

Therefore w is a single-valued and continuous transformation mapping M^* onto M and $\tilde{M}^* - M^*$ onto $\tilde{M} - M$. Incidentally we have proved the following theorem:

(9.21) *If M is any connected set having property S, then M is homeomorphic with a uniformly locally connected subset M^* of a compact locally connected continuum $\tilde{M}^*$.*

(9.3) THEOREM. *If M is connected and uniformly locally connected, w maps $\tilde{M}^*$ topologically onto $\tilde{M}$, carrying M^* onto M and A onto $B' = B$.*

For by virtue of (7) and (8) f and f^{-1} are uniformly continuous. Consequently, f and f^{-1} can be extended to the points of $\tilde{M} - M$ and $\tilde{M}^* - M^*$, respectively, in a unique manner so that the resulting transformations are single-valued and continuous on $\tilde{M}$ and $\tilde{M}^*$, respectively, and so that their values on M and M^* are unaltered. Also it is seen at once that under the extended transformations f and f^{-1} it is true that for each point x in $\tilde{M}$ (whether x be in M or not), $f^{-1}[f(x)] = x$, so that f^{-1} is the true inverse of f over the completed spaces $\tilde{M}$ and $\tilde{M}^*$. Therefore, the extended transformation f is a topological transformation throwing $\tilde{M}$ into $\tilde{M}^*$, M into M^*, and $\tilde{M} - M$ into $\tilde{M}^* - M^*$. Thus we have shown that $\tilde{M}$ and $\tilde{M}^*$ are homeomorphic and also that $\tilde{M} - M$ and $\tilde{M}^* - M^*$ are homeomorphic.

DEFINITION. *A metric space X is said to be locally complete provided every point x has a neighborhood U such that $\bar{U}$ is complete.*

(9.4) THEOREM. *If M is locally complete (or, in particular, locally compact), M is open in $\tilde{M}$; thus B' is closed in $\tilde{M}$ and M^* is open in $\tilde{M}^*$ because $M^* = w^{-1}(M)$. Further, every point of A is regularly accessible from M^* and B is identically the set of all points of B' which are accessible from M.*

The first statement follows because clearly the completion process does not alter M in any neighborhood U such that $\bar{U}$ is already complete. Also, since by (6) of (9.1) M^* is uniformly locally connected, $M^* + p$, for any $p \in A$, is a locally connected G_δ-set in $\tilde{M}^*$; and hence any $p \in A$ is regularly accessible from M^*.

To prove the final statement first let $q \in B$. Then $q = w(p)$ for some $p \in A$. By the above proof there exists an arc rp with $rp - p \subset M^*$. By continuity, $w(rp)$ is a locally connected continuum in $M + q$ with $w(rp) - q \subset M$. Hence $w(rp)$ contains an arc sq with $sq - q \subset M$.

On the other hand, let $q \in B'$ be accessible from M. Let sq be an arc with $sq - q \subset M$. Let (x_i) be a sequence of points on $sq - q$ with $x_i \to q$. Since $\lim_{m,n\to\infty} \delta(x_m x_n) = 0$ ($x_m x_n$ being the subarc of sq from x_m to x_n) it follows that (x_i^*) is a fundamental sequence in M^*, $[x_i^* = f(x_i)]$. Accordingly there exists a point $p \in A$ with $x_i^* \to p$. By continuity we have $w(x_i^*) = x_i \to w(p) = q$. Hence $q \in B$.

(9.41) COROLLARY. *If every point of B' is accessible from M,*

$$w(A) = B = B'.$$

(b) APPLICATIONS TO PLANE REGIONS

Note. A region is always connected.

(9.5) *If M is a plane region with connected boundary B', for any non-cut point p of B' in B, $w^{-1}(p)$ is single-valued.*

For let $a, b \in w^{-1}(p)$ and let ab be an arc with $ab \subset M^* + a + b$ [see (9.4)]. Since $w(a) = w(b) = p$ and w is (1-1) on $ab - (a + b)$, $w(ab) = J$ is a simple closed curve with $J - p \subset M$. Now since for $x, y \in ab$, any connected subset of M containing $w(x) + w(y)$ is of diameter as large as $\rho^*(x, y)$, it results that both regions of the plane complementary to J must contain points of B. Accordingly, p cuts B'.

(9.51) *If B' has no cut point, w is (1-1) on A and on $\tilde{M}^*$.*

(9.6) *Let M be a plane region with a connected boundary. If $x, y \in A$ and xy is an arc in M^* with $xy \cdot A = x + y$, then there exist separations:*

$$\tilde{M}^* - xy = M_1^* + M_2^*,$$

$$A - (x + y) = A_1 + A_2, \qquad A_i = M_i^* \cdot A.$$

The first of these is unique and the second is of course determined by the first. Further, if $w(x) = w(y)$, we have

$$w(A_1) \cdot w(A_2) = w(x),$$

so that w is "non-alternating" on A.

For let

$$M - w(xy - x - y) = M_1 + M_2,$$

where M_1 and M_2 are regions [see VI (3.5)],

$$N_i = w^{-1}(M_i), \qquad i = 1, 2,$$

then

$$M_i^* = \bar{N}_i - xy;$$

$$\tilde{M}^* - xy = M_1^* + M_2^*.$$

Suppose $\overline{M}_1^* \cdot M_2^* \supset z$. Then $z \subset A - (x + y)$. Let $z_i^1 \to z$, $z_i^2 \to z$, where $z_i^1 \epsilon M_1^* \cdot M^*$, $z_i^2 \epsilon M_2^* \cdot M^*$. Since $\rho^*(z_i^1, z_i^2) \to 0$ we have $z_i^1 + z_i^2 \subset C_i$, where C_i is connected, $\delta(C_i) \to 0$, and $C_i \subset M^*$. Clearly this is impossible, since $C_i \cdot xy \neq 0$ for each i, whereas $z \cdot xy = 0$. Accordingly M_1^* and M_2^* are separated. Obviously also A_1 and A_2 are separated, where $A_i = M_i^* \cdot A$ $(i = 1, 2)$.

Now suppose $w(x) = w(y) = p$. Then $w(xy) = J$ is a simple closed curve having just p in common with B'. Furthermore, $w(M_1^*)$ and $w(M_2^*)$ lie in different regions complementary to J. Accordingly

$$w(A_1) \cdot w(A_2) \subset p = w(x),$$

because $w(A_1) \subset w(M_1^*) \cdot B'$, $w(A_2) \subset w(M_2^*) \cdot B'$.

(9.61) *Any compact connected subset A_0 of A is either an arc or a simple closed curve.*

This follows directly from

(9.62) *The sets $A_1 + x + y$ and $A_2 + x + y$ in* (9.6) *are naturally ordered and dense, i.e., between any two points there is a third point.*

To prove (9.62), let $p \epsilon A_2$. Then for $a, b \epsilon A_1 + x + y$, define $a < b$ provided a and x lie in the same summand of the separation of A given in (9.6) by the pair b, p. This is readily seen to be a natural ordering of $A_1 + x + y$. Obviously, $A_2 + x + y$ is similarly ordered. That these sets are dense results from (9.6).

To prove (9.61) directly without (9.62), suppose A_0 is not a simple closed curve. Then it contains two points a and b such that $C = A_0 - (a + b)$ is connected. Then if $x \epsilon C$, $y \epsilon A - A_0$, x and y separate a and b in A in the sense of (9.6). Hence x must separate a and b in A_0. Accordingly, A_0 is an arc from a to b.

(9.7) THEOREM. *If M is a plane region with connected boundary, there exists a* (1-1), *continuous, and "order"-preserving mapping* ϕ *of A onto a subset A' of the unit circle C such that A' is dense in C, i.e.,*

$$\phi(A) = A' \subset C, \qquad \bar{A}' = C.$$

Proof. This is readily proved with the aid of (9. 62). For if $x, y \in A$, the sets $A_1 + x + y$ and $A_2 + x + y$ can readily be mapped onto a dense subset of the unit interval I—hence into a dense subset of a semicircle on C—by a (1-1), continuous, and order-preserving transformation. To do this for $A_1 + x + y$, say, we need only choose a countable sequence $p_1, p_2, \cdots$ in A_1 such that every point of A_1 is a point or a limit point of $P = \sum p_i$. Then set $\phi(x) = 0$, $\phi(y) = 1$, $\phi(p_1) = 1/2$, $\phi(p_{n_1}) = 1/4$ [where p_{n_1} is the first point in (p_i) preceding p_1], $\phi(p_{n_2}) = 3/4$ [where p_{n_2} is the first point in (p_i) following p_1], $\phi(p_{n_3}) = 1/8$ [where p_{n_3} is the first point in (p_i) preceding p_{n_1}], $\phi(p_{n_4}) = 3/8$ [where p_{n_3} is the first point in (p_i) between p_{n_1} and p_1], and so on indefinitely. Clearly ϕ maps $P + x + y$ in a (1-1) and order-preserving way onto the dyadic rationals on I. Thus if $p \in A_1$ and $p_{k_1} < p_{k_2} < p_{k_3} < \cdots \to p$, $\phi(p_{k_1}) < \phi(p_{k_2}) < \cdots$ so that $\phi(p_{k_i})$ converges to a point which we call $\phi(p)$ and which must be $\phi(p)$ in case $p \in P$. Hence as thus defined ϕ maps $A_1 + x + y$ in the required way into I. Obviously, two such mappings ϕ on $A_1 + x + y$ and $A_2 + x + y$ can be combined to yield the required mapping of A into C.

(9.71) COROLLARY. *Let M be a plane region with connected boundary. Then if the set A is compact it is a simple closed curve.*

For in this case, $A' = \phi(A)$ is also compact and hence must coincide with C since $A' = \bar{A}' = C$. Also ϕ is topological. Accordingly A is a simple closed curve.

It is interesting to note that even if A is not compact, on account of the order relations in A given by (9.62), A can be "compactified" by adjoining a new point to A corresponding to every class of equivalent monotone sequences in $A_1 + x + y$ or $A_2 + x + y$ which does not converge to a point of A, in an obvious way. The resulting space $\hat{A}$ will be compact and the mapping ϕ of (9.7) extends to yield a topological mapping $\phi(\hat{A}) = C$, so that $\hat{A}$ is a simple closed curve. This curve $\hat{A}$ may naturally be thought of as the curve of "boundary elements" of the region M. Thus we have

(9.72) *For any plane region M with connected boundary, the set A may be "compactified" to give a simple closed curve* $\hat{A}$ *of boundary elements of M.*

If M is bounded, the space $M^* + \hat{A}$ can be similarly compactified. For let K be the space consisting of the points of $M^* + \hat{A}$ and in which a neighborhood of p is (1) the set in K corresponding to any neighborhood of p in M^* when $p \in M^*$, (2) the set in K corresponding to any set M_1, where $(M^* + \hat{A}) - xy = M_1 + M_2$ is a separation given by an arc xy with $x, y \in \hat{A}$ and with $p \in M_1$. Using the compactness of $\hat{A}$ and the boundedness of M, it is readily shown that K is a compact space; and, further, the set M in K corresponding to M^* has property S and is homeomorphic with M^* (and hence also with M) and the set in K corresponding to $\hat{A}$, which we also denote by $\hat{A}$, is homeomorphic with $\hat{A}$. Since thus K is a locally connected continuum and every point of $\hat{A}$ is regularly accessible from M it readily follows from (9.6) and [VI, (5.3)] that K is a 2-cell with boundary curve $\hat{A}$. Whence

(9.73) *Any bounded plane region with a connected boundary is homeomorphic with the interior of the unit circle.*

(9.8) THEOREM. *Let M be a plane region with connected boundary B'. If M has property S, A is a simple closed curve and w maps A onto B' and the mapping $w(A) = B'$ is non-alternating. Furthermore, M^* is a 2-cell with boundary curve A.*

For by (9.2), M and M^* are compact and $w(A) = B'$. Thus A $[= w^{-1}(B')]$ is compact and hence by (9.7) it is a simple closed curve. That the mapping $w(A) = B'$ is non-alternating now results from (9.6). Finally, that M^* is a 2-cell with boundary curve A is an immediate consequence of (9.6) and [VI, (5.3)].

(9.81) Compare with [VI, (4.3)]. *If the plane region R with a connected boundary B' has property S, B' is a locally connected continuum.*

It should be noted that this conclusion is reached independently of [VI, (4.2)].

Now—continuing with a plane region M with connected boundary B'—if M is bounded and uniformly locally connected, it must have property S. Hence A is a simple closed curve and by (9.3), the mapping $w(A) = B'$ is a homeomorphism. Accordingly, B' is a simple closed curve. Conversely, if M is bounded and B' is a simple closed curve, by [VI, (4.2)], M has property S. Hence $\tilde{M}$ and $\tilde{M}^*$ are compact. Now by (9.5), w^{-1} is single-valued on B'. Accordingly w is a topological mapping, since $\tilde{M}^*$ is compact. But on M, w^{-1} is identical with f. Accordingly, f is uniformly continuous on M. Therefore, by (9.1), (8), M is uniformly locally connected. Thus we have

(9.82) THEOREM. *In order that a bounded plane region with connected boundary have a simple closed curve for its boundary it is necessary and sufficient that it be uniformly locally connected.*

We continue to let M be a plane region with connected boundary B' and let $B = w(A)$ so that by (9.4), B consists of exactly the points of B' which are accessible from M. Now for each $p \in B$, let $\alpha(p)$ be the number of components of $B' - p$ and let $m(p)$ be the number of points in $w^{-1}(p)$. We have

(9.9) THEOREM. *If either of the numbers $\alpha(p)$ and $m(p)$ is finite, the other is finite and the two are equal.*

First let $x_1, x_2, \cdots, x_m \in w^{-1}(p)$, the notation being chosen so that these points are cyclically ordered as in (9.7). Take $a \in M^*$ and let $D = \sum_1^m ax_i$ be a dendrite made up of arcs ax_i where $ax_i \subset M^* + x_i$ and $ax_i \cdot ax_j = a$, $i \neq j$. Now $M^* - (D - \sum x_i)$ consists of m regions $R_1, R_2, \cdots, R_m$ where $F(R_i) \supset ax_i + ax_{i+1}$, $(x_{m+1} = x_1)$. Thus $M - [w(D) - p]$ consists of m regions $w(R_1), w(R_2), \cdots, w(R_m)$ and $F[w(R_i)] \supset w(ax_i + ax_{i+1}) = J_i$ and J_i is a simple closed curve. Furthermore, if S_i denotes the region in the plane complementary to J_i which contains $w(R_i)$, $S_i \cdot S_j = 0$ for $i \neq j$. Now for each i, $S_i \cdot B \neq 0$. Hence there are at least m components of $B' - p$. Thus, in particular, if $\alpha(p)$ is finite, $m(p) \leqq \alpha(p)$.

Now on the other hand, if $m(p)$ is finite, let us take $m = m(p)$ in the above. Then $S_i \cdot B'$ must be connected for each i. For, if not, we could get a simple arc qp in $S_i + q + p$, where $q \in J$, separating $S_i \cdot B'$ in S_i and not intersecting $S_i \cdot B'$. But if $y_1, y_2, \cdots$ is a monotone sequence on qp converging to p, by the proof of (9.4), $[w^{-1}(y_j)]$ converges to a point y of $w^{-1}(p)$; and since $w^{-1}(y_j) \subset R_i$ and $x_i \neq y \neq x_{i+1}$, we get $y \cdot \sum_1^m x_i = y \cdot w^{-1}(p) = 0$, which is absurd.

10. Irreducibility of transformations. If A is a continuum, a continuous transformation $T(A) = B$ is *irreducible* provided no proper subcontinuum of A maps onto all of B under T. Similarly, merely assuming A compact, T is said to be *strongly irreducible* provided no closed proper subset of A maps onto all of B under T.

First certain auxiliary non-negative real-valued functions will be defined for such a transformation T on a compact set A and application will be made of the continuity properties of these functions to yield certain conclusions about T, in particular about irreducibility properties of T.

Let A be compact and metric and let $T(A) = B$ be continuous. We define

$f(x) = \delta(T^{-1}T(x))$, $\qquad x \in A$;

$g(x) = \delta(C_x)$, where C_x is the component of $T^{-1}T(x)$ containing x;

$h(x) = \text{l.u.b.}\ [\delta(C)]$, where C is a component of $T^{-1}T(x)$.

It is readily seen that the functions f, g and h are upper semi-continuous. Since they are non-negative, they therefore are continuous on the sets D_f, D_g, D_h, respectively, where they vanish. Now if E_f, E_g, E_h denote the sets on which f, g, and h, respectively, are continuous, we have that D_f, D_g, D_h are relatively closed subsets of E_f, E_g, E_h, respectively. Thus since E_f, E_g, E_h are G_δ-sets, so also are D_f, D_g, and D_h.

Thus we have

(10.1) THEOREM. *If A is compact and $T(A) = B$ is continuous, the sets*
D_f of all points of A on which T is (1-1),
D_g of all points x such that x is a component of $T^{-1}T(x)$,
D_h of all points where T is light (i.e., *all points x such that $T^{-1}T(x)$ is 0-dimensional) are G_δ-sets.*

In a similar way, if we define $n(x)$ to be the number of components of $T^{-1}T(x)$ and assume T interior, it follows that $n(x)$ is lower semi-continuous on A. Thus for any integer n, the set E_n of all points x of A with $n(x) \leqq n$ is closed. In particular, since $n(x) \geqq 1$ everywhere on A, the set E_n of all points x of A with $n(x) = 1$, i.e., the subset of A on which T is monotone, is closed. Of course, this is a conclusion easily arrived at directly without the use of the function $n(x)$.

(10.2) THEOREM. *In order that a continuous transformation $T(A) = B$ on a compact set A be strongly irreducible it is necessary and sufficient that the set D_f of all points $x \in A$ with $x = T^{-1}T(x)$ be dense in A.*

Proof. The condition clearly is sufficient, since obviously any subset of A which maps onto all of B must contain D_f. To establish the necessity of the condition, let O be any open subset of A. Then since T is strongly irreducible there must exist at least one point x such that $T^{-1}T(x) \subset O$, because $T(A - O) \neq B$. Thus for any $\epsilon > 0$, the set E_ϵ of all points $x \in A$ with $f(x) < \epsilon$ is dense in A; and since $f(x)$ is upper semi-continuous, it follows that the set D_f of all points $x \in A$ where $f(x) = 0$ is also dense in A.

COROLLARY. *If $T(A) = B$ is strongly irreducible and interior, where A is compact, then T is topological.*

(10.3) THEOREM. *Let A be a semi-locally-connected continuum. In order that a continuous transformation $T(A) = B$ be irreducible it is necessary and sufficient that the set D_f of all points $x \in A$ with $x = T^{-1}T(x)$ be dense on the set N of all non-cut points of A.*

Proof. The condition is sufficient. For suppose on the contrary that some proper subcontinuum K of A maps onto all of B under T. Then clearly we have $D_f \subset K$. But this gives $N \subset K$, since $\bar{D}_f \supset N$. Thus any $x \in A - K$ is a cut point; and this is impossible since any component of $A - x$ necessarily contains at least one non-cut point of A. To prove the necessity of the condition we again make use of the function $f(x)$. For any open set O in A intersecting $\bar{N}$, there exists at least one x such that $T^{-1}T(x) \subset O$, because about a point of $N \cdot O$ we can get an open set O_1 such that $O_1 \subset O$ and $A - O_1$ is a continuum and $T(A - O_1) \neq B$. Thus for any $\epsilon > 0$ the set E_ϵ of all points $x \in A$ with $f(x) < \epsilon$ is dense in $\bar{N}$; and since $f(x)$ is upper semi-continuous, the set D_f of all points $x \in A$ with $f(x) = 0$ must be dense in $\bar{N}$.

(10.4) THEOREM. *If A is a semi-locally-connected continuum, a necessary and sufficient condition that every irreducible transformation on A be strongly irreducible is that the non-cut points of A be dense in A.*

The sufficiency of the condition follows at once from (10.2) and (10.3) To see that the condition is necessary we need only note that if the non-cut points of a continuum A are not dense in A, then A contains a free arc, say $axuvyb$ of cut points; and if we fold this arc so that v goes into x, u into y, and xu, vu and vy go into an arc xy and all other points of A go into themselves, we set up a transformation on A which is irreducible but not strongly irreducible.

(10.5) THEOREM. *If A is a semi-locally connected continuum, any irreducible interior transformation $T(A) = B$ is a homeomorphism.*

Proof. Suppose, on the contrary, that for some $b \in B$, $T^{-1}(b) \supset x + y$ where $x \neq y$. Then x must be a cut point of A. For if not, let V_x and V_y be disjoint neighborhoods of x and y, respectively, so chosen that $T(V_x) \subset T(V_y)$;

there exists a continuum N such that $A - V_x \subset N \subset A - x$; whence $T(N) \supset T(V_y) \supset T(V_x)$ so that $T(N) = T(A)$, contrary to the irreducibility of T. Now let R be the component of $A - x$ containing y. Then R must map onto all of B. For since R is open and T is interior, $T(R)$ is open in B. Further, since $R + x$ is compact and $T(R + x) = T(R)$, $T(R)$ is compact and hence closed. Accordingly $T(R) = B$. But $R + x$ is a proper subcontinuum of A. Hence we have a contradiction and our theorem is proved.

REFERENCES

§§1–3. Whyburn [22]. For extensions of the notion of a non-alternating transformation, see Hall [3] and Schweigert, Wallace [4], White [1].

§4. Whyburn [22, 26], Eilenberg [5].

§5. Whyburn [22].

§6. Whyburn [22]. Result (6.23) in the case of locally connected continua was first proved by G. E. Schweigert.

§7. Whyburn [26, 27, 31].

§8. Wallace [1]. For the definition of the number $r(X)$ and the result (8.6), see Eilenberg [3]. The proof of (8.6) given here is due to the author and has not previously been published.

§9. Whyburn [28]. For the original relative distance definition and for properties (1) and (3) in (9.1) see Mazurkiewicz [1]. For property (4) of (9.1) see Urysohn [2]. Compare (9.72) and (9.73) with well known results of Carathéodory [1]. (9.81) and (9.82) are due to R. L. Moore, see Moore [5, 12]. Also see Newman [1]. In connection with (9.9) see Nöbeling [2].

§10. Whyburn [29, 30]. In this connection, see also Rozanska [1], Simond [1] and Harrold [1].

CHAPTER IX

APPLICATIONS OF MONOTONE AND NON-ALTERNATING TRANSFORMATIONS

1. **Non-alternating transformations on boundary curves.** We first treat some special cases which serve as a basis for handling the general situation. For simplicity we shall regard a set consisting of a single point as being a (degenerate) simple arc, simple closed curve, dendrite, sphere, and so on.

(1.1) *The property of being a* $\left\{\begin{matrix}\textit{simple arc} \\ \textit{simple closed curve}\end{matrix}\right\}$ *is invariant under monotone transformations.*

First let A be a simple arc from a to b and let $f(A) = B$ be monotone. Then B is a simple arc from $f(a)$ to $f(b)$. For let $x \in B - f(a) - f(b)$. Then $f^{-1}(x)$ separates a and b in A, and hence x separates $f(a)$ and $f(b)$ in B. Thus B is an arc $f(a)f(b)$.

Now let A be a simple closed curve and again suppose $f(A) = B$ is monotone. Let x and y be distinct points (if any such exist) of B. If $B - (x + y)$ were connected, $A - f^{-1}(x) - f^{-1}(y)$ would be connected. Obviously this is impossible. Hence B is a simple closed curve.

Note. It may be remarked that (1.1) states that *if A is an arc or a simple closed curve then B either is a single point or is homeomorphic with A*, even though f need not be a homeomorphism. Further, the condition employed in [VII, §7] that for each $b \in B$ there exist a point $a \in A$ such that $A - a$ and $A - f^{-1}(b)$ be homeomorphic clearly is satisfied for *any monotone transformation* in each of these cases. In fact a continuous transformation on an arc or a simple closed curve is monotone *if and only if* it satisfies this condition.

(1.2) *If A is a simple closed curve and B is cyclic, any non-alternating transformation $f(A) = B$ is monotone.*

For let $x \in B$. Since $B - x$ is connected it follows from [VIII, (2.3)] that $A - f^{-1}(x)$ $[= f^{-1}(B - x)]$ is connected. Since A is a simple closed curve, it results that $f^{-1}(x)$ must be connected.

Now it has already been shown [VIII, (6.21)] that the property of being a dendrite is invariant under non-alternating transformations. In connection with (1.2) we recall also [VIII, (2.41)] that *any non-alternating transformation on a dendrite is necessarily monotone.*

(1.3) Theorem. *The property of being a boundary curve is invariant under non-alternating transformations.*

For let A be a boundary curve and let $f(A) = B$ be non-alternating. Then if E_b is any true cyclic element of B it follows by [VIII, (6.5)] that there exists a true cyclic element E_a of A and a non-alternating transformation $z(E_a) = E_b$.

Since E_a is a simple closed curve, it follows from (1.2) that z is monotone. Hence by (1.1), E_b is a simple closed curve.

Thus in particular, any non-alternating image of a circle is a boundary curve. Now we wish to show, conversely, that any boundary curve is the image of a circle under some non-alternating transformation. This is a ready consequence of the results obtained on the relative distance transformation in [VIII, §9].

(1.4) THEOREM. *If B is any boundary curve, there exists a non-alternating light transformation $f(A) = B$ where A is a simple closed curve.*

For by [IV, (7.31)], B is homeomorphic with the boundary of a plane (or spherical) bounded region R. Hence we may suppose $B = F(R)$ on a sphere. Now if we apply the relative distance transformation d to R obtaining a set R^* and complete R^* by adding to it the set A, then since R has property S it follows by [VIII, §9] that A is a simple closed curve. Furthermore, if d^{-1} be extended to A we obtain a non-alternating transformation $w(A) = B$ as desired. That this transformation is non-alternating follows either directly from [VIII, (9.8)] or from [VIII, (9.9)], together with (1.7) below.

Thus if $[B]$ is the class of all boundary curves, we can derive any element of $[B]$ and only elements of $[B]$ from a circle by applying a non-alternating transformation. Further, a non-alternating transformation applied to any element of $[B]$ always yields an element of $[B]$.

Now it is clear that if B is any boundary curve which is not a dendrite and B' is any element whatever of $[B]$, *we can map B onto B' by a non-alternating transformation.* For if C is a true cyclic element of B we have only to map C onto B' by a non-alternating transformation $g(x)$, let $r(x)$ be the monotone transformation retracting B into C and define

$$f(x) = gr(x).$$

Then by [VIII, (3.5)], f is non-alternating and clearly it maps B onto B'. Thus any non-dendritic element B of $[B]$ may be taken as a "generator" and any element of $[B]$ may be obtained from B by a non-alternating transformation.

(1.51) LEMMA. *If $P = \sum_1^m p_i$ is any finite subset of a boundary curve B such that for $i, j \leqq m$, $E(p_i, p_j) \subset P$, we have*

$$\text{(i)} \qquad \alpha(P) = \sum_1^m \alpha(p_i),$$

where for any subset X of B, $\alpha(X)$ denotes the number of components of $B - X$.

Proof. If R is any component of $B - P$, the points of $F(R)$ are mutually conjugate. Hence $F(R)$ consists of at most two points; and since any given pair of points of P can be the boundary of at most two components of $B - R$, it results that if either side of (i) is infinite so also is the other. Thus we may assume both sides finite.

Let $R_1, R_2, \cdots, R_{\alpha(P)}$ be the components of $B - P$, where $F(R_i)$ consists of two points for $1 \leqq i \leqq k$ and of one point for $i > k$. Thus if for each $i \leqq \alpha(P)$ $n(R_i)$ denotes the number of points of $F(R_i)$, and for each $j \leqq m$, $m(p_j)$ denotes the number of regions R_i with $F(R_i) \supset p_j$, $i \leqq \alpha(P)$,

(ii) $\sum_1^{\alpha(P)} n(R_i) = \sum_1^k n(R_i) + \sum_{k+1}^{\alpha(P)} n(R_i) = 2k + \alpha(P) - k = \alpha(P) + k$,

(iii) $\alpha(p_j) = m(p_j) - (1/2)q_j$, $j \leqq m$, where q_j is the number of regions R_i with $i \leqq k$ and $F(R_i) \supset p_j$. Relation (iii) holds because in a given component of $B - p_j$ there must be either 0 or 2 regions R_i with $i \leqq k$ and $F(R_i) \supset p_j$. Summing (iii), we get

$$\begin{aligned} \sum_1^m m(p_j) &= \sum_1^m \alpha(p_j) + (1/2) \sum_1^m q_j \\ &= \sum_1^m \alpha(p_j) + (1/2) \sum_1^k n(R_i) \qquad \text{(iv)} \\ &= \sum_1^m \alpha(p_j) + k. \end{aligned}$$

Now clearly

$$\sum_1^{\alpha(P)} n(R_i) = \sum_1^m m(p_i).$$

Hence (ii) and (iv) give (i).

(1.52) THEOREM. *If A is a boundary curve and $f(A) = B$ is non-alternating and light, for each $p \in B$ we have*

$$(*) \qquad \alpha(p) = \sum_{x \in f^{-1}(p)} \alpha(x).$$

Proof. Let $P = f^{-1}(p)$. Since f is non-alternating, the components of $A - P$ map onto the components of $B - p$ in (1-1) way under f. Hence

$$\alpha(p) = \alpha(P).$$

Now if $\alpha(p)$ is finite, since each component of $A - P$ has at most two boundary points it results that P is finite, because each point of P is a boundary point of at least one of the $\alpha(P)$ components of $A - P$. Accordingly (*) follows from the lemma. On the other hand, if the right member of (*) is finite, again P must be finite since $\alpha(x) \geqq 1$ for each $x \in P$. Hence again the lemma applies.

COROLLARIES. (1) *If no point of $f^{-1}(p)$ is a cut point of A, $\alpha(x) = 1$ for each $x \in f^{-1}(p)$ so that $\alpha(p) = \sum \alpha(x) = m(p)$, where $m(p)$ is the number of points of $f^{-1}(p)$.*

(2) *If A is a simple closed curve, $\alpha(p) = m(p)$ for all $p \in B$.*

(3) *If p is a non-cut point of B, $1 = \alpha(p) = \sum \alpha(x)$, so that $f^{-1}(p)$ is a single non-cut point of A.*

(4) *If $\alpha(p) = 2$, $f^{-1}(p)$ consists either of two conjugate non-cut points of A or of a single point cutting A into just two components.*

(1.61) LEMMA. *If $P = \sum p_i$ is any finite subset of a boundary curve B, we have*

$$(\ddagger) \qquad \alpha(P) \leqq \sum_1^m \alpha(p_i).$$

Proof. Since any component of $B - P$ having just one boundary point p_i is a component of $B - p_i$ and since there are only a finite number of components of $B - P$ having more than one boundary point, it follows that if either side of ($\ddagger$) is infinite, so also is the other. Hence we may assume both sides finite.

Let $R_1, R_2, \cdots, R_{\alpha(P)}$ be the components of $B - P$ where $F(R_i)$ consists of more than one point for $1 \leqq i \leqq k$ and of just one point for $k < i \leqq \alpha(P)$. Then if for each i, $n(R_i)$ denotes the number of points of $F(R_i)$ we have

$$\text{(i)} \qquad \begin{aligned} \sum_1^{\alpha(P)} n(R_i) &= \sum_1^k n(R_i) + \sum_{k+1}^{\alpha(P)} n(R_i) \\ &= \sum_1^k n(R_i) + \alpha(P) - k \geqq \alpha(P) + (1/2) \sum_1^k n(R_i). \end{aligned}$$

Also if for each $j \leqq m$, $m(p_j)$ denotes the number of regions R_i with $F(R_i) \supset p_j$ and q_j denotes the number of such regions with $i \leqq k$, then since in any given component of $B - p_j$ there are at most two regions R_i with $F(R_i) \supset p_j$ and when there are two, both must have subscript less than or equal to k, we have

$$\text{(ii)} \qquad m(p_j) \leqq \alpha(p_j) + (1/2)q_j .$$

Whence

$$\text{(iii)} \qquad \sum_1^m m(p_j) \leqq \sum_1^m \alpha(p_j) + (1/2) \sum_1^m q_j \leqq \sum_1^m \alpha(p_j) + (1/2) \sum_1^k n(R_i).$$

Thus since $\sum_1^{\alpha(P)} n(R_i) = \sum_1^m m(p_j)$, from (i) and (iii) we get

$$\alpha(P) + (1/2) \sum_1^k n(R_i) \leqq \sum_1^m \alpha(p_j) + (1/2) \sum_1^k n(R_i)$$

which is equivalent to ($\ddagger$).

(1.62) THEOREM. *If A is a boundary curve, $f(A) = B$ is light and, for each $p \in B$, $\alpha(p)$ is finite and*

$$\alpha(p) = \sum_{x \in f^{-1}(p)} \alpha(x),$$

then f is non-alternating.

Proof. Let $P = f^{-1}(p)$. Since $\alpha(x) \geqq 1$ for all x, it follows from our hypothesis that P is a finite set. Now by continuity of f we have

$$\alpha(P) \geqq \alpha(p)$$

and by the lemma ($\ddagger$),

$$\alpha(P) \leqq \sum_{x \in P} \alpha(x) = \alpha(p).$$

Whence

$$\alpha(P) = \alpha(p).$$

Thus if $q \in B$, $q \neq p$, $f^{-1}(q)$ must lie in a single component of $A - P$, because otherwise two components of $B - P$ would map into a single component of $B - p$ giving $\alpha(P) > \alpha(p)$. Accordingly f is non-alternating.

(1.7) THEOREM. *If A and B are boundary curves and $f(A) = B$ is light, then in order that f be non-alternating it is necessary and sufficient that*

(*) $$\alpha(y) = \sum_{x \in f^{-1}(y)} \alpha(x)$$

for each $y \in B$ for which $\alpha(y) = 1$ or 2.

Proof. The condition is sufficient. For, let $p, q \in B$. Since B is a boundary curve, there exists a set Y consisting either of two non-cut points of B or of a single point of B which cuts B into just two components such that Y separates p and q in B. In either case $\sum_{y \in Y} \alpha(y) = 2$ and hence by our hypothesis $\sum_{x \in f^{-1}(Y)} \alpha(x) = 2$. Thus $f^{-1}(Y) = X$ consists either of two non-cut points of A or a single point of A which cuts A into just two regions. Since in any case, by continuity of f, X must separate $f^{-1}(p)$ and $f^{-1}(q)$ in A, it results that $A - X$ consists of exactly two regions (because A is a boundary curve). Hence one of these regions contains $f^{-1}(p)$ and the other contains $f^{-1}(p)$. Accordingly $f^{-1}(p)$ does not separate $f^{-1}(q)$ in A. Thus f is non-alternating.

The necessity of the condition is a consequence of (1.52).

(1.71) THEOREM. *If A and B are boundary curves, in order that a continuous transformation $f(A) = B$ be non-alternating it is necessary and sufficient that*

(*) $$\alpha(y) = \sum_{X \text{ a component of } f^{-1}(y)} \alpha(X),$$

for each $y \in B$ for which $\alpha(y) = 1$ or 2.

For let $f(A) = g\phi(A) = g(A') = B$ be the monotone-light factorization of f. Since ϕ is monotone, for each $x \in A'$ we have

$$\alpha(x) = \alpha[\phi^{-1}(x)] = \alpha(X),$$

where $X = \phi^{-1}(x)$ is a component of $f^{-1}g(x)$.

Whence, for any $y \in B$

(†) $$\sum_{x \in g^{-1}(y)} \alpha(x) = \sum_{x \in g^{-1}(y)} \alpha[\phi^{-1}(x)] = \sum_{X \text{ a component of } f^{-1}(y)} \alpha(X).$$

Thus if f is non-alternating, since then g is non-alternating and light it follows that when $\alpha(y)$ is 1 or 2 it equals the left member of (†) and hence (*) holds. Conversely, if (*) holds for $\alpha(y) = 1$ or 2, $\alpha(y)$ equals the left member of (†) and hence g—and therefore also f—is non-alternating.

(1.8) THEOREM. *If A, B and C are boundary curves and the transformations* $f_1(A) = B$ *and* $f_2(B) = C$ *are non-alternating and light, so also is the transformation* $f(A) = f_2f_1(A) = C$.

Proof. Let $y \in B$ such that $\alpha(y) = 1$ or 2. Then since

$$\text{(i)} \qquad \alpha(y) = \sum_{x \in f_2^{-1}(y)} \alpha(x),$$

we have $\alpha(x) = 1$ or 2 for all $x \in f_2^{-1}(y)$. Hence

$$\text{(ii)} \qquad \alpha(x) = \sum_{t \in f_1^{-1}(x)} \alpha(t) \qquad \text{for each } x \in f_2^{-1}(y).$$

Whence

$$\alpha(y) = \sum_{x \in f_2^{-1}(y)} \sum_{t \in f_1^{-1}(x)} \alpha(t) = \sum_{t \in f^{-1}(y)} \alpha(t).$$

Accordingly, f is non-alternating.

2. Monotone transformations on spheres, cactoids, planes, and 2-cells. In this section we shall have occasion to use the readily established fact that *if M is any continuum on the surface S of a sphere such that* $S - M$ *is connected, then* $S - M$ *is homeomorphic with* $S - p$ *where* $p \in S$. This results at once from the fact that since both $S - p$ and $S - M$ are simply connected regions on S, each of them is homeomorphic with the set $|z| < 1$ in the complex plane [see VIII, (9.73)]. It could be proved directly by taking a sequence of regions bounded by "concentric" simple closed curves (polygons) on S closing down on M and mapping the annular regions between these in a sense-preserving way topologically onto the annular regions between a sequence of concentric circles on S closing down on p. (For a detailed proof, see for example, Newman [1].)

(2.1) THEOREM. *If A is a topological sphere, B is cyclic and* $f(A) = B$ *is monotone, B is a topological sphere.*

Proof. We set aside the trivial case where B reduces to a single point. To establish our theorem, by virtue of [VI, §5] we have to prove two things, namely,

(i) *no simple arc in B separates B*;

(ii) *every simple closed curve in B separates B.*

To prove (i) suppose, on the contrary, that some arc α in B separates two points a and b in B. Then α contains an arc uv which separates a and b and is such that if R_a and R_b denote the components of $B - uv$ containing a and b, respectively,

$$(*) \qquad F(R_a) \cdot F(R_b) \supset u + v.$$

For we have only to let G_a be the component of $B - \alpha$ containing a, let α' be the least subarc of α containing $F(G_a)$, let R_b be the component of $B - \alpha'$ containing B, let uv be the least subarc of α' containing $F(R_b)$ and let R_a be the component of $B - uv$ containing a.

Let $x \in uv - (u + v)$ and $X = f^{-1}(x)$. Decompose A into the set X and the individual points of $A - X$, let S be the hyperspace of this decomposition and let $g(A) = S$ be the associated continuous transformation. Then g is monotone and since $A - X$ is connected it follows by the above remarks and [VII, (7.1)], that S is a topological sphere.

Now since f and g are monotone it follows that the point $p = g(X) = gf^{-1}(x)$ separates the continuum $K = gf^{-1}(uv)$ into two connected sets $U = gf^{-1}(ux - x)$ and $V = gf^{-1}(xv - x)$. Hence by the plane separation theorem [VI, (3.1)] there exists on S a simple closed curve J separating U and V on S and so that $K \cdot J = p$. Now from (*) it follows that both U and V contain a limit point of each of the sets $gf^{-1}(R_a)$ and $gf^{-1}(R_b)$; and since each of these sets is connected each must intersect $J - p$. But again by monotoneity, $fg^{-1}(J - p)$ is connected and does not intersect uv. This contradicts the fact that uv separates R_a and R_b in B. Hence (i) is proved.

To establish (ii), suppose there exists a simple closed curve C on B such that $B - C$ is connected. Let x, y be points of C decomposing C into arcs xuy and xvy and let $X = f^{-1}(x)$, $Y = f^{-1}(y)$. Decompose A into the continua X and Y and the individual points of $A - (X + Y)$, let S be the hyperspace of this decomposition and let $g(A) = S$ be the associated monotone transformation. Since neither X nor Y separates A (because B is cyclic) it follows by [VII, (7.1)], that S is a topological sphere. For we can factor g into the form $g(x) = g_2g_1(x)$ where $g_1(A) = A'$ and $g_2(A') = S$ are such that g_1 maps X into a single point of A' and is otherwise topological and g_2 maps $g_1(Y) = Y'$ into a single point and is otherwise topological. Then since Y' does not separate A', it follows by successive application of [VII, (7.1)] that A' and S are topological spheres.

Now since f and g are monotone, it follows that the points $p = g(X) = gf^{-1}(x)$ and $q = g(Y) = gf^{-1}(y)$ together separate the continuum $K = gf^{-1}(C)$ into two connected sets $U = gf^{-1}(xuy - x - y)$ and $V = gf^{-1}(xvy - x - y)$. Hence by the Plane Separation Theorem there exists on S a simple closed curve J separating U and V and such that $J \cdot K = p + q$. Now since $B - C$ is connected and f and g are monotone, the set $Q = gf^{-1}(B - C)$ is connected. Since $Q = S - K$ it follows that there exists a simple arc rt in Q such that r is on one of the arcs of J from p to q and t is on the other and $rt \cdot J = r + t$. Thus if $\theta = J + rt = rpt + rqt + rt$, θ is a θ-curve with vertices r and t and $\theta \cdot K = p + q$ so that $\theta \cdot (U + V) = 0$. But since J separates U and V and both p and q are limit points of both U and V, $U + V$ intersects all three regions of $S - \theta$. Hence either U or V, say U, must intersect two regions of $S - \theta$ and hence must intersect θ since U is connected. This contradiction proves (ii). Hence our theorem is established.

(2.1)′ *The hyperspace of any upper semi-continuous decomposition of a sphere into continua not separating the sphere is a topological sphere.*

(2.2) THEOREM. *The property of being a cactoid is invariant under monotone transformations.*

For let A be a cactoid and let $f(A) = B$ be monotone. Then if E_b is any true cyclic element of B it follows by [VIII, (5.5)] that there exists a true cyclic element E_a of A and a monotone transformation $z(E_a) = E_b$. Since E_a is a topological sphere it follows by (2.1) that E_b is a topological sphere. Accordingly B is a cactoid.

(2.2)′ *The hyperspace of any upper semi-continuous decomposition of a cactoid into continua is a cactoid.*

DEFINITION. *A mapping is strongly monotone provided the inverse of every continuum is a continuum.*

(2.3) THEOREM. *If A is a plane and $f(A) = B$ is strongly monotone, B is topologically equivalent to a set obtained by removing some non-cut point from a cactoid.*

Proof. Clearly there is no loss in generality in supposing $A = S - p$ where S is a topological sphere and $p \in S$. Now on S the collection $[f^{-1}(y)]_{y \in B}$ is semi-closed. Hence if we add p to this collection we obtain an upper semi-continuous decomposition of S, and if K is the hyperspace of this decomposition and $g(S) = K$ is the associated transformation, g is monotone and hence K is a cactoid. Now since $S - p$ is connected and $p = g^{-1}g(p)$, $K - g(p)$ is connected. Finally, if for each $x \in K - g(p)$ we define

$$h(x) = fg^{-1}(x)$$

it is clear that since $g^{-1}(x) = f^{-1}h(x)$, h is a homeomorphism mapping $K - g(p)$ onto B.

(2.31) COROLLARY. *If A is a plane, B is cyclic and $f(A) = B$ is strongly monotone, B is a topological plane.*

(2.31)′ COROLLARY. *The hyperspace of an upper semi-continuous decomposition of a plane A into continua not separating A is itself a topological plane.*

DEFINITIONS. *A locally connected continuum which is homeomorphic with a plane continuum not separating the plane is called a base set. In other words, a base set is homeomorphic with a plane locally connected continuum every true cyclic element of which is a 2-cell. Hence any base set β is the sum of a boundary curve B and a countable set of disjoint open 2-cells each bounded by a true cyclic element of β.*

A locally connected continuum which is the sum of a base set β and a null sequence of disjoint cactoids each intersecting β in a single point is called a *hemi-cactoid.*

If we think of the base set β of a hemi-cactoid H as being on a sphere S such that $S \cdot H = \beta$, it is clear that $S - \beta$ is an open 2-cell R bounded by the boundary curve B of β. Hence we see that *a hemi-cactoid is a set H of the form*

$$H = K - R$$

where K is a cactoid and R is an open 2-cell having property S.

(2.4) THEOREM. *If A is a 2-cell with boundary J and $f(A) = H$ is monotone, H is a hemi-cactoid with boundary curve $f(J)$.*

Proof. There is no loss of generality in supposing that A is on a sphere S. Decompose S into the sets $[f^{-1}(y)]_{y \in H}$ and the individual points of $S - A$ and let $g(S) = K$ be the associated transformation. Since g is monotone, K is a cactoid. Let $g(S - A) = U$, $g(A) = B$, $g(J) = C$. Since $S - A$ is an open 2-cell and g is topological on $S - A$, there exists a true cyclic element E of K containing $U + C$. Thus E is a topological sphere and $B \cdot E$ is a locally connected continuum on E not separating E (because $E - E \cdot B = U$). Hence $B \cdot E$ is a base set, and thus B is a hemi-cactoid with base set $B \cdot E$ and boundary curve C. Now for each $x \in B$ let

$$h(x) = fg^{-1}(x).$$

Then clearly h maps B topologically onto H and maps C onto $f(J)$. Hence H is a hemi-cactoid with boundary curve $f(J)$

(2.41) Corollary. *If A is a 2-cell, B is cyclic and $f(A) = B$ is monotone, B is either a sphere or a 2-cell. If B is a sphere, the boundary J of A maps into a single point; if B is a 2-cell with boundary C, $f(J) = C$.*

(2.5) Theorem. *The property of being a hemi-cactoid is invariant under monotone transformations.*

Proof. Let A be a hemi-cactoid with base set B and boundary curve C and let $f(A) = H$ be monotone. Since B is homeomorphic with a subset of a sphere, clearly there is no loss of generality in supposing A is a subset of a cactoid K such that $K - A$ is a region on a cyclic element E of K such that $K \cdot E = B$. Let us decompose K into the sets $[f^{-1}(y)]_{y \in H}$ together with the individual points of $K - A$. Let N be the hyperspace of this decomposition and let $g(K) = N$ be the associated transformation. Since g is monotone, N is a cactoid. Furthermore, since g is topological on $K - A$, there exists a true cyclic element F of N containing $g(K - A)$. Thus $g(A) \cdot F$ is a locally connected continuum on F not separating F (because $F - g(A) \cdot F = g(K - A)$) and hence is a base set. Accordingly $g(A)$ is a hemi-cactoid. Then, just as in previous proofs,

$$h(x) = fg^{-1}(x), \qquad x \in g(A),$$

is a homeomorphism mapping $g(A)$ onto H.

(2.51) *If A is a hemi-cactoid with boundary curve C and $f(A) = H$ is monotone, H is a hemi-cactoid with boundary curve $f(C)$.*

It will be noted from this that the boundary curve of a hemi-cactoid maps onto a boundary curve when the hemi-cactoid undergoes a monotone transformation. In this connection the following theorem which also yields this conclusion is of interest.

(2.52) Theorem. *If H is a hemi-cactoid with boundary curve C, any monotone transformation on H is non-alternating on C.*

For let $f(H) = K$ be monotone and let $f(C) = B$. If $y \in B$, and $a, b \in C \cdot f^{-1}(y)$, the monotoneity of f on H gives $E(a, b) \subset f^{-1}(y)$ where $E(a, b)$ is taken in H. Furthermore, since any point which separates a and b in C must separate a

and b also in H, we have $E(a, b) \subset C$. Hence $E(a, b) \subset f^{-1}(y)\cdot C$. Hence by [VIII, (2.4)] it remains only to be shown that f is non-alternating on each cyclic element J of C. If this were not so, there would exist two points x, $y \in B$ and points $x_1, x_2 \in J\cdot f^{-1}(x)$ and $y_1, y_2 \in J\cdot f^{-1}(y)$ such that x_1 and x_2 separate y_1 and y_2 on J. But now if E is the cyclic element of H containing J, E is a 2-cell; and since $E\cdot f^{-1}(x)$ and $E\cdot f^{-1}(y)$ are continua containing $x_1 + x_2$ and $y_1 + y_2$, respectively, these continua would have to intersect. This is impossible since one of them maps into x and the other into y. This completes the proof.

3. **Existence theorems.** A sequence of sets $X_1, X_2, X_3, \cdots$ is said to *converge 0-regularly* to a set X, written "$X_n \to X$ 0-regularly," provided $[X_n]$ converges to X and for any $\epsilon > 0$ a $\delta > 0$ and an integer $N > 0$ exist such that if $n > N$, any two points x, y of X_n with $\rho(x, y) < \delta$ lie together in a connected subset of X_n of diameter less than ϵ. It is clear that 0-regularity of convergence is a sort of "equi-uniform local connectedness" restriction on the sequence $[X_n]$. Further, a sequence of the type $X, X, X, \cdots$, where X is a closed set, clearly converges 0-regularly to X if and only if X is uniformly locally connected.

(3.1) THEOREM. *Let A be a continuum, let $f_n(A) = B_n$ be a sequence of monotone transformations converging uniformly to a continuous transformation $f(A) = B$ and let $B_n \to B$ 0-regularly. Then f is monotone.*

Proof. Let $y \in B$ and $a, b \in f^{-1}(y)$. For each n let $f_n(a) = p_n$, $f_n(b) = q_n$ and let L_n be a continuum in B_n of minimum diameter containing $p_n + q_n$, and let $K_n = f_n^{-1}(L_n)$. Then K_n is a continuum containing $a + b$. Let $[K_{n_i}]$ be a subsequence of $[K_n]$ converging to a limit continuum K. Then $K \supset a + b$. We shall show that $K \subset f^{-1}(y)$. In the first place, since $p_n \to y$ and $q_n \to y$ it follows by 0-regular convergence of B_n to B that $L_n \to y$. Suppose there were a point $x \in K$ with $f(x) \neq y$. Then if $3d = \rho[y, f(x)]$, there exists an N such that for $n \geqq N$, $\rho[L_n, f(x)] > 2d$, and for any $t \in A$, $\rho[f_n(t), f(t)] < d$; but since $K_{n_i} \to K \supset x$, there exists an $n_j > N$ and a point $t \in K_{n_j}$ such that

$$\rho[f(t), f(x)] < d;$$

and since

$$\rho[f_{n_j}(t), f(t)] < d,$$

this gives

$$\rho[f_{n_j}(t), f(x)] < 2d$$

contrary to $f_{n_j}(t) \in L_{n_j}$ and $\rho[L_{n_j}, f(x)] > 2d$. Therefore $K \subset f^{-1}(y)$ and hence $f^{-1}(y)$ is connected.

(3.11) COROLLARY. *If A and B are continua, B is locally connected and $f_n(A) = B$ is a sequence of monotone transformations converging uniformly to a transformation $f(A) = B$, then f is monotone.*

(3.12) LEMMA. *If* $[H_n]$ *is any sequence of H-sets in a locally connected continuum* M *converging to a limit set* L, *the convergence* $H_n \rightarrow L$ *is* 0-*regular.* (*Incidentally,* L *is an A-set.*)

For if $\epsilon > 0$, we have only to choose δ so that any two points in M at a distance apart less than δ lie in a connected subset of M of diameter less than ϵ and choose $N = 1$. Then if $x, y \in H_n$, $\rho(x, y) < \delta$ there exists a connected subset Q of M with $x + y \subset Q$ and $\delta(Q) < \epsilon$. Hence $H_n \cdot Q$ is a connected subset of H_n of diameter less than ϵ containing $x + y$.

This lemma together with the preceding theorem yields at once the

(3.2) THEOREM. *Let* A *be a continuum and let* $f_n(A) = B_n$ *be a sequence of monotone transformations converging uniformly to a transformation* $f(A) = B$, *where for each* n, B_n *is an A-set in a locally connected continuum* M. *Then* f *is monotone.*

(3.3) THEOREM. *If* B *is any cactoid, there exists a monotone transformation of a sphere onto* B.

Proof. We first establish the following.

(*) *If* $C(a, b)$ *is any cyclic chain in* B, *there exists a monotone transformation* $\phi(S) = C(a, b)$, *where* S *is any sphere, such that for each point* y *of* $E(a, b) + a + b = K$ *which is* a, b *or a point of order greater than* 2 *in* B, $\phi^{-1}(y)$ *contains a* 2-*cell on* S.

To prove (*) we suppose, without loss of generality, that S is made up of the segment $0 \leqq z \leqq 1$ of the cylinder $r = 1$ in 3-space with cylindrical coordinate system (r, θ, z) together with the two "ends;" $z = 0, r \leqq 1$ and $z = 1, r \leqq 1$. Furthermore, we may suppose that $E(a, b) + a + b = K$ is situated on the unit interval I: $0 \leqq z \leqq 1$ so that $I \subset C(a, b)$ and $a = (0, 0, 0)$, $b = (1, 0, 0)$ and so that every point y of the set Y of all points of $E(a, b) + a + b$ of order greater than 2 in B together with a and b has a coordinate z of the form $m/2^n$. [For we can take an arc ab in $C(a, b)$ and map this topologically onto I so that these conditions hold.]

Now let $g(z)$ be the mapping of the Cantor ternary set C on I onto I described in [II, §4], and let us suppose $g(z)$ extended to a mapping $g(I) = I$ by defining $g(z)$ to be constant on each segment of $I - C$. Now for each $x \in S$, $x = (r, \theta, z)$, let us define $z(x) = z$ and

$$q(x) = g(z) = g[z(x)].$$

Now let us decompose S into the sets $q^{-1}(z)$, $z \in K$, together with the individual points of $S - q^{-1}(K)$. Since K is compact, this decomposition is upper semi-continuous. Furthermore, if D is the hyperspace and $\phi(x)$ the associated transformation, ϕ is monotone and D is homeomorphic with $C(a, b)$. Finally, since any point $y \in Y$ is of the form $m/2^n$, $g^{-1}(y)$ contains an open segment of I and hence $z^{-1}g^{-1}(y) = q^{-1}(y) = \phi^{-1}(y)$ contains an open cylinder on S and thus surely a 2-cell. Accordingly (*) is proved.

Now to prove the theorem, we take a cyclic chain development of B:

$$B = H + \sum_{1}^{\infty} H_n = H + \sum_{1}^{\infty} C(p_i, q_i),$$

$$H_n = \sum_{1}^{n} C(p_i, q_i)$$

satisfying conditions laid down in [IV, (7.1)]. For each i, let $K_i = E(p_i, q_i) + p_i + q_i$. Let A be a fixed sphere. Let $f_1(A) = C(p_1, q_1)$ be a mapping satisfying (*).

Now if $q_2 \in K_1$, $f_1^{-1}(q_2)$ contains a 2-cell $E_1 = R_1 + J_1$ so that $f_1^{-1}(q_2) - E_1$ contains a 2-cell. Consider a transformation $g_1(x)$ which is identical with $f_1(x)$ on $A - E_1$ and (1-1) in R_1. The set $g_1(E_1)$ is then a sphere S_1 containing the point $f_1(J_1) = q_2$. Now let us apply (*) to obtain a mapping

$$\phi_1(S_1) = C(p_2, q_2)$$

satisfying all conditions in (*) and so that $\phi_1(q_2) = q_2$. Then let

$$f_2(x) = \phi_1 g_1(x), \qquad x \in A.$$

In case q_2 non $\in K_1$, we have $q_2 \in W$, where W is a sphere in $C(p_1, q_1)$ and $f_1^{-1}(q_2)$ reduces to a single point x_2. Let J_1 and J_2 be concentric circles on A with center x_2 and interiors R_1 and R_2 with $R_2 + J_2 \subset R_1$, $\delta[f_1(R_1 + J_1)] < 1/2, f_1(R_1 + J_1) \subset W - W \cdot K_1$. Let g_1 be a mapping which is identical with f_1 on $A - (R_1 + J_1)$ maps J_2 into q_2 and the ring $R_1 - (R_2 + J_2)$ topologically onto $f_1(R_1 + J_1) - q_2$ and is (1-1) on R_2. Then $g_1(R_2 + J_2)$ is a sphere S_1 containing q_2. Now defining ϕ_1 and f_2 as in the preceding case, we obtain our desired mapping $f_2(A) = H_2$.

It will be noted that for any $x \in A$,

(i) $$\rho[f_1(x), f_2(x)] < 1/2 + \text{l.u.b. } [\delta(Q_i^1)]$$

where $[Q^1_i]$ is the collection of components of $B - H_1$.

We next define a mapping

$$f_3(A) = H_3$$

by modifying f_2 on the points of a 2-cell whose image under f_2 is of diameter less than $1/2^2$ using the point q_3 just as q_2 was used above in obtaining f_2 by modifying f_1 and so that for each $x \in A$

(ii) $$\rho[f_2(x), f_3(x)] < 1/2^2 + \text{l.u.b. } [\delta(Q_i^2)]$$

where $[Q_i^2]$ is the collection of components of $B - H_2$. Similarly we obtain mappings

$$f_4(A) = H_4,$$

$$\cdots\cdots\cdots\cdots$$

$$f_n(A) = H_n$$

such that at any stage n for any $y \in \sum_1^n K_i$ which either is a p_i or a q_i for $i \leqq n$ or is a point of order greater than 2 in B, $f_n^{-1}(y)$ contains a 2-cell. Thus the process may be continued indefinitely unless the series $\sum H_n$ is finite. Hence we get a sequence of mappings $f_1, f_2, \cdots$.

This sequence of mappings has the property that for any $x \in A$ and any two integers m and n $(n > m)$

(iii) $$\rho[f_n(x), f_m(x)] < 1/2^{m-1} + \text{l.u.b. } \delta[Q_i^m]$$

where $[Q_i^m]$ is the collection of components of $B - H_m$.

This property yields at once the uniform convergence of the sequence. For let $\epsilon > 0$. Choose N so that for $m \geqq N$, $1/2^m < \epsilon/2$ and each component of of $B - H_m$ is of diameter less than $\epsilon/2$. Then if $m, n > N$ and $x \in A$ we have, by (iii),

$$\rho[f_n(x), f_m(x)] < \epsilon/2 + \epsilon/2 = \epsilon.$$

Thus the sequence $[f_n(x)]$ converges uniformly to a limit transformation $f(x)$. Clearly $f(A) = B$. Furthermore, since each f_n is monotone and each H_n is an A-set in B it results from (2.63) that f is monotone. This completes the proof of our theorem.

Our next objective will be to establish a similar existence theorem for the representability of an arbitrary hemi-cactoid as the monotone image of a 2-cell. To this end we first prove

(3.31) Lemma. *If B is a base set on a sphere S, $E = G + J$ is a closed 2-cell lying in $S - B$ and $A = S - G$, there exists a monotone retraction $\phi(A) = B$.*

Proof. Let $R = S - B$ and let B' denote the boundary of R. Then if $F = H + C$ is the unit circle C plus its interior H it follows by [VIII, (9.8)] that there exists a continuous transformation $w(F) = R + B'$ which maps H onto R topologically and C onto B' in a non-alternating way.

Let $Q = w^{-1}(G)$. Since $F - Q$ is a closed circular ring, there exists a monotone retraction $\phi(F - Q) = C$.

Now let us define

$$\phi(y) = w\phi w^{-1}(y), \qquad y \in R + B' - G.$$

$$\phi(y) = y, \qquad y \in A - R,$$

Since ϕ is a retraction it results at once that ϕ retracts A into B. To see that ϕ is monotone, let $z \in B$. If z non $\in B'$, $\phi^{-1}(z) = z$. If $z \in B'$,

$$\phi^{-1}(z) = w\phi^{-1}w^{-1}(z);$$

and since ϕ is monotone, every component of $\phi^{-1}w^{-1}(z)$ intersects $w^{-1}(z)$. Accordingly every component of $w\phi^{-1}w^{-1}(z) = \phi^{-1}(z)$ must contain z, and hence $\phi^{-1}(z)$ is connected.

(3.4) Theorem. *If H is any hemi-cactoid, there exists a monotone transformation $f(A) = H$, where A is a 2-cell.*

Proof. We may suppose, without loss of generality, that H is a subset of a cactoid K such that $K - H$ is a region R on a true cyclic element S of K and where $B = S - R$ is the base set of H.

By (3.3) there exists a monotone transformation $g(N) = K$ where N is a sphere. Furthermore, since R is an open 2-cell which is open in K it is clear from the proof of (3.3) that g can be chosen so that on a closed 2-cell $E = G + J$ lying in R, g is (1-1).

Let A denote the closed 2-cell $N - g^{-1}(G)$ and let A' be the set $S - G$. By the preceding lemma, there exists a monotone retraction

$$\phi(A') = B.$$

Clearly if for each $y \in K - B$ we define

$$\phi(y) = y$$

we get a monotone retraction

$$\phi(K - G) = H.$$

Now define

$$f(x) = \phi g(x), \qquad x \in A.$$

Then

$$f(A) = \phi g(A) = \phi g[N - g^{-1}(G)] = \phi(K - G) = H,$$

and since ϕ and g are monotone, so also is f.

Thus we have that any cactoid can be represented as the monotone image of a sphere and any hemi-cactoid can be represented as the monotone image of a 2-cell.

We note next that, using these existence theorems, we can obtain from the special cases of a sphere or 2-cell the results (2.2) and (2.5) to the effect that the monotone image of a cactoid or of a hemi-cactoid is necessarily a cactoid or a hemi-cactoid, respectively, i.e.,

The property of being a $\left\{\begin{matrix}\textit{cactoid}\\ \textit{hemi-cactoid}\end{matrix}\right\}$ *is invariant under monotone transformations.*

For let K be a cactoid and let $f(K) = K'$ be monotone. We can write

$$K = g(S)$$

where S is a sphere and g is monotone. Whence

$$K' = fg(S).$$

Now since f and g are monotone, so also is fg. Thus K' is a cactoid.

Similarly, let $f(H) = H'$ where H is a hemi-cactoid and f is monotone. Writing H in the form

$$H = g(E)$$

where E is a 2-cell, we have

$$H' = fg(E).$$

Since again fg is monotone it follows that H' is a hemi-cactoid.

It will now be noted that we have established properties for the classes $[K]$ and $[H]$ of cactoids and hemi-cactoids, respectively, relative to monotone transformations similar to those of the boundary curves relative to non-alternating transformations. That is, any element of either of these classes maps onto another element in the same class under a monotone transformation; any cactoid is representable on a sphere and any hemi-cactoid on a 2-cell by a monotone transformation. Thus any cactoid K with a nondegenerate cyclic element may be taken as a "generator" of the class $[K]$ and any hemi-cactoid H with a non-dendritic base set may be taken as a generator of $[H]$ in the sense that *any* element of $[K]$ or $[H]$, respectively, may be obtained from this generator by applying a monotone transformation.

4. **Extension theorems.** If $A \subset X$, $B \subset Y$ and $f(A) = B$ and $g(X) = Y$ are continuous, g is said to be an *extension* of f to X provided

$$g(x) = f(x)$$

for all $x \in A$.

If $A \subset X, f(A) = B$ and $g(X) = Y$ are continuous and $B' = g(A)$, g is said to be an extension of f provided the transformations

$$f(A) = B \text{ and } g(A) = B'$$

are topologically equivalent, i.e., provided there exists a homeomorphism

$$h(B) = B'$$

such that

$$g(x) = hf(x), \qquad x \in A.$$

In other words, g is an extension of f provided we can imbed B topologically in Y so that

$$g(x) = f(x) \qquad \text{for all } x \in A.$$

Now it was shown in (2.52) that if H is a hemi-cactoid with boundary curve B, any monotone transformation on H is non-alternating on B. Thus when a monotone transformation on H is *reduced* to B, it retains the important property of being non-alternating. We shall establish next a converse result to the effect that if we start with a non-alternating transformation on the boundary curve B of a hemi-cactoid H, we can *extend* this transformation to H in such a way that the extended transformation is monotone on H.

To that end we prove first

(4.1) LEMMA. *If C is a circle in a plane, E is C plus its interior and $f(C) = B$ is non-alternating, there exists a monotone extension $g(E) = Q$ of f to E where Q is a base set with boundary curve $B' = g(C)$.*

Proof. Let us decompose E into the sets $[f^{-1}(y)_c]$, $y \in B$, together with the individual points of $E - \sum_{y \in B} f^{-1}(y)_c$, where $f^{-1}(y)_c$ denotes the convex closure of $f^{-1}(y)$, i.e., the common part of all convex sets in E containing $f^{-1}(y)$. Call this collection of sets G. Clearly every element of G is a continuum (convex), and since f is non-alternating it results at once that the elements of G are disjoint. For if $p \in f^{-1}(y_1)_c \cdot f^{-1}(y_2)_c$, y_1, $y_2 \in B$, p is on a linear segment x_1x_1' in $f^{-1}(y_1)_c$ and on a linear segment x_2x_2' in $f^{-1}(y_2)_c$ where x_1, x_1', x_2, $x_2' \in C$. But then x_1 and x_1' separate x_2 and x_2' on C contrary to the fact that f is non-alternating.

We prove next that G is upper semi-continuous. To this end it suffices to show that if g_1, g_2, $\cdots$ is a convergent sequence of distinct nondegenerate elements of G with limit L, then L is contained in some single element of G. If L is a single point, this is evident. Hence we suppose L nondegenerate. Then L contains a point p lying within C. Furthermore, there exist points a_i, $b_i \in g_i \cdot C$ and an infinite sequence n_i such that $a_{n_i} \rightarrow a$, $b_{n_i} \rightarrow b$ where $a \neq b$. Since for each i, $f(a_i) = f(b_i)$, it follows that $f(a) = f(b) = y$. We shall show that L is identical with the linear segment ab. Clearly $L \supset ab$. Further, since infinitely many of the sets g_i are in one of the two segments into which ab divides E we may suppose all of them are in one such segment S. Then since the linear segments $a_{n_i}b_{n_i}$ lie in g_{n_i} and converge to ab, no point not on ab can be on L. Thus $L = ab \subset f^{-1}(z)_c$. Accordingly, G is upper semi-continuous.

Now let Q be the hyperspace of the decomposition G and let $g(E) = Q$ be the associated (monotone) transformation. Since g is monotone, Q is a hemicactoid. But since for any $q \in Q$, $g^{-1}(q)$ separates E if and only if it separates two points of C in E, it follows that Q reduces to a base set. Further, if $B' = g(C)$ and if for each $y \in B$ we define

$$h(y) = gf^{-1}(y),$$

clearly h is a homeomorphism mapping B onto B' and is such that

$$g(x) = hf(x), \qquad x \in C.$$

Accordingly, g is an extension of f to E.

Now since a given boundary curve can be the boundary curve B for one and only one topological type of base set Q, i.e., Q consists of B plus an open 2-cell fitted into each true cyclic element of B, we have

(4.2) THEOREM. *If $E = C + I$ is a 2-cell, Q is any base set with boundary curve B and $f(C) = B$ is non-alternating, there exists a monotone extension $g(E) = Q$ of f.*

(4.3) THEOREM. *If H is a hemi-cactoid with boundary curve A and $f(A) = B$ is non-alternating, there exists a monotone extension $g(H) = Q$ of f to H where Q is a base set with boundary curve $B' = g(A)$.*

Proof. Let W denote the base set of H. As each true cyclic element E of W is a 2-cell bounded by a simple closed curve C of A on which f is non-alternating, by (4.1) f can be extended monotonically to each such E. Now let us decompose W into the sets $f^{-1}(y)$, $y \in B$, + all points x in W which map into the same point as some point of $f^{-1}(y)$ under the extension of f in some cyclic element of W, together with the individual points of W not in any set of the first type. Let G denote this collection of sets. Since the extensions of f in each cyclic element of W are monotone and since f is non-alternating so that any element of G contains all points of W separating two points of that element, it results that the elements of G are disjoint continua. If $g_1, g_2, \cdots$ is any sequence of elements of G converging to a limit L, there exists a cyclic element E of W so that

$$L = \lim (E \cdot g_i).$$

Thus since the decomposition of E was upper semi-continuous, we have $L \subset$ some element of G. Hence G is upper semi-continuous.

Now let Q be the hyperspace of the decomposition G and let

$$\phi(W) = Q$$

be the associated (monotone) transformation. It follows just as in the proof of (4.1) that ϕ is an extension of f to W and Q is a base set with boundary curve $B' = \phi(A)$.

Finally, let $r(H) = W$ be the monotone retraction of H onto W and let us define

$$g(x) = \phi r(x).$$

Since ϕ and r are monotone, so also is g. Also g maps H onto Q, and since r is a retraction, clearly g is an extension of f to H.

REFERENCES

§1. Whyburn [22].

§2. (2.1) Moore [13, 9]. (Proof in text new.)

(2.2), (2.3) Whyburn [22], Moore [9].

(2.4) Morrey [1]. (Proof in text new.)

In connection with §2 see also Puckett [2] and Roberts [2] and Steenrod.

§3. For a treatment of regular convergence, in all dimensions, as well as for results related to (3.1) and (3.2) but in some cases more general, see Whyburn [32, 33]. In this connection see also Arnold [1], Puckett [1], Wallace [2], and Whitney [2].

(3.3) Moore [9]. (Proof in text new.)

(3.4) Morrey [1]. The proof given and the lemma (3.41) are believed original and published here for the first time.

CHAPTER X

INTERIOR TRANSFORMATIONS

1. Action on linear graphs. We begin by analyzing the action of interior transformations on linear graphs, arcs and simple closed curves. We recall that a set A is a (linear) graph provided it is the sum of a finite set V of points, called *vertices*, and a finite number of open free arcs $\alpha_1, \alpha_2, \cdots, \alpha_n$, called *edges*, so that the two end points of each edge α_i are distinct and belong to V. Such a decomposition of a graph A into vertices and edges is called a *subdivision* of A.

It is verified at once that either of the following conditions is necessary and sufficient in order that a continuum A be a graph:

(1) almost all points of A (i.e., all save a finite number) are of order 2;

(2) every point of A is of some finite order and almost all points of A are of order less than or equal to 2.

If A and B are graphs and K_a and K_b are subdivisions of A and B, respectively, so that K_a and K_b are finite 1-dimensional complexes in the sense of combinatorial topology, a mapping $f(A) = B$ which maps vertices (0-simplexes) and edges (1-simplexes) of K_a onto vertices and edges, respectively, of K_b is said to be a *simplicial transformation* on K_a. In the next theorem to be established, these and some other elementary notions of combinatorial topology are used. For adequate definitions and developments of these concepts the reader is referred to Alexandroff [6]-Hopf, Lefschetz [1], or Veblen [3]. Since the results of this nature which we obtain are so few and are generally end-results or applications and hence not used in what succeeds them, it seems hardly justifiable to take the space to develop them here. We use R to denote the rational coefficient field, $B^n_R(K)$ to denote the n-dimensional Betti group of a complex K, and $p^n(K)$ to denote its rank which is the n-dimensional Betti number of K.

(1.1) THEOREM. *Let A be a linear graph and let $f(A) = B$ be interior. Then B is a linear graph and there exist subdivisions of A and B, respectively, into finite complexes K_a and K_b such that f maps each edge of K_a topologically into an edge in K_b. Thus $f(K_a) = K_b$ is a simplicial transformation. Furthermore, if J is any simple closed curve in B, there exists a simple closed curve C in A such that $f(C) = J$ and, on C, f is topologically equivalent to the transformation $w = z^k$ on $|z| = 1$ (k an integer). Consequently $B^1_R(K_a)$ maps homomorphically onto $B^1_R(K_b)$ under f.*

Proof. Since A has only a finite number of points of order greater than 2 and has no point of increasing order, and since by [VIII, (7.31)], the order of no point can be increased under f, it follows that B has these same properties. Accordingly, B is a linear graph.

Furthermore, for each $b \in B, f^{-1}(b)$ must be a finite set of points. For otherwise there would exist infinitely many components of $A - f^{-1}(b)$, and clearly this is impossible, since by [VIII, (7.41)], the inverse of each one of the finite number of components of $B - b$ is a finite set of components of $A - T^{-1}(b)$.

Now let E_a and E_b, respectively, denote the sets of points in A and B of order different from 2. Let $D_b = E_b + f(E_a) +$ one point from each component of $B - [E_b + f(E_a)]$ which has at most one limit point in $E_b + f(E_a)$; let $D_a = f^{-1}(D_b)$. The sets D_a and D_b subdivide A and B into complexes K_a and K_b, respectively. Let xy be any edge of K_a and let $f(x) = x', f(y) = y', f(xy) = x'y'$. Then $x' \neq y'$, since otherwise $f(xy - x - y)$ would be a free arc in B having just one end point, which is impossible by the choice of D_b. Furthermore, for any $p \in x'y' - x' - y', f^{-1}(p) \cdot xy$ reduces to just one point. For if not, there would exist an open arc $\widehat{p_1p_2}$ in $xy - xy \cdot f^{-1}(p)$ with $p_1 + p_2 \subset f^{-1}(p)$; this would give $f(\widehat{p_1p_2}) \supset x'$ or $f(\widehat{p_1p_2}) \supset y'$; and both are impossible, since $f^{-1}(x') \cdot xy = x$, $f^{-1}(y') \cdot xy = y'$. Accordingly, $f(xy) = x'y'$ is topological and $f(K_a) = K_b$ is simplicial.

Now let J be any simple closed curve in B, let the vertices of K_b on J be ordered cyclically $p_0, p_1, \cdots, p_n, p_0$ and let $q_0 \in f^{-1}(p_0)$. There exists an edge q_0q_1 of K_a which is contained in $f^{-1}(p_0p_1)$ and such that $q_1 \in f^{-1}(p_1)$. Similarly there is an edge q_1q_2 with $q_1q_2 \subset f^{-1}(p_1p_2)$ and $q_2 \in f^{-1}(p_2)$, and so on to p_n. There is an edge $q_nq_0^1$ with $q_nq_0^1 \subset f^{-1}(p_np_0)$ and $q_0^1 \in f^{-1}(p_0)$. If $q_0^1 = q_0$, we can bring our selection to a close. If not, we choose an edge $q_0^1q_1^1$ in $f^{-1}(p_0p_1)$. Again if $q_1^1 = q_1$, we may stop. Otherwise we choose an edge $q_1^1q_2^1$ in $f^{-1}(p_1p_2)$, and so on. Continuing this process, since $f^{-1}(b)$ is a finite set for each $b \in B$, after a finite number of steps we must eventually reach, for the first time, a point q_i^j $(i \leqq n)$ such that $q_i^j = q_i^m$ $(m < j)$. Then clearly the edges

$$q_0^{m+1}q_1^{m+1}, q_1^{m+1}q_2^{m+1}, \cdots, q_{n-1}^{m+1}q_n^{m+1}, q_n^{m+1}q_0^{m+2}, \cdots, q_{i-1}^jq_i^j, q_i^jq_{i+1}^m,$$
$$q_{i+1}^mq_{i+2}^m, \cdots, q_n^mq_0^{m+1}$$

fit together to form a simple closed curve C which maps onto J exactly $j - m$ times; and since $q_s^rq_{s+1}^r$ maps topologically onto p_sp_{s+1}, it follows at once that if we set $j - m = k$, the transformation $f(C) = J$ is topologically equivalent to the transformation $w = z^k$ on the circle $|z| = 1$ of the complex z-plane.

Since, as shown above, f can be considered as a simplicial transformation of K_a onto K_b, it follows that f generates a homomorphism, which we shall also call f, of $B_R^1(K_a)$ into $B_R^1(K_b)$. To see that actually $f[B_R^1(K_a)] = B_R^1(K_b)$, let us take any rational 1-cycle z in K_b. As is well known, z can be expressed as a linear form

$$z = a_1z_1 + a_2z_2 + \cdots + a_nz_n, \qquad a_i \text{ rational},$$

where z_i is a simply oriented simple closed polygon. By the above, we can find a simple closed curve C_i in A such that $f(c_i) = k_i z_i$ ($i \leqq n$, k_i an integer), where c_i is an oriented fundamental cycle on C_i. Accordingly if we set

$$\gamma = \frac{a_1}{k_1} c_1 + \frac{a_2}{k_2} c_2 + \cdots + \frac{a_n}{k_n} c_n,$$

then γ is a rational 1-cycle in K_a and $f(\gamma) = z$. Thus the homology class of γ in $B^1_R(K_a)$ maps onto the homology class of z in $B^1_R(K_b)$ and our result follows.

(1.11) COROLLARY. $p^1(K_b) \leqq p^1(K_a)$ [*where* $p^r(X)$ *denotes the r-dimensional Betti number of* X].

(1.2) *If* A *is a simple closed curve and* $f(A) = B$ *is interior, then* B *is either a simple closed curve or a simple arc. If* B *is a simple closed curve, there exists an integer* K *such that* f *is topologically equivalent to the transformation* $w = z^k$ *on the circle* $|z| = 1$. *If* B *is an arc, there exists an integer* k *such that* f *is equivalent to the transformation* $f(1, \theta) = \sin k\theta/2$ *of the circle* $\rho = 1$ *into the interval* $(-1, 1)$.

Proof. It follows from [VIII, (7.31)] that every point of B is of order less than or equal to 2. Accordingly, B is either a simple closed curve or a simple arc.

If B is a simple closed curve, it follows from (1.1) by taking $J = B$ that f is equivalent to $w = z^k$ on $|z| = 1$ for some integer k.

If B is a simple arc xy, referring back to the proof of (1.1), we see that $D_b = x + y$ and hence $D_a = f^{-1}(x) + f^{-1}(y)$. Thus if we order the points of D_a cyclically on A: $p_1, p_2, \cdots, p_k, p_1$, where $p_1 \in f^{-1}(x)$, then by (1.1), p_1p_2 maps topologically onto xy, p_2p_3 maps topologically onto yx, p_3p_4 onto xy and so on to p_kp_1, which maps onto yx. (Note that k must be even.) Hence f is equivalent to the transformation $f(1, \theta) = \sin k\theta/2$ of $\rho = 1$ into $(-1, 1)$.

Remark. In each case the integer k is exactly the maximum power among the sets $f^{-1}(b)$ for all $b \in B$. If k is odd, B is necessarily a simple closed curve, f is completely alternating, and all the sets $f^{-1}(b)$ are of power k. If k is even, B may be either a simple closed curve or an arc xy; and in the latter case $f^{-1}(x)$ and $f^{-1}(y)$ are each of power $k/2$, while for every other point b of B, $f^{-1}(b)$ is of power k.

(1.3) *If* A *is a simple arc* pq *and* $f(A) = B$ *is interior,* B *is a simple arc and there exists an integer* k *such that* f *is topologically equivalent to the transformation* $f(x) = \cos k\pi x$ *of the interval* $(0, 1)$ *into the interval* $(-1, 1)$.

Proof. Since, by [VIII, (7.3)], B must have at least two points of order 1 but no point of order greater than 2, it follows that B is a simple arc uv.

Referring again to the proof of (1.1), we see that in this case $D_b = u + v \supset f(p) + f(q)$ and $D_a = f^{-1}(u) + f^{-1}(v)$. Hence if we order the points of D_a on A from p to q: $p = p_0, p_1, \cdots, p_k = q$ and suppose the notation chosen so that $f(p) = u$, then by (1.1), p_0p_1 maps topologically onto uv, p_1p_2 maps topologically onto uv, p_2p_3 onto uv, and so on to $p_{k-1}p_k$, which may map either onto uv or vu. Hence it is seen at once that f is equivalent to the transformation $f(x) = \cos k\pi x$ of $(0, 1)$ into $(-1, 1)$.

DEFINITION. *For any transformation* $f(A) = B$ *and any* $x \in B$, *let the multiplicity* $k(x)$ *of* x *be the number of points in* $f^{-1}(x)$ *if this number is finite, otherwise we define* $k(x) = \infty$. *Also for each integer* $n > 0$, *let* $B(n)$ *denote the set of all* $x \in B$ *with* $k(x) = n$.

We prove next

(1.4) THEOREM. *Let* $f(A) = B$ *be interior, where* A *is a locally connected continuum. If* x *is a point of* B *such that* $B - x$ *has just a finite number of components* $R_1, R_2, \cdots, R_k$, *and if for each* i, n_i *is the number of components of* $f^{-1}(R_i)$, *then* $f^{-1}(x)$ *contains less than* $n = \sum_1^k n_i$ *cut points of* A.

Proof. Since any set of n cut points of a connected set A cuts A into at least $n + 1$ mutually separated (open) sets and since no open set can map into x, it follows that if $f^{-1}(x)$ contained as many as n cut points of A, we would have

$$A - f^{-1}(x) = A_1 + A_2 + \cdots + A_{n+1},$$

where the sets A_i are disjoint and open. But clearly this is impossible, since each A_i must contain at least one component of $A - f^{-1}(x)$ and there are only n such components because each one is a component of $f^{-1}(R_i)$ for some i.

We next prove a theorem concerning the action of interior transformations on dendrites. We recall from [VIII, (7.7)] that the property of being a dendrite is invariant under interior transformations.

(1.5) THEOREM. *Let* $f(A) = B$ *be interior, where* A *is a dendrite. For any two points* $a, b \in B$ *and any point* x *on the arc* ab *of* B *we have*

$$\text{(i)} \qquad k(x) \leqq k(a) + k(b) - 1.$$

Proof. Suppose, first, that x is a point of order 2, interior to ab. Then there are just two components R_a and R_b of $B - x$ and these contain a and b, respectively. Now it follows by a former theorem [VIII, (7.41)] that each component of $A - f^{-1}(x)$ maps onto either R_a or R_b. Hence there can be at most $k(a)$ of these components mapping onto R_a and at most $k(b)$ mapping onto R_b. Accordingly there are at most $k(a) + k(b)$ components of $A - f^{-1}(x)$; and since each point of $f^{-1}(x)$ must be a point of order greater than or equal to 2 and hence be a cut point of A, it follows by (1.4) that $k(x) \leqq k(a) + k(b) - 1$.

Thus every interior point x of ab of order 2 satisfies (i). Since such points are dense on ab and the points x with $k(x)$ less than or equal to any integer form a closed set, it follows that (i) holds for any x on ab.

We have the following corollaries.

(1.51) *If* $k(a) = k(b) = 1$, *then* $k(x) = 1$ *for each* $x \in ab$. *Accordingly the set* $B(1)$ *is a continuum, as is also* $f^{-1}[B(1)]$.

(1.52) *In general, for any* m, *there exists a dendrite* K *in* B *such that*

$$B(m) \subset K \subset B(2m - 1).$$

(1.53) *If* $k(a) = 1$ [*or* $k(b) = 1$], *then* $k(x) \leqq k(b)$ [*or* $k(x) \leqq k(a)$]. *Thus if* $B(1) \neq 0$, *then for each* m, $\sum_1^m B(i)$ *is a dendrite.*

(1.6) THEOREM. *The property of being a boundary curve is invariant under interior transformations.*

Proof. Let A be a boundary curve, let $f(A) = B$ be interior, and let B_0 be any true cyclic element of B. If A_0 is any component $f^{-1}(B_0)$, then $f(A_0) = B_0$, and on A_0, f is interior because by [VIII, (7.5)], there can be only a finite number of components of $f^{-1}(B_0)$. Hence by [VIII, (7.61)] any node E_0 of A_0 maps onto all of B_0 under f. Since E_0 is a simple closed curve, B_0 can have at most one point of order different from 2; and hence B_0 is also a simple closed curve.

2. Inversion of simple arcs and dendrites.

(2.1) THEOREM. *Let $T(A) = B$ be a light interior transformation, where A is compact. Then if pq is any simple arc in B and p_0 is any point in $T^{-1}(p)$, there exists a simple arc p_0q_0 in A such that $T(p_0q_0) = pq$ and T is topological on p_0q_0.*

Our proof for this theorem will make use of the following definition and lemma.

DEFINITION. *If $\sigma: p = x_0, x_1, \cdots, x_n = q$ (x_i precedes x_{i+1} in the order p, q) is a subdivision of pq, a continuum K in $T^{-1}(p)$ will be said to proceed directly from $T^{-1}(p)$ to $T^{-1}(q)$ relative to the subdivision σ provided $T(K) = pq$ and $K = \sum_1^n K_i$, where K_i is a continuum in $T^{-1}(x_{i-1}x_i)$. By the norm of such a continuum K will be meant* max $\delta(K_i)$.

LEMMA. *Given any subdivision σ' of pq and any $\epsilon > 0$, there exists a subdivision σ of pq containing σ' and a continuum K of norm less than ϵ in $T^{-1}(pq)$ which contains p_0 and which proceeds directly from $T^{-1}(p)$ to $T^{-1}(q)$ relative to σ.*

Proof of Lemma. Let S denote the set of all points q' on pq for which such a subdivision exists on pq' containing all points of σ' on pq'. Let x be any point of pq.

If $x \in S$, let us take the subdivision $\sigma = \sigma_x$ on px and the corresponding continuum $K = K_x$. Let $z \in K \cdot T^{-1}(x)$. There exists an ϵ-neighborhood U of z such that $F(U) \cdot T^{-1}(x) = 0$. There exists a point v on xq such that if w is any point whatever of xv, $F(U) \cdot T^{-1}(w) = 0$ and w does not belong to σ' unless $w = x$. It follows that $Q = T^{-1}(xv) \cdot U$ is both open and closed in $T^{-1}(xv)$. Accordingly $T(Q) = xv$ is an interior transformation. Let K_{n+1} be the component of Q containing z, let σ_v be the subdivision $p = x_0, x_1, \cdots, x_n, x_{n+1} = v$ of pv and let $K_v = \sum_1^{n+1} K_i$. Then clearly K_v is a continuum of norm less than ϵ which proceeds directly from $T^{-1}(p)$ to $T^{-1}(v)$ relative to σ_v, so that v belongs to S. Thus S is open in pq.

Now if $x \in pq - S$, we can choose x to be the first point of $pq - S$ on pq. Hence $px - x \subset S$. If we take a sequence $y_1, y_2, \cdots$ on px so that $y_i \to x$, and choose points $z_i \in K_{y_i} \cdot T^{-1}(y_i)$, where K_{y_i} is a continuum of norm less than ϵ in $T^{-1}(py_i)$ proceeding directly from $T^{-1}(p)$ to $T^{-1}(y_i)$ relative to a subdivision σ_{y_i} of py_i which contains all points of σ' on py_i, the sequence $z_1, z_2, \cdots$ will have a limit point z on $T^{-1}(x)$, and we may suppose $z_i \to z$. Let us choose the ϵ-neighborhood U of z exactly as before. There exists an i such that for any $w \in y_ix$, $T^{-1}(w) \cdot F(U) = 0$ and such that no point of y_ix, except possibly x, belongs to σ'. Hence if $Q = T^{-1}(y_i x) \cdot U$, Q is both open and closed in $T^{-1}(y_i x)$ and $T(Q) = y_i x$ is interior. Now if σ_{y_i} is denoted by $p = x_0, x_1, \cdots, x_n = y_i$ and $K_{y_i} = \sum_1^n K_m$, let σ_x be the subdivision $p = x_0, x_1, \cdots, x_n, x_{n+1} = x$ of px, let K_{n+1} be the component of Q containing z_i and $K_x = \sum_1^{n+1} K_j$. Then clearly K_x is a continuum of norm less than ϵ which proceeds directly from $T^{-1}(p)$ to $T^{-1}(x)$ relative to σ_x, and σ_x contains all points of σ' on px. But this gives $x \in S$, contrary to our supposition. Accordingly $S = pq$, and the lemma is proved.

Proof of the Theorem. There exists a subdivision $\sigma_1 : p = x_0^1, \cdots, x_{n_i}^1 = q$ of pq and a continuum $K^1 = \sum_1^{n_1} K_i^1$ of norm less than 1 which contains p_0 and proceeds directly from $T^{-1}(p)$ to $T^{-1}(q)$ relative to σ_1.

In general, for each k there exists a subdivision

$$\sigma_k : p = x_0^k, \cdots, x_{n_k}^k = q,$$

which contains σ_{k-1}, and a continuum $K^k = \sum_1^{n_k} K_i^k$ of norm less than $1/k$ which contains p_0 and proceeds directly from $T^{-1}(p)$ to $T^{-1}(q)$ relative to σ_k.

Since the sequence $K^1, K^2, \cdots$ contains a convergent subsequence, we may suppose, without loss of generality, that the whole sequence converges to a limit which we shall call E. Since each K^i is a continuum with $T(K^i) = pq$, it follows that E is a continuum and $T(E) = pq$. To prove that E is an arc and $T(E) = pq$ is topological, it suffices to prove that for each $y \in pq$, $T^{-1}(y) \cdot E$ reduces to a single point.

Now since for each k, i, j ($i < j$, $i, j \leqq n_k$) there exists a continuum $K_{ij}^k = \sum_{m=i+1}^{j} K_m^k$ in K^k which contains $K^k \cdot T^{-1}(\widehat{x_i^k x_j^k})$, it follows that for each k, i, j there exists a continuum $E_{ij}^k = \lim_{(m)} K_{i_m j_m}^{k+m}$ in E which contains $E \cdot T^{-1}(\widehat{x_i^k x_j^k})$. (Note: i_m and j_m are integers such that $x_{i_m}^{k+m} = x_i^k$, $x_{j_m}^{k+m} = x_j^k$.)

Now for any $y \in pq$ and any k let us choose i and j so that (i) y is interior to the arc $x_i^k x_j^k$ if $p \neq y \neq q$, (ii) $y = x_i^k = x_0^k = p$ if $y = p$, $y = x_j^k = q$ if $y = q$, and (iii) $0 < j - i \leqq 2$. Then since in all cases $E \cdot T^{-1}(y)$ is interior to E_{ij}^k (rel. E), it follows that $E \cdot T^{-1}(y) = \lim_{(k \to \infty)} E_{ij}^k$. Accordingly, since each E_{ij}^k is a continuum, $E \cdot T^{-1}(y)$ is a continuum. Therefore $E \cdot T^{-1}(y)$ reduces to a single point, since $T^{-1}(y)$ is totally disconnected, and our theorem is proved.

Remarks. It is of interest to note that it results at once from this theorem that the property of containing no arc is invariant for compact sets under light interior transformations. It may be noted further that, whereas the initial point p_0 of the arc $p_0 q_0$ can be chosen arbitrarily in $T^{-1}(p)$, the terminal point q_0 is not subject to choice but may easily depend on the choice of p_0.

The methods of proof used in (1.1) together with the theorem just proved yield at once

(2.2) THEOREM. *Under the conditions of* (2.1), *if J is any simple closed curve in B, then for any integer $n > 0$ either there exists a simple closed curve C in A and an integer $k \leqq n$ such that $T(C) = J$ and, on C, T is topologically equivalent to the transformation $w = z^k$ on $|z| = 1$, or there exists a simple arc X in A such that*

$T(X) = J$ and, on X, T is equivalent to the transformation $w = e^{yi}$ on the interval $-(n+1)\pi \leqq y \leqq (n+1)\pi$.

(2.3) THEOREM. *Let $T(A) = B$ be interior, where A is compact. For any simple arc pq in A there exists a continuum E in A such that $T(E) = pq$ and, on E, T is monotone.*

Proof. Let us factor T into the form T_2T_1 where T_1 is monotone and T_2 is light. Let $A' = T_1(A)$. Then $T_2(A') = B$ is interior and light. Hence by (2.1) there exists an arc p_0q_0 in A' such that $T_2(p_0q_0) = pq$ is topological. Let $E = T_1^{-1}(p_0q_0)$. Then E is a continuum since T_1 is monotone, and since

$$T(E) = T_2T_1(E) = T_2T_1T_1^{-1}(p_0q_0) = T_2(p_0q_0) = pq$$

and T_2 is topological on $T_1(E) = p_0q_0$, it follows that $T(E) = pq$ is monotone.

Note. Theorem (2.1) also follows directly from this theorem.

The following extension to arbitrary dendrites (that is, locally connected continua containing no simple closed curves) in B will be made.

(2.4) THEOREM. *Let $T(A) = B$ be interior and light, where A is compact. For any dendrite D in B and any $x_0 \in T^{-1}(D)$ there exists a dendrite E in A containing x_0 which maps topologically onto D under T.*

Proof. Since the transformation $TT^{-1}(D) = D$ is interior, clearly there is no loss of generality in assuming that $D = B$. Now let $H = \sum p_iq_i$ be an arc-development of B; that is, for each $i > 0$, $p_iq_i \cdot \sum_{j=1}^{j=i-1} p_jq_j = q_i$, and p_i ($i = 1, 2, \cdots$) and all points of $\bar{H} - H$ are end points of B, where we may suppose that $p_0 = T(x_0)$. Now by the result quoted above, there exists an arc x_0y_0 such that $T(x_0y_0) = p_0q_0$ and T is topological on x_0y_0. Likewise there exists an arc x_1y_1 in A such that $T(x_1y_1) = p_1q_1$, T is topological on x_1y_1, and furthermore so that $y_1 = T^{-1}(q_1) \cdot x_0y_0$. Similarly there is an arc x_2y_2 in A so that $T(x_2y_2) = p_2q_2$, T is topological on x_2y_2, and $y_2 = T^{-1}(q_2) \cdot \sum_0^1 x_iy_i$. Continuing this process indefinitely we obtain $K = \sum x_iy_i$, so that $T(K) = H$, and T is topological on K. Clearly $\bar{K}$ is a continuum and $T(\bar{K}) = B$. Hence our proof will be complete as soon as we show that T is (1-1) on $\bar{K}$, or, what amounts to the same thing, that for each $p \in B$, $T^{-1}(p) \cdot \bar{K}$ reduces to a single point.

Now there exists a monotone decreasing sequence of connected neighborhoods, $V_1, V_2, V_3, \cdots$ of p in B with $\delta(V_i) \to 0$. Furthermore, since $\bar{H} - H$ contains only end points of B, it follows that $H \cdot V_1, H \cdot V_2, \cdots$ are connected sets which, of course, are open in H. Hence $K \cdot T^{-1}(V_1)$, $K \cdot T^{-1}(V_2)$, $\cdots$ is a monotone decreasing sequence of connected sets in K which are open in K. Let $L = \lim K \cdot T^{-1}(V_i)$. Then, since $V_i \to p$, we have $L \subset T^{-1}(p)$. But, since L is necessarily connected and $T^{-1}(p)$ is totally disconnected, L must reduce to a single point $q \in T^{-1}(p)$. Hence $T^{-1}(p) \cdot \bar{K} = q$, and our theorem is established.

That a similar conclusion cannot be obtained, if we permit D to contain simple closed curves, is readily seen, since it fails to hold even in the simple transformation $w = z^2$ of the circle $|z| = 1$ into the circle $|w| = 1$. Still more striking, however, is the following:

EXAMPLE. *There exist two connected graphs A and B and an interior transformation $T(A) = B$, where A contains no subset homeomorphic with B.*

Let $J = a_1b_1f_1c_1a_2b_2f_2c_2$ be a simple closed curve in a plane, where the points $a_1, b_1, \cdots$ are cyclically ordered on J as indicated. Let $a_1d_1f_2$ and $b_1e_1c_1$ be disjoint arcs lying within J except for their end points, and let d_1e_1 be an arc within J having only its end points in common with $a_1d_1f_2 + b_1e_1c_1$. Similarly, let $f_1d_2a_2$ and $b_2e_2c_2$ be disjoint arcs lying, except for their end points, without J, and let d_2e_2 be an arc without J having just its end points in common with $f_1d_2a_2 + b_2e_2c_2$. Let A be the graph thus constructed, that is, let

$$\begin{aligned} A &= J + a_1d_1f_2 + b_1e_1c_1 + d_1e_1 + f_1d_2a_2 + b_2e_2c_2 + d_2e_2 \\ &= a_1b_1 + b_1f_1 + f_1c_1 + c_1a_2 + b_1e_1 + e_1c_1 + a_1d_1 + d_1f_2 + d_1e_1 \\ &\quad + a_2b_2 + b_2f_2 + f_2c_2 + c_2a_1 + b_2e_2 + e_2c_2 + a_2d_2 + d_2f_1 + d_2e_2. \end{aligned}$$

Then A is a graph of eighteen edges as indicated in the latter sum.

We now construct a graph B in 3-space as follows: Let θ be a θ-curve $bac + bec + bfc$ in a plane π. Let d be a point not in π, and let da, de, and df be arcs having just d in common and having just a, e, and f, respectively, in common with θ. Let

$$B = \theta + da + de + df = ab + bf + fc + ca + be + ec + ad + df + de.$$

Then B is a graph of nine edges as here indicated.

Now let $T(A) = B$ be the transformation which maps a_1b_1, a_2b_2 topologically into ab, b_1f_1 and b_2f_2 topologically into bf, and so on. In general T maps any edge x_iy_i of A topologically, preserving end points, into the edge xy of B. Then T is an interior transformation. In fact T is a local homeomorphism which is 2-to-1.

Since A is a planar graph, whereas B is non-planar (B is, in fact, one of the two well known Kuratowski primitive skew curves), clearly A contains no subset homeomorphic with B. Incidentally, this example shows that *planarity is not an interior property* (that is, it is not invariant under interior transformations).

3. Inversion of local connectedness and the finite-to-one property on 2-manifolds.

(3.1) THEOREM. *If A is a locally connected continuum, $f(A) = B$ is interior and light and Y is a locally connected continuum in B whose interior is dense in Y, $f^{-1}(Y)$ is locally connected.*

Proof. Let $\epsilon > 0$. Choose $\delta > 0$ so that if N is a continuum in B of diameter less than δ, each component of $f^{-1}(N)$ is of diameter less than ϵ.

Since Y has property S we can write

$$Y = \sum_1^n K_i$$

where each K_i is the closure of a region in Y of diameter less than δ. Each K_i therefore has a nonvacuous interior in B. Hence, by [VIII, (7.41), (7.5)] it follows that $f^{-1}(K_i)$ has just a finite number of components; and as each of these is of diameter less than ϵ and $f^{-1}(Y) = \sum_1^n f^{-1}(K_i)$, it follows that $f^{-1}(Y)$ has property S. **(See Note 2, p. 279.)**

Essentially the same proof clearly gives

(3.11) *If A is a locally connected continuum, $f(A) = B$ is interior and light and Y is an open set in B having property S, $f^{-1}(Y)$ has property S.*

DEFINITION. *A point p of a locally connected continuum M will be called a branch point of M provided there exist three arcs ap, bp, cp in M forming a "triod" with vertex at p, i.e., $ap \cdot bp = bp \cdot cp = ap \cdot cp = p$. Actually p will be a branch point if and only if it is a point of order greater than 2 of M (see Menger [1]), though no use will be made of this equivalence in what follows.*

(3.2) THEOREM. *Let A be a locally connected continuum on a plane or sphere or 2-cell π, and let $f(A) = B$ be interior and light. For any branch point p of B, $f^{-1}(p)$ is a finite set of points.*

Proof. There exist S-regions (i.e., having property S) R_1', R_2', R_3' in B with closures K_1, K_2, K_3, respectively, such that $K_1 \cdot K_2 = K_1 \cdot K_3 = K_2 \cdot K_3 = p$. Let $K = K_1 + K_2 + K_3$. By the preceding theorem, $f^{-1}(K)$ is locally connected. Now if $f^{-1}(p)$ were infinite, there would exist a component X of $f^{-1}(K)$ such that $f^{-1}(p) \cdot X$ is infinite. The transformation $f(X) = K$ is interior and light. There exist components R_1, R_2, R_3 of $f^{-1}(K_1 - p), f^{-1}(K_2 - p), f^{-1}(K_3 - p)$, respectively, such that $I = F(R_1) \cdot F(R_2) \cdot F(R_3)$ is infinite. But clearly this is impossible, since every point of I is accessible from each of the regions R_i.

(3.21) COROLLARY. *The same conclusion as in (3.2) holds where π is any 2-dimensional manifold* (compact or not).*

To prove this, let Y be an S-region about p so small that any component X of $f^{-1}(\bar{Y})$ is in a closed 2-cell E of π. Then apply (3.2) to the transformation $f(X) = \bar{Y}$.

(3.3) THEOREM. *Let A be a locally connected continuum on any 2-manifold* (compact or not) and let $f(A) = B$ be interior and light. For any nondegenerate locally connected continuum K in B, $f^{-1}(K)$ is locally connected.*

Proof. For any $\epsilon > 0$ choose $\delta > 0$ such that if N is a continuum in B of diameter less than δ, any component of $f^{-1}(N)$ is of diameter less than ϵ. Write

$$K = \sum_1^n K_i$$

* For a definition of this term, see §4 below.

where K_i is a locally connected continuum of diameter less than δ. Each K_i contains either interior points relative to B or branch points of B. In either case it follows from the preceding theorems that there is only a finite number of components of $f^{-1}(K_i)$ for each i. Since each component of $f^{-1}(K_i)$ is of diameter less than ϵ and $f^{-1}(K) = \sum_1^n f^{-1}(K_i)$, it results that $f^{-1}(K)$ has property S.

(3.4) Theorem. *Let M be a 2-manifold and suppose A is a locally connected subcontinuum of M which is locally a 2-dimensional Cantorian manifold (in particular A may $= M$). Any light interior transformation $f(A) = B$ is finite-to-one, i.e., $f^{-1}(y)$ is a finite set for each $y \in B$.*

This results from (3.21) together with the fact that every point of B must be a branch point of B. To prove the latter fact, we recall from [VIII, (7.83)] that B is locally a 2-dimensional Cantorian manifold. Thus in particular B has no local separating point. Hence if $p \in B$, there exists an arc apd in B. Let R be an S-region in B containing $pd - p$ and such that $\bar{R} \cdot ap = p$. Then $\bar{R}$ must be cyclic, since R has no cut point. Hence if $b, c \in \bar{R}$, $b \neq c$, there exists an arc bpc in $\bar{R}$. Then $ap + bp + pc$ is a triod in B with vertex p.

(3.41) Corollary. *Any light interior transformation on a compact 2-manifold is finite-to-one.*

(3.42) Corollary. *If M is any 2-manifold (compact or not) and $f(M) = N$ is interior and light, for each $y \in N$, any $x \in f^{-1}(y)$ is an isolated point of $f^{-1}(y)$.*

For let E be a closed 2-cell neighborhood of x on M and let $f(E) = F$. Since the mapping $f(E) = F$ is light and E is compact, there exists a closed 2-cell neighborhood A_0 of x such that if $B = f(A_0)$ and A is the component of $f^{-1}(B)$ containing x then A is interior to E. Since, by (3.3), $f^{-1}(B)$ is locally connected, the mapping $f(A) = B$ is interior. Furthermore, by [VIII, (7.82)], no point of B is a local separating point of B and hence as in the above proof every point of B is a branch point. Thus $A \cdot f^{-1}(y)$ is a finite set; and as x is interior to A, it must be an isolated point of $f^{-1}(y)$.

4. **Invariance of the 2-manifold property.** We begin with a result on the mapping of 2-cells which will be used not only in the main theorem of this section but also in the main theorem of §5.

(4.1) Theorem. *Let $f(A) = B$ be interior and light where A is a 2-cell with boundary J and suppose B contains a simple closed curve C such that $f^{-1}(C) = J$. Then any arc in B spanning C irreducibly separates B into two components each containing an open segment of J. Furthermore,*

(i) *if f is (1-1) on J it is (1-1) on A;*

(ii) *if there is a point q within J such that $p = f(q)$, $q = f^{-1}(p)$, then on A, f is equivalent to $w = z^k$ on $|z| \leqq 1$ for some integer k.*

Proof. Let ab be an arc in B spanning C. We first show that ab separates in B the two open arcs of C from a to b. If this were not so, there would exist an arc xy in B spanning C such that x and y separate a and b on C.

Now let us decompose A into the components of the sets $f^{-1}(ab)$ and $f^{-1}(xy)$ together with the individual points of A not on $f^{-1}(ab)$ or $f^{-1}(xy)$. Since the transformation associated with this decomposition is monotone, it therefore generates a non-alternating mapping

$$g(J) = M$$

on the boundary J of A such that two points of J have the same image in M if and only if they lie together in a component of $f^{-1}(ab)$ or $f^{-1}(xy)$. Since there is only a finite number of such components, M can have only a finite number of cut points. Thus if N is a node of M and $q = N \cdot \overline{M - N}$, N is a non-degenerate simple closed curve and $g^{-1}(N - q)$ is a component R of $J - g^{-1}(q)$. Also

$$(N - q) \cdot g[f^{-1}(ab + xy) \cdot J] = 0$$

so that

$$(*) \qquad g^{-1}(N - q) \cdot f^{-1}(ab + xy) = R \cdot f^{-1}(ab + xy) = 0.$$

Thus since $g^{-1}(q)$ is contained wholly either in $f^{-1}(ab)$ or in $f^{-1}(xy)$, say $f^{-1}(ab)$, it follows that R is a component of $J - J \cdot f^{-1}(ab)$. But since the transformation $f(J) = C$ is interior and x and y separate a and b on C, $f(R)$ must contain either x or y, contrary to (*). Hence ab separates in B the two open arcs of C from a to b.

Clearly no proper subarc of ab can separate x and y in B. Thus since B is unicoherent, no proper subset of ab can separate x and y in B. Now if there were a third component of $B - ab$, its boundary would be a subarc α of ab; and we could then find a subcontinuum β of $f^{-1}(\alpha)$ and components R_1, R_2, R_3, respectively, of the inverse of these components of $B - ab$ each having β on its boundary. Since $f^{-1}(\alpha)$ is hereditarily locally connected, clearly this is impossible. Accordingly ab separates B irreducibly into two components and each of these contains a segment of $J - a - b$.

Note. It results from what we have just shown that B is a 2-cell. For it is readily shown by argument similar to that just given to show that there are only two components of $B - ab$ that if $R = B - C$, then R is a connected region and any spanning arc γ of C in B separates R into two regions bounded by $\gamma + C_1$ and $\gamma + C_2$, respectively, where C_1 and C_2 are the two arcs of C into which the end points of γ divide C. Hence [VI, (5.2)] applies. See also Zippin [2]. However, use will not be made of this fact (i.e., of the 2-cell characterization) in what follows. By avoiding the use of the 2-cell characterization at this point we are able to prove the invariance of the 2-manifold property by direct analysis of the action of the transformation rather than by leaning on this characterization.

Now to prove (i), suppose, on the contrary, that there exist two points x

and y in A such that $f(x) = f(y) = p$. Let apb be an arc in B such that $apb \cdot f(J) = a + b$. There exist arcs $a'xb'$ and $a'yb'$ in A which map topologically onto apb under f. Since these two arcs are different, there exists a component S of $A - (a'xb' + a'yb')$ such that $S \cdot J = 0$ and hence so that $S \subset A - J$. Clearly this is impossible, since S contains a component U of $A - f^{-1}(apb)$ and U necessarily maps onto one of the components of $B - apb$ and each of these contains points of $f(J)$.

Finally, to prove (ii), let apb be an arc in B such that $apb \cdot C = a + b$. Now since the mapping $f(J) = C$ is interior, it is equivalent to $w = z^k$ on $|z| = 1$ for some integer k. Let $f^{-1}(a) = a_1 + a_2 + \cdots + a_k$, $f^{-1}(b) = b_1 + b_2 + \cdots + b_k$, where we have the cyclic order $a_1, b_1, a_2, b_2, \cdots, a_k, b_k, a_1$ on C. There exist arcs a_iq and b_iq $(1 \leqq i \leqq k)$ in A which map topologically onto ap and bp, respectively.

Now since $a_iq + a_jq$ separates b_i and b_j in A and $b_iq + b_jq$ separates a_i and a_j in A whereas

$$(a_iq + a_jq)(b_iq + b_jq) = q,$$

it follows that

$$a_iq \cdot a_jq = q = b_iq \cdot b_jq, \qquad i \neq j.$$

Thus any pair of the arcs intersect in just q.

We now show that $\sum (a_iq + b_iq) = f^{-1}(apb)$. If this were not so, then since $f^{-1}(apb) + J = f^{-1}(apb + C)$ is cyclic (because any node of it would have to contain J), there would exist a simple closed curve D in $f^{-1}(apb)$. The interior I of D would contain no point of J but would contain a component Q of $A - f^{-1}(apb)$. This is impossible since $f(Q)$ maps onto one of the two components into which apb separates B and both of these contain points of C.

Now let $J_i = a_iq + b_iq +$ arc a_ib_i of J which contains no other a_j or b_j (if $k = 1$, take either arc). Let R_i be the interior of J_i (considering A as in a plane). Similarly let $K_i = b_iq + a_{i+1}q +$ arc b_ia_{i+1} of J and let S_i be the interior of K_i $(i \leqq k,\ k + 1 \equiv 1)$. Let R and S be the two components of $B - apb$ where $R \supset f(a_1b_1) - a - b$, $S \supset f(b_1a_2) - a - b$. We then have

$$(\dagger) \qquad f(R_i + J_i) = R + apb,$$

$$f(S_i + K_i) = S + apb.$$

These mappings are interior, $R_i + J_i$ and $S_i + K_i$ are 2-cells with boundaries J_i and K_i, respectively, and $f(a_ib_i) + apb = f(J_i)$ is a simple closed curve on which f is (1-1) as is also $f(b_ia_{i+1}) + apb$. Accordingly, by (i), the mappings (†) are homeomorphisms. Hence f is equivalent to the mapping $w = z^k$ on $|z| \leqq 1$.

Definition. *A separable metric space M is said to be a 2-dimensional manifold provided that for any $x \in M$ there exists a neighborhood U of x such that $\overline{U}$ is a 2-cell, say $\overline{U} = R + J$ where J is the edge or boundary curve of $\overline{U}$ and R is* *the interior. If $x \in R$, it is an ordinary or regular point of M whereas if $x \in J$, it is an edge point (or boundary or singular point) of M. If there are no edge points the manifold is said to be an ordinary manifold or a "closed" manifold, whereas if there are edge points it is a manifold with boundary.*

If β denotes the set of all edge points of a manifold M, since each point of β is a point of order 2 of β and each component of β is open in β while β itself is a closed set, it results that each component of β is either a simple closed curve or an open curve (topological line or open segment). Thus if M is compact, β consists of a finite number (possibly 0) of simple closed curves.

(4.2) THEOREM. *If A is a compact 2-dimensional manifold (with or without boundary curves) and $f(A) = B$ is interior and light, then B is a 2-dimensional manifold.*

For the purpose of this proof a point x will be called a *regular point* of A or B provided there exists a neighborhood U of x (in A or B) such that $\overline{U}$ is a 2-cell whose boundary curve does not contain x. Any non-regular point will be called *singular*.

The proof will be given in the form of a series of statements.

(i) *The boundary of any component of the complement of a compact locally connected set in B is locally connected.*

To prove this, let R be a component of $B - M$ where M is a compact locally connected set in B and let $F = F(R)$. Let $f^{-1}(M) = N$, let S be a component of $f^{-1}(R)$ and let $E = F(S)$. Then by (3.21) N is locally connected. Also S is a component of $A - N$. Accordingly E is locally connected. Now since $f(S) = R$ it follows that $f(E) = F$. Therefore F is locally connected, because f is continuous.

(ii) *For any $b \in B$ and any $\epsilon > 0$ there exists a simple closed curve or an arc X in B which irreducibly separates B into just two components and so that the component of $B - X$ containing b is of diameter less than ϵ.* (See Note 3, p. 279.)

Proof of (ii). Let $a \in B - b$, let $b' \in f^{-1}(b)$ and let U be a 2-cell neighborhood of b' with boundary curve J such that the topological boundary $F(U)$ of U is either J or an arc cvd of J and such that $\overline{U} \cdot f^{-1}(b) = b'$, $\overline{U} \cdot f^{-1}(a) = 0$. [Note: $f^{-1}(b)$ is finite by (3.4).] Now since, by [VIII, (7.83)], B has no local separating point, there exists a locally connected subcontinuum M of B which ϵ-separates b in B and is such that $f^{-1}(M) \cdot U \neq 0$, $f^{-1}(M) \cdot F(U) = 0$, and such that the component S^* of $B - M$ containing a also contains $f[F(U)] + B - V_{\epsilon/2}(b)$. Let R be the component of $B - \bar{S}^*$ containing b, let S be the component of $B - \bar{R}$ containing a and let X be the common boundary of R and S. Then X is a compact locally connected subset of M [by (i)] which irreducibly separates a and b in B and $\delta(R) < \epsilon$. Now if R' and S' denote the components of $f^{-1}(R)$ and $f^{-1}(S)$ containing b' and $F(U)$, respectively, and if $Z = U \cdot f^{-1}(X)$, it follows that since $b' = U \cdot f^{-1}(b)$ and $f^{-1}(a) \cdot \overline{U} = 0$ there can be no other components of either $f^{-1}(R)$ or $f^{-1}(S)$ intersecting $\overline{U}$. Accordingly $Z = F(R')$ and $Z \subset F(S')$. Hence Z is a continuum and since it is locally connected (because X is locally connected) it must be either a simple closed curve or a simple arc according as b' is a regular point or a singular point of A. Furthermore, since $Z = F(R')$ and $f(R') = R$ it follows that $f(Z) = X$, and since the transformation $f(Z) = X$ is interior on Z, X is either an arc or a simple closed curve. Finally, since there can exist no component of $A - f^{-1}(X)$ other than R' and S' with boundary points in Z, clearly $R + S = B - X$ so that X irreducibly separates B.

Now continuing with the same notation we next prove

(iii) *If X is a simple closed curve, $R + X = \bar{R}$ is a 2-cell with boundary curve X; and thus b is a regular point.*

For if X is a simple closed curve, Z must be a simple closed curve and $R' + Z = \bar{R}'$ is a 2-cell with boundary curve Z. Furthermore the transformation $f(\bar{R}') = \bar{R}$ is interior, since $\bar{R}' = U \cdot f^{-1}(\bar{R})$. Thus since $Z = \bar{R} \cdot f^{-1}(X)$ it follows by [(4.1), (ii)] that $\bar{R}$ is a 2-cell with boundary curve X.

(iii′) *Every singular point of A maps into a singular point of B.*

This is a corollary to (iii).

Still retaining the notation of (ii) we next establish (see Note 4, p. 279)

(iv) *A point x of X is a singular point of B if and only if it is an end point of X.*

First let x be a non-end point of X and let $q \in f^{-1}(x)$. Let σ be arbitrarily small. Since x is not a cut point of B, there exists a subcontinuum N of B such that $B - x \supset N \supset B - V_{\sigma/2}(x)$ and $N \supset f(u) + f(v)$, where u and v are points of Z so chosen that the arc uqv of Z maps topologically into an arc exf of X under f. Now we can construct in A an arbitrarily small simple closed curve $C = rhs + rks$ such that rhs and rks are arcs in $R' + r + s$ and $S' + r + s$, respectively, $r + s = C \cdot Z$, and $f(C) \cdot N = 0$. Now since C separates q and u in A it follows that $f(C)$ must separate x and N in B. Hence by the argument used in proving (ii), $f(C)$ must contain a set X' irreducibly separating x and N in B and X' is either an arc or a simple closed curve. But now since the set $f^{-1}(X')$ must separate x and u in A and f is topological on uqv, it follows that $f^{-1}(X')$ contains continua H and K such that $r + s \subset H \subset R' + r + s$, $r + s \subset K \subset S' + r + s$. Accordingly X' contains the continua $f(H)$ and $f(K)$. Also $f(H) \cdot f(K) = f(r) + f(s)$ and $f(r) \neq f(s)$. Hence X' cannot be an arc and thus it is a simple closed curve. Therefore, by (iii), x is a regular point.

Now if x is an end point of X, since clearly x can have no 2-cell neighborhood not having x on its boundary curve it follows that x is a singular point.

Now let K denote the set of all singular points of B. Clearly K is closed. It may be vacuous in case A has no singular points, but not otherwise, by (iii′).

(v) *Every point of K is of order* 2 (*rel. K*). (See Note 5, p. 279.)

Proof. By (iv) it follows that every point of K is of order less than or equal to 2. Now if there were a point x of K of order less than 2, there would exist an arbitrarily small locally connected continuum X in B such that $X \cdot K$ contains less than 2 points and X irreducibly ϵ-separates x in B. By (iii) it follows that X is an arc and by (iv) that both its end points must belong to K, contrary to supposition.

(v′) *Each component of K is a simple closed curve.*

(vi) *$B - K$ is connected.*

If this were not so we could find a subset K' of K and two distinct components R and S of $B - K'$ with a boundary point p in common, $p \in K'$. But by (ii) and (iv) there exists an arbitrarily small arc ab in B with $ab \cdot K = a + b$ and so that ab separates p from both r and s where $r \in R$, $s \in S$. Clearly this is impossible, since we would then have $ab - (a + b) \subset R \cdot S$.

(vii) *K has only a finite number of components.*

To prove this we note first that by (vi) it follows that there is only a finite number of components of $A - f^{-1}(K)$. Since by (v) for any $e > 0$, almost all components of K are of diameter less than e and f is light it follows that the same is true of $f^{-1}(K)$. But each component of $f^{-1}(K)$ must contain a simple closed curve since its image is a simple closed curve. As any sufficiently small simple closed curve on A disconnects A it results that $f^{-1}(K)$ and hence also K can have only a finite number of components.

(viii) *Every singular point of B has a 2-cell neighborhood.* (See Note 6, p. 279.)

Proof. Let $p \in K_i \subset K$ where K is the set of all singular points of B and K_i is a component of K. Let u and v be points of $K_i - p$, let q be a point of $f^{-1}(p)$. Since $f^{-1}(K)$ is a finite graph, there exists a 2-cell neighborhood U of q with boundary J (which may or may not contain q) such that $\bar{U} \cdot f^{-1}(K)$ consists of a finite number of arcs qx_0, qx_1, $\cdots$, qx_n such that (a) $qx_i \cdot qx_j = q$, (b) x_0, x_1, $\cdots$, x_n are cyclically ordered on J and the arc x_0x_n of J not containing q is the set-theoretic boundary of U in case q is on J, (c) qx_0 maps topologically into a subarc either of pu or pv, say pu, qx_1 maps topologically onto a subarc of pv, qx_2 into a subarc of pu, and so on to qx_n, where pu and pv are the arcs of K_i not containing v and u, respectively, (d) $\bar{U} \cdot f^{-1}(p) = q$.

Now let ab be an arc in B irreducibly ϵ-separating p in B where $a \in pu$, $b \in pv$ and where ϵ is so chosen that the component R of $B - ab$ containing p satisfies $\bar{R} \cdot f(\bar{U} - U) = 0$. Let S be the other component of $B - ab$. Since $f^{-1}(S) \supset x_0x_1$ (arc of J) and $f^{-1}(R) \supset q$, it follows that the open 2-cell V in U bounded by the simple closed curve $G = qx_0 + x_0x_1 + qx_1$ contains a component R' of the inverse of the region $Q = R - \widehat{apb}$ (open arc of K_i). Since $x_0x_1 \subset f^{-1}(S)$, it follows at once that the boundary of R' is a simple closed curve $W = a'q + qb' + a'b'$, where $a'q$ and qb' are arcs on qx_0 and qx_1, respectively, and where $a'b'$ is an arc which maps into ab and lies except for a' and b' in V. Since $a'b'$ is the common boundary of R' and the component S' of $f^{-1}(S)$ containing x_0x_1 and $a' = f^{-1}(ab) \cdot qx_0$, $b' = f^{-1}(ab) \cdot qx_1$, it follows that $a'b' = f^{-1}(ab) \cdot \bar{V}$ and hence that $a'b'$ maps interiorly onto ab. Thus since $f^{-1}(a) \cdot a'b' = a'$, $f^{-1}(b) \cdot a'b' = b'$ it follows that $a'b'$ maps topologically onto ab.

Therefore we have shown that the boundary W of the open 2-cell R' maps topologically into the boundary C of the region Q while R' maps onto Q. Hence surely the transformation $f(\bar{R}') = \bar{Q}$ is interior, and since $f^{-1}(C) \cdot \bar{Q} = W$, it follows by [(4.1), (i)] that $\bar{Q}$ is a 2-cell with boundary C.

Clearly the statements (i)–(viii) yield our theorem.

As a direct consequence of the arguments given under (i) and (ii) we get

(4.3) THEOREM. *If M is any 2-dimensional manifold (compact or not) and $f(M) = N$ is interior and light, for any $x \in M$ there exists an arbitrarily small closed 2-cell neighborhood E of x on which f is interior.*

For let C be a closed 2-cell neighborhood of x so chosen that $C \cdot f^{-1}f(x) = x$ [see (3.42)]. Then since C is compact and $y = f(x)$ is interior to $D = f(C)$ and the mapping $f(C) = D$ is light, the argument given under (i) and (ii) of (4.2) applies to yield a closed 2-cell neighborhood $R' + Z = E$ of x with set-theoretic boundary Z which is either a simple closed curve or a simple arc according as x is an ordinary point or an edge point of M and such that the mapping f is interior on E.

This together with (4.2) yields at once

(4.4) THEOREM. *If M is any 2-dimensional manifold (compact or not) and $f(M) = N$ is light and interior, N is a 2-dimensional manifold.*

For if $y \in N$, $y = f(x)$, $x \in M$, applying (4.3) we find E so that $f(E) = Q$ is interior. By (4.2) Q is a 2-manifold; and since y is interior to Q, our theorem follows.

EXAMPLES AND APPLICATIONS. It is readily seen that the sphere may be mapped onto either a 2-cell or a projective plane by a light interior transformation. For we have only to map the sphere $x^2 + y^2 + z^2 = 1$ by the transformation $f(x, y, z) = (x, y, 0)$ to obtain the 2-cell; and the identification of pairs of diametrically opposite points on a sphere produces a projective plane by an interior mapping. [Since the sphere is unicoherent, note that in view of [VIII, (8.61)] this proves the unicoherence of the projective plane and of the 2-cell.] In this connection we note that (i) *if A is a sphere or projective plane and $f(A) = B$ is interior and light, B is necessarily a sphere, 2-cell, or projective plane; if A is a 2-cell, so also is B.*

For in any case by (4.2) and [VIII, (8.61)], B must be a unicoherent 2-dimensional manifold and hence is necessarily either a sphere, projective plane or 2-cell (since these are the only unicoherent compact 2-manifolds). Further, if A is a 2-cell, by (iii′) under (4.2), B necessarily has edge points and hence must be a 2-cell. [Concerning the projective plane, see further the application at end of (7.2).]

We note further that (ii) *if every cyclic element of the locally connected continuum A is a sphere or 2-cell and $f(A) = B$ is any interior transformation, then every cyclic element of B is either a sphere, a projective plane, or a 2-cell.*

For let $f(x) = f_2f_1(x)$, $x \in A$, be the monotone-light factorization of f and let $f_1(A) = A'$. Then by [VIII, (6.51)] every true cyclic element of A' is either a sphere or a 2-cell. Thus since $f_2(A') = B$ is light and interior it results from [VIII, (7.92)] that every true cyclic element E of B is the image of some cyclic element of A' under a light interior mapping and hence by the above statement (i), E is either a sphere, projective plane, or 2-cell.

5. Analysis in the small on 2-manifolds.

(5.1) THEOREM. *Let $f(A) = B$ be interior and light, where A is any 2-dimensional manifold. For any point q of A there exists a closed 2-cell neighborhood E of q in A and a positive integer k such that* (a) *if $f(q)$ is a regular point of B, then on E, f is topologically equivalent to the transformation $w = z^k$ on $|z| \leqq 1$;* (b) *if $f(q)$ is a singular point of B, then on E, f is topologically equivalent to the transformation*

$$\phi(z) = \rho\,(\cos k\theta/2 + i\,|\sin k\theta/2\,|), \qquad k \text{ even},$$

on $|z| \leqq 1$ when q is a regular point of A and to this same transformation (for a different value of k) on $|z| \leqq 1$, $y \geqq 0$ when q is a singular point.

We note first that by virtue of (4.3) and (4.4) we may suppose without loss of generality that A is compact.

Now in case $f(q)$ is a regular point of B, it will be noted that the 2-cell neighborhood $E = R + Z$ of q (which must be a regular point of A) was chosen in the proof of (ii) part of (4.2) [substituting q for b', $f(q)$ for b], so that its edge Z not only maps onto the simple closed curve X in $f(E) = W$, but also in this case $E \cdot f^{-1}(X) = Z$. Since the mapping $f(E) = W$ is interior and $E \cdot f^{-1}f(q) = q$, our conclusion follows from (4.1), (ii).

Finally, consider the case where $f(q)$ is a singular point of B. Then, referring back to the proof of (viii) in (4.2), it follows by (4.1) that the set $\bar{R}'$ there defined maps topologically under f onto the set $\bar{Q}$. In that proof let us set $V = V_0$, $a' = a_0$, $b' = a_1$, $R' = R_0$. Let V_i be the open 2-cell in U bounded by the simple closed curve $qx_i + x_ix_{i+1} + qx_{i+1}$ for $i < n$ if q is on J and for $i \leqq n$ $(n + 1 \equiv 0)$ if q is within J. Then in exactly the same way it follows that there exist arcs $a_1a_2, a_2a_3, \cdots, a_ia_{i+1}, \cdots$ in $\bar{V}_1, \bar{V}_2, \cdots, \bar{V}_i, \cdots$ so that if R_i is the open 2-cell in V_i bounded by the simple closed curve $qa_i + a_ia_{i+1} + qa_{i+1}$ (where $0 \leqq i \leqq n - 1$ for q on J and $0 \leqq i \leqq n$, $n + 1 \equiv 0$, if q is within J), $\bar{R}_i$ maps topologically onto $\bar{Q}$ under f. Hence if we call E the closed 2-cell $\sum \bar{R}_i$, it follows that, on E, f is topologically equivalent to the transformation $\phi(z) = \rho\,(\cos n\theta + i\,|\sin n\theta\,|)$ on $|z| \leqq 1$, $y \geqq 0$, if q is on J and to the transformation

$$\phi(z) = \rho\,[\cos\,(n + 1)\theta/2 + i\,|\sin\,(n + 1)\theta/2\,|], \qquad n + 1 \text{ even},$$

on $|z| \leqq 1$ if q is within J. Thus (b) is established.

Let $f(A) = B$ be interior and finite-to-one, where A is any 2-dimensional manifold (compact or not). For each $y \in B$ let $k(y)$ be the number of points in $f^{-1}(y)$. For each n $(n = 1, 2, \cdots)$ let Y_n be the set of all $y \in B$ such that $k(y) = n$. As a consequence of (5.1) we have

(5.2) THEOREM. *If B is connected and the decomposition G of A generated by f is upper semi-continuous, there exists one and only one of the sets $Y_1, Y_2, \cdots$ which has interior points. If this is Y_k we have*

$$B = \sum_1^k Y_i \quad \text{and} \quad Y_n = 0 \qquad \text{for } n > k.$$

Proof. Since for each n the set $\sum_1^n Y_i$ is closed and $B = \sum Y_i$, at least one Y_n must have interior points (rel. B). Let k be the least index such that Y_k has interior points and let K be the set of all interior points of $H = \sum_1^k Y_i$ (rel. B). Then K is open in B. But K is also closed in B. For if p is a limit point of K, $p \in H$ and by (5.1) it follows that there exists a neighborhood V of p in B such that $k(y) \leqq k$ for all $y \in V$, because for any $q \in f^{-1}(p)$ any neighborhood of q contains interior points x (rel. A) such that $k[f(x)] \leqq k$ and hence, since G is upper semi-continuous, in the whole neighborhood of q given by (5.1), $k[f(x)] \leqq k$. Thus K is both open and closed in B, which gives $K = B$.

(5.21) COROLLARY. *If $f(A) = B$ is interior and finite-to-one and the decomposition of A generated by f is upper semi-continuous where A is any 2-manifold and B is connected, there exists an integer k such that f is at most k-to-1 on A and is exactly k-to-1 on $A - G$ where G is locally a finite linear graph.*

6. **Degree. Local homeomorphisms.** Let $f(A) = B$ be continuous. For any $y \in B$, let $k(y)$ be the number of points in $f^{-1}(y)$ if this number is finite; otherwise let $k(y) = \infty$. In case $k(y)$ has a finite maximum k we define the *degree* of f to be k. Clearly we may regard $k(x)$ as being defined on A by

$$k(x) = k[f(x)] = \text{number of points in } f^{-1}f(x).$$

It results at once from §5 that any light interior mapping $f(A) = B$ on a compact 2-manifold A has a finite degree k. However, it will be of interest to determine k in another way by means of local homeomorphisms.

A continuous transformation $f(A) = B$ is said to be *locally topological* or to be a *local homeomorphism* at a point $x \in A$ provided there exists an open set U in A containing x which maps topologically under f onto an open set in B. If this condition is satisfied at all points of A, f is said to be a local homeomorphism on A. Also f is said to be a local homeomorphism on a subset A_0 of A provided the mapping of A_0 onto $f(A_0)$ is a local homeomorphism.

Obviously any local homeomorphism is interior and light. Furthermore, any local homeomorphism on a compact set is finite-to-one.

(6.1) THEOREM. *If A is compact, $f(A) = B$ is continuous and is a local homeomorphism on $f^{-1}(C)$ where C is any arcwise connected subset of B, then $k(y)$ is constant on C.*

Proof. Let $x, y \in C$ and let xy be a simple arc in C. Clearly our theorem will follow as soon as we prove $k(x) = k(y)$. Let $f^{-1}(x) = x_1 + \cdots + x_{k(x)}$. By (2.1) there exist arcs $x_1y_1, x_2y_2, \cdots, x_{k(x)}y_{k(x)}$ in $f^{-1}(xy)$ each mapping topologically onto xy. Since f is a local homeomorphism, no two of these arcs can intersect. Accordingly, $k(y) \geqq k(x)$ because $y_1 + y_2 + \cdots + y_{k(x)} \subset f^{-1}(y)$. Similarly, $k(x) \geqq k(y)$ and our result follows.

COROLLARY. *Let A be compact and suppose $f(A) = B$ is a local homeomorphism. If D is any dendrite in B, $f^{-1}(D)$ is the sum of a finite number of disjoint dendrites each of which maps topologically onto D. In particular, if B is a dendrite and A is a continuum, A is a dendrite and f is a homeomorphism.*

The following considerably more general result is readily established.

(6.2) *If A is compact and* $f(A) = B$ *is continuous,* $k(y)$ *is continuous on any subset C of B such that f is a local homeomorphism on* $f^{-1}(C)$.

For let $y \in C$ and let U be an open subset of C containing y. Let $f^{-1}(y) = x_1 + x_2 + \cdots + x_{k(y)}$. Choose disjoint open subsets U_i of $f^{-1}(C)$ about x_i ($i = 1, 2, \cdots, k(y)$) such that each U_i maps topologically onto an open subset $f(U_i)$ of C. Choose an open subset V of C, $y \subset V \subset \prod f(U_i)$, so that $f^{-1}(V) \subset U_1 + U_2 + \cdots + U_{k(y)}$. Then for any $y' \in V$ we have

$$k(y') \geqq k(y), \qquad \text{since } V \subset \prod f(U_i);$$

and

$$k(y') \leqq k(y) \qquad \text{since } f^{-1}(V) \subset U_1 + U_2 + \cdots + U_{k(y)};$$

whence

$$k(y') = k(y) \qquad \text{for all } y' \in V.$$

(6.21) COROLLARY. *On any connected subset of C,* $k(y)$ *is constant.*

Note. To see that the condition in the above results that f be defined and continuous on a compact set is essential, we have only to note that the mapping $w = z^2$ on the set $A: |z| = 1, z \neq 1$ is a local homeomorphism mapping A onto the set $B: |z| = 1$, but $k(1) = 1$, $k(z) = 2$ for $z \neq 1$.

Now to find the degree of a light interior mapping $f(A) = B$ on a compact 2-dimensional manifold A, let β denote the boundary (edges) of B. By §5 it follows that there can exist only a finite set Z of points in $B - \beta$ whose inverse contains a point where f is not locally topological. Hence if $C = B - \beta - Z$, C is an arcwise connected subset of B and on $f^{-1}(C)$, f is locally topological. Accordingly, by (6.1) or (6.21), the multiplicity $k(y)$ is constant on C. Let $k(y) = k$ on C. Since, by interiority, the set of points on B where $k(y) > k$ is open, it follows that this set is empty because C is dense in B. Accordingly k is the maximum value of $k(y)$, $y \in B$. Hence

(6.3) THEOREM. *Any light interior transformation* $f(A) = B$ *on a compact 2-manifold A has a finite degree k which is exactly the multiplicity of f on the set* $C = B - \beta - Z$, *where* β *is the edge graph of B and Z is the (at most finite) set of* $y \in B - \beta$ *such that f fails to be locally topological at some point of* $f^{-1}(y)$.

7. Analysis in the large on 2-manifolds. We consider a light interior transformation $f(A) = B$ where A, and hence also B, is a compact 2-dimensional manifold. Let α and β denote the boundaries (if any) of A and B, respectively, and, just as in §§5 and 6, let k denote the degree of f and let Z be the (finite) set of points y of $B - \beta$ such that f fails to be locally topological at some point of $f^{-1}(y)$. We shall have occasion to use here the fact, which we do not prove (see Cairns [1]), that any compact 2-manifold can be "triangulated," i.e., subdivided so as to form a (finite) Euclidean 2-dimensional complex.

Throughout this section the letters used above will retain their significance as there defined. Also $\chi(X)$ will stand for the *Euler characteristic* of the complex or surface X, i.e., $\chi(X) = \alpha^2 - \alpha^1 + \alpha^0$, where α^i $(i = 0, 1, 2)$ is the number of i-dimensional simplexes in a subdivision of X. We proceed to establish two lemmas.

(7.1) LEMMA 1. *If $N \subset B$ is a graph dividing B into only a finite number of components, then $f^{-1}(N)$ is a graph.*

Proof. Since all but a finite number of points of N are of order 2, $f^{-1}(N)$ has at most a finite number of end points. Since $A - f^{-1}(N)$ has only a finite number of components (each of which maps onto a component of $B - N$), it follows that $f^{-1}(N)$ contains only a finite number of simple closed curves. Thus each cyclic element of a component C of $f^{-1}(N)$ is a graph, and as C can have only a finite number of nodes, C is a graph and so is $f^{-1}(N)$.

In particular, $f^{-1}(\beta)$ is a graph, since $B - \beta$ is connected.

(7.2) LEMMA 2. *There exist simplicial subdivisions K and H of A and B, respectively, such that each simplex of K maps topologically onto a simplex of H.*

Proof. Let Z be the set of all points of $B - \beta$ whose inverse contains a point at which f is not locally topological. For each point $q \in f^{-1}(Z)$, where f is not locally topological, let E_q be a 2-cell neighborhood of q on which f is equivalent to the transformation $w = z^k$ on $|z| \leqq 1$, and let G_q be the graph consisting of the part in E_q of the inverse of the set corresponding to $v = 0$, $-1 \leqq u \leqq 1$, where $w = u + iv$. As the number of points in $f^{-1}(Z)$ is finite, we may suppose the graphs G_q are disjoint and do not intersect $f^{-1}(\beta)$. Let G be the graph consisting of $f^{-1}(\beta)$, $f^{-1}(Z)$ and all the graphs G_q. Let $W = f^{-1}(\beta) + f^{-1}(Z)$.

(i) There exists a $d > 0$ such that if E is any region in A not intersecting G and such that $\rho(E, W) < d > \delta(E)$, then f is topological on $\bar{E}$.

Proof of (i). Let $Q = q_1 + q_2 + \cdots + q_m$ be the points of $f^{-1}(Z)$ where f is not locally topological. Now the set $V = f^{-1}(Z) - Q + f^{-1}(\beta)$ can be covered by a finite number of 2-cells $D_1 + D_2 + \cdots + D_p$ such that the part, if any, of $f^{-1}(\beta)$ within D_i divides D_i into a finite number of 2-cells on each of which f is topological. Now there exists a $d_0 > 0$ such that any subset A' of A such that $\delta(A') < d_0$ and $\rho(V, A') < d_0$ lies wholly in some one of the sets D_i. For each $j \leqq m$, let d_j be less than half the distance from q_j to the boundary of E_{q_j}.

Finally, let $d = \min d_i$ $(0 \leqq i \leqq m)$. Then if E is any region in A of diameter less than d not intersecting G and such that $\rho(E, W) < d$, it is seen at once that E lies wholly in some one of the sets D_i or in one of the sets E_{q_j}; and, in either case, since the interior of E does not intersect G, it follows that f is topological on $\bar{E}$.

(ii) There exists a number $e > 0$ such that if K is any subdivision (simplicial or cellular) of A of norm less than e whose 1-dimensional structure includes all points of the graph G, then on any simplex of K, f is topological.

For if we take a sufficiently small neighborhood U of $\beta + Z$ in B we will have $W \subset f^{-1}(U) \subset V_d(W)$. Hence on the set $f^{-1}(B - U)$, f is locally topological. Accordingly, there exists a number $c > 0$ such that on any subset of $f^{-1}(B - U)$ of diameter less than c, f is topological. Let $e = \min(d, c)$. Then if K is any subdivision of A of norm less than e whose 1-dimensional structure includes G and if E is any 2-cell of K; we have either $E \subset f^{-1}(B - U)$ or $\rho(E, W) < d$; and since the interior of E cannot intersect G, it follows in either case that f is topological on E. Hence surely f is topological on any cell of K.

Since f is light, we have

(iii) There exists a number $h > 0$ such that if H' is any graph in B including $\beta + Z$ and such that every component of $B - H'$ is of diameter less than h, the diameter of every component of $f^{-1}(B - H') = A - f^{-1}(H')$ is less than e.

Let H be a simplicial subdivision of B whose 1-dimensional structure H' is such a graph.

Since every point of Z is a vertex of H, it follows that $f^{-1}(H') = K'$ contains a graph G_q about each point q_i of q. Thus by (iii) and (ii) the closure of each component of $A - K'$ is a 2-cell E mapping topologically onto a simplex F of H. Thus if for each such E we select edges and vertices by the (topological) transformation $f^{-1}(F) = E$, it is clear that we obtain a simplicial subdivision K of A with 1-dimensional structure K' such that f is topological on each simplex of K.

COROLLARY. *If B is a closed orientable* surface, so also is A.*

For we have only to assign to each 2-simplex in K the orientation of its image simplex in H, where H is "coherently" oriented.* Then since each 1-simplex in K is on exactly two 2-simplexes and these have distinct images, it follows that K is coherently oriented.

Thus non-orientability is invariant under light interior transformations between closed surfaces. That this is not true for surfaces with boundary will be seen below in example (iii).

APPLICATION. We note that this corollary exhibits the impossibility of mapping the projective plane onto a sphere by a light interior transformation. Accordingly, in view of the applications given in §5 we can now assert that any interior light image of the projective plane is either a projective plane or a 2-cell. Further, both image sets are possible as will be seen in example (iii) below.

(7.3) THEOREM. *If A is a 2-dimensional manifold and $f(A) = B$ is a light interior transformation of degree k, then*

$$k\chi(B) - \chi(A) = kr - n - m$$

where r and n are the numbers of points in Y and $f^{-1}(Y)$, respectively, where Y is the set of all $y \in B$ such that $f^{-1}(y)$ contains either a branch point of $f^{-1}(\beta)$ or a

* For definitions of these terms see Alexandroff [6]-Hopf.

point of $A - f^{-1}(\beta)$ *at which* f *is not locally topological, and where* m *is the number of components of* $f^{-1}(\beta) - f^{-1}(Y)\cdot f^{-1}(\beta) - \alpha$ *which are not simple closed curves.*

Proof. Let K and H be subdivisions of A and B given by Lemma 2. Let α^i and β^i ($i = 0, 1, 2$) be the number of i-dimensional simplexes in K and H, respectively.

Since f is k-to-1 on the set $A - f^{-1}(Z) - f^{-1}(\beta)$ we have

$$\text{(i)} \qquad \alpha^2 = k\beta^2.$$

Since for any 1-simplex x^1 of K which is in $f^{-1}(\beta)$ but not in α the two 2-simplexes on x^1 have the same image, it follows that if t is the total number of such 1-simplexes x^1 in K, we have

$$\text{(ii)} \qquad \alpha^1 = k\beta^1 - t.$$

For the same reason, if s is the number of vertices of K on

$$f^{-1}(\beta) - f^{-1}(Y)\cdot f^{-1}(\beta) - \alpha,$$

we have

$$\text{(iii)} \qquad \alpha^0 = k\beta^0 - (kr - n) - s.$$

Now let $E_1, E_2, \cdots, E_m, E_{m+1}, \cdots, E_q$ be the components of $f^{-1}(\beta) - f^{-1}(Y)\cdot f^{-1}(\beta) - \alpha$, where $E_1, E_2, \cdots, E_m$ are open arcs and $E_{m+1}, E_{m+2}, \cdots, E_q$ are simple closed curves.

Let t_i and s_i be the number of edges and vertices, respectively, of K on E_i ($0 \leqq i \leqq q$). Then

$$\sum_1^q t_i = t, \qquad \sum_1^q s_i = s.$$

Further, for $i > m$, $t_i = s_i$; and for $i \leqq m$, $t_i = s_i + 1$. Whence

$$\text{(iv)} \qquad t - s = \sum_1^q (t_i - s_i) = \sum_1^m (t_i - s_i) + \sum_{m+1}^q (t_i - s_i)$$

$$= \sum_1^m 1 + 0 = m.$$

Now, by (i)–(iii), we have

$$\alpha^2 - \alpha^1 + \alpha^0 = k(\beta^2 - \beta^1 + \beta^0) - (kr - n) + t - s.$$

Whence, by (iv),

$$\chi(A) = k\chi(B) - (kr - n) + m,$$

or

$$k\chi(B) - \chi(A) = kr - n - m.$$

EXAMPLES. (i) Let A be a sphere with center at the origin in 3-space with a cylindrical coordinate system (ρ, θ, z). Let f be the transformation of A into a sphere B effected by identifying the points (ρ, θ, z) and $(\rho, \theta + 2\pi/k, z)$, where k is a positive integer. We have $k = k$, $r = n = 2$, $m = 0$, so that both sides of the equation in our theorem reduce to $2k - 2$.

(ii) Let f_1 be the transformation of a torus A into a circular ring R effected by merely flattening the torus onto a plane which cuts it into two rings. Let f_2 be the transformation of R into a 2-cell B, effected by folding R across a line cutting R into two 2-cells. Finally, let $f = f_2 f_1$. For f_1 we have: $k = 2$, $r = n = m = \chi(A) = \chi(R) = 0$. For $f_2 : k = 2$, $r = n = 4$, $m = 2$, $\chi(R) = 0$, $\chi(B) = 1$. For f: $k = 4$, $r = n = 4$, $m = 8$, $\chi(A) = 0$, $\chi(B) = 1$.

(iii) Let f be the transformation of a projective plane A into a 2-cell B, effected as follows. Cut A into a Möbius band M and a 2-cell E, each bounded by a simple closed curve J. Regard M as generated by an interval ab moving "parallel" to itself with its ends on J. Now in M identify points on ab equidistant from the midpoint of ab; and on E let f be equivalent to the transformation $w = z^2$ on $|z| \leqq 1$. We have: $k = 2$, $r = n = 1$, $m = 0$, $\chi(A) = \chi(B) = 1$.

(iv) Let f be a transformation of a torus A into a sphere B effected as follows. Cut A into a 2-cell E and a "handle" H each bounded by a simple closed curve J. On E let f be equivalent to $w = z^2$ on $|z| \leqq 1$. Let H be mapped into a 2-cell F as follows. On B select three points u', v', w' inside F and join each to the edge of F by 2 intervals, thus dividing F into 4 regions. On H select three points u, v, w which will be branch points of order 2; for convenience these may be taken with u and v opposite points on one generator of the torus and w opposite v on the other generator. Now join each of these points on H to J by 4 intervals so that H is divided into 8 regions. If selected properly, f can be defined so as to map these in pairs topologically onto the four regions of F. For this transformation we have: $k = 2$, $r = n = 4$, $m = 0$, $\chi(A) = 0$, $\chi(B) = 2$.

APPLICATIONS. (i) *If* $m = 0$ [*as for example when* $\alpha = f^{-1}(\beta)$] *and* $\chi(A) = \chi(B)$, *either* f *is topological or* $r \geqq \chi(A)$.

For in this case our equation in (7.3) takes the form

$$(k - 1)(\chi(A) - r) = r - n;$$

and since the right-hand member is less than or equal to 0 we have either $k = 1$ or $r \geqq \chi(A)$.

Thus, for example, any non-topological light interior transformation of a sphere into a sphere has at least two singular points in the image sphere [see example (i)]; and any such transformation of a projective plane into a 2-cell or another projective plane or of a 2-cell into a 2-cell [see example (iii)] has at least 1 singular point in the image set.

(ii) *If* $\chi(A) = \chi(B) < 0$ *and* $m = 0$, *the transformation is necessarily topological.*

For our equation takes the form

$$(k - 1)\chi(A) = kr - n;$$

and as the right member is greater than or equal to 0 and $\chi(A) < 0$, it follows that $k = 1$. Obviously this conclusion does not hold for surfaces of characteristic greater than or equal to 0.

(iii) *If $kr = n$ and $m = 0$ (as for example, when f is locally topological),*

$$k\chi(B) = \chi(A).$$

Thus $\chi(A)$ and $\chi(B)$ vanish or fail to vanish together and if $\chi(A) = \chi(B) \neq 0$, f is necessarily topological.

It is interesting to note that the conditions imposed here do not restrict f to a local homeomorphism—see the transformation f_1 under example (ii) above.

For the surfaces of positive characteristic we have the following striking consequences. If A is a sphere, $\chi(A) = 2$ and hence $k = 1$ or 2 and $\chi(B) = 1$ or 2. Thus B is either a sphere, projective plane or 2-cell and all are possible under such a transformation. If B is a sphere, $k = 1$ and the transformation is topological; if B is a projective plane, then $k = 2$ and f is locally topological and exactly (2, 1); if B is a 2-cell, $k = 2$ and f is not locally topological. Thus *the only exactly $(k, 1)$ interior transformations on a sphere are the topological transformation into another sphere and the (2, 1) nonsingular transformation into a projective plane* (identifying diametrically opposite points). For no value of k other than 1 and 2 does there exist a $(k, 1)$ interior transformation on a sphere.

If A is a projective plane or a 2-cell, $\chi(A) = 1$ and hence $\chi(B) = k = 1$. Thus the only transformation satisfying our conditions on A is a homeomorphism. In particular, it follows that *there exists no exactly $(k, 1)$ interior transformation on a projective plane or a 2-cell other than a homeomorphism.*

For surfaces of characteristic less than or equal to 0, of course, no such conclusions hold. For example, it is easy to define transformations of a torus into a torus or of a circular ring into a circular ring which are locally topological and exactly $(k, 1)$ for every positive integral value of k. Also, it is easily seen that for any $n = 0, 1, 2, \cdots$ and any $k = 1, 2, \cdots$ we can map the closed orientable surface of characteristic $-2kn$ into the one of characteristic $-2n$ by a $(k, 1)$ local homeomorphism.

In case of a 2-cell or the projectve plane, even without the assumption that $kr = n$, we have $k - 1 = kr - n$; thus $r = 1$ gives $n = 1$ so that if there is only 1 branch point in B, the same holds for A; and in the case of a 2-cell, f is equivalent to $w = z^k$ on $|z| \leqq 1$.

8. **Extension to pseudo-manifolds.** A set of points P is called a 2-dimensional pseudo-manifold (see Alexandroff [6] and Hopf) provided it admits of subdivision so as to form a 2-dimensional Euclidean complex K such that each 1-simplex is on exactly two 2-simplexes and any two 2-simplexes can be joined by a finite chain of 2-simplexes in K such that successive links in the chain have a 2-simplex in common. In this section it will be shown that any light interior transformation on a 2-dimensional pseudo-manifold reduces essentially to a light interior transformation defined on a manifold. This reduction will be made with the aid of a transformation studied in Chapter VIII which makes use of

the relative distance space introduced by Mazurkiewicz. If M is a connected metric space of finite diameter, the relative distance space M^* of M consists of the same points as M but the metric $\rho^*(x, y)$ is defined, for $x, y \in M^*$, as the greatest lower bound of $[\delta(C)]$, where C is any connected subset of M containing $x + y$.

It was shown in [VIII, §9] that if M has property S (i.e., M is the sum of a finite number of arbitrarily small connected sets), and if, in general for any metric space X, $\tilde{X}$ denotes the space obtained by "completing" X, then the spaces $\tilde{M}$ and $\tilde{M}^*$ are compact and the change of metric from M^* to M generates a continuous transformation

$$w(\tilde{M}^*) = \tilde{M}$$

which maps M^* onto M topologically and uniformly continuously and maps $\tilde{M}^* - M$ onto $\tilde{M} - M$.

(8.1) THEOREM. *Let $f(A) = B$ be interior and light where A is a compact 2-dimensional pseudo-manifold. There exist compact 2-dimensional manifolds A' and B' and continuous transformations $w(A') = A, f'(A') = B', z(B') = B$ such that*

(1) *w is topological except at g points which map into s points of A,*
(2) *z is topological except at e points which map into t points of B,*
(3) *f' is interior and light, and*
(4) $zf'w^{-1} \equiv f$.

Proof. Let P denote the (finite) set of all local separating points of A and let $R = A - P$. Then R is connected and has property S. Let R^* be the relative distance space for R and set $\tilde{R}^* = A'$. Then since $\tilde{R} = A$, we have

$$w(A') = A.$$

Furthermore, since for each $p \in P$ there exists a closed neighborhood of p which is the sum of a finite number n_p of 2-cells intersecting by pairs only in p, it follows by [VIII, proof of part (ii) of (9.1)] that there are exactly n_p points in $w^{-1}(p)$ and each of these points has a closed 2-cell neighborhood. Thus since w is topological on $A' - w^{-1}(P)$, it follows that A' is a 2-dimensional manifold and (1) holds.

The transformation $f'(A') = B'$ is given by the following continuous decomposition G of A' into disjoint closed sets G_x. For any $x \in A'$ such that $w(x)$ is not a local separating point of A, let

$$G_x = w^{-1}f^{-1}fw(x).$$

For an $x \in A'$ such that $w(x)$ is a local separating point of A, let E_x be a 2-cell in A' containing x in its interior rel. A' and so small that w is topological on E_x. Let $x_1, x_2, \cdots$ be a sequence of points in E_x converging to x, and let $G_x = \lim (G_{x_i})$. Clearly G_x is independent of the sequence x_i, since it consists merely of all $y \in A'$ such that

$$fw(y) = fw(x) = u$$

and $fw(E_y)$ is essentially the same as $fw(E_x)$, i.e., u is interior to the common part of these sets relative to their sum. For the same reason it follows that any two distinct sets G_x are disjoint and that the decomposition G of A' into the sets G_x is continuous. Let B' be the hyperspace of this decomposition and let

$$f'(A') = B'$$

be the associated (light interior) transformation. Then B' is a 2-dimensional manifold, since f' is interior and light.

Finally, if for each $x \in B'$ we define

$$z(x) = fwf'^{-1}(x),$$

clearly we get $z(B') = B$ and (4) holds. Also, from the definitions of z and f' it follows that if $x, y \in B' - f'w^{-1}(P)$, $x \neq y$, we have

$$z(x) \neq z(y).$$

Hence (2) holds and our theorem is proved.

(8.11) Corollary. *The image of a 2-dimensional pseudo-manifold under any light interior transformation is itself a 2-dimensional pseudo-manifold.*

Now to extend our theorem of §7 to pseudo-manifolds, we let $f(A) = B$ be a light interior transformation where A (hence also B) is a 2-dimensional pseudo-manifold. Let α, β, k, Y, r, n, m be defined for f exactly as in §7 where now the s and t points of A and B mentioned in the preceding theorem are included in the n and r points of A and B, respectively, given by $f^{-1}(Y)$ and Y. (Note that the existence of the integers k, r, n, m follows from their existence in the manifold case by virtue of the preceding theorem.)

Now from the transformations w and z we have by virtue of (1) and (2)

$$\text{(i)}\qquad \begin{aligned} \chi(A') &= \chi(A) + g - s, \\ \chi(B') &= \chi(B) + e - t. \end{aligned}$$

By §7, the transformation $f'(A') = B'$ gives

$$\text{(ii)}\qquad \begin{aligned} k\chi(B') - \chi(A') &= k(r - t + e) - (n - s + g) - m \\ &= k(r - t) - (n - s) + ke - g - m. \end{aligned}$$

Now, by (i),

$$k\chi(B') - \chi(A') = k\chi(B) - \chi(A) + ke - g - kt + s,$$

whence, by (ii),

$$\begin{aligned} k\chi(B) - \chi(A) &= k(r - t) - (n - s) - m + kt - s \\ &= kr - n - m. \end{aligned}$$

This is identically the same relation as in the theorem of §7.

Thus we have the

(8.2) THEOREM. *If $f(A) = B$ is interior and light where A is a compact 2-dimensional pseudo-manifold, then B is a 2-dimensional pseudo-manifold, and if k, r, n, m are defined as in* §7 *we have*

$$k\chi(B) - \chi(A) = kr - n - m.$$

REFERENCES

§§1, 2. Whyburn [24, 26, 35]. In connection with (2.1) see Stoïlow [1] and Montgomery [1].

§§3, 4. Whyburn [26, 27].

§5. Whyburn [27]. In connection with (5.1) see Stoïlow [1].

§6. Compare results with Eilenberg [6].

§§7, 8. Whyburn [36]. Compare Stoïlow [2].

CHAPTER XI

EXISTENCE THEOREMS. MAPPINGS ONTO THE CIRCLE

a. Existence theorems

It is well known that any topological n-cell, and hence certainly any simple arc, is a retract of any containing topological space; and any topological polyhedron, in particular any simple closed curve, is a retract of some of its suitably chosen neighborhoods in any containing space. When anything more than continuity is required of the "retracting" transformation, however, the situation is obviously markedly altered and yet it seems not to have been the subject of much investigation.

In the first part of this chapter conditions will be developed under which a locally connected continuum M can be retracted onto an arc or a simple closed curve by non-alternating interior transformations. The principal results are: (1) for M to be retractable into an arbitrarily given arc axb by such a transformation it is necessary and sufficient that M be identical with the cyclic chain $C(a, b)$ in M; and (2) for M to be retractable into a given simple closed curve J by such a transformation it is necessary and sufficient that M be cyclic and not unicoherent about J. In the proofs, no use will be made of the facts concerning retracts mentioned above.

It will be recalled that a continuous transformation $f(S) = R$ is said to be *retracting* (and R is then a *retract* of S) provided $R \subset S$ and for each $x \in R$, $f(x) = x$. Also, a continuous transformation $T(A) = B$ is said to be *interior* provided sets open in A map into sets open in B, and T is said to be *non-alternating* provided that for no two points $x, y \in B$ does $T^{-1}(x)$ separate two points of $T^{-1}(y)$ in A.

Throughout §§1–4 M will denote a locally connected continuum. For any two points a and b of M, the *cyclic chain* $C(a, b)$ in M is the set consisting of a, b, the set K of all points separating a and b in M, and all cyclic elements of M containing two points of $K + a + b$. More simply, $C(a, b)$ consists of all simple arcs in M of the form axb.

1. Separation and subdivision.

(1.1) Lemma. *Let M be a locally connected continuum such that $M = C(a, b)$ for $a, b \in M$. Let $A \supset a$, $B \supset b$ be continua in M not separating M. Then for any $x \in M - (A + B)$, there exists an irreducible continuum AxB in M between A and B containing x which is locally an arc at all points $AxB - (A + B)$.*

Proof. Let G be the upper semi-continuous decomposition of M into the sets A, B and points of $M - (A + B)$. Let $T(M) = M'$ be the associated (monotone) transformation. Let $T(A) = a'$, $T(B) = b'$. Then M' is a locally connected continuum; and since a' and b' are non-cut points of M' and T is monotone, it readily follows that $M' = C(a', b')$. Hence M' contains an arc $a'xb'$ and we have only to set

$$AxB = T^{-1}(a'xb').$$

(1.2) LEMMA. *Let $M = C(a, b)$, let A and B be disjoint continua in M containing a and b, respectively, and let $a'xb'$ be any arc in M with $a'xb' \cdot A = a'$, $a'xb' \cdot B = b'$. There exists a set X with $X \cdot a'xb' = x$ which irreducibly separates M between A and B into just two components.*

Proof. If x separates A and B in M, we have only to set $X = x$. If not, then $M - x$ is connected and locally connected. Hence by a result previously established [III, (2.4)] we can find a decomposition

$$M - x = R_a + F + R_b,$$

where R_a and R_b are disjoint connected open sets containing $A + a'x - x$ and $B + xb' - x$, respectively, and where $F = F(R_a) = F(R_b)$ (boundaries taken relative to $M - x$). Thus if we set $F + x = X$ we have

$$M - X = R_a + R_b,$$

where $F(R_a) = F(R_b) = X$ (boundaries taken relative to M), and our lemma is satisfied.

DEFINITIONS. *Let the locally connected continuum M be a cyclic chain $C(a, b)$. By a subdivision σ of M will be meant a finite, linearly ordered set of disjoint irreducible cuttings of M between a and b*

$$a = X_0, X_1, X_2, X_3, \cdots, X_n, X_{n+1} = b,$$

where X_i $(1 \leqq i \leqq n)$ cuts M into two connected sets $M_a(X_i)$, $M_b(X_i)$ so that $M_a(X_i) \supset a + \sum_1^{i-1} X_j$ and $M_b(X_i) \supset b + \sum_{i+1}^{n} X_j$. The set

$$X_{i-1} + X_i + M_a(X_i) \cdot M_b(X_{i-1}) = I_i$$

will be called the interval from X_{i-1} to X_i ($a = X_0$, $b = X_{n+1}$). A set of the form $M_a(X_i) \cdot M_b(X_{i-1})$ will be called an open interval of σ. A subdivision σ' will be called a refinement of σ provided it is obtained from σ by inserting additional elements. (The sets X_i are called elements of the subdivision.)

Still assuming $M = C(a, b)$, let axb be any arc in M from a to b. We proceed to prove

(1.3) LEMMA. *Given any subdivision σ_0 of M each element of which contains just one point of axb and any $\epsilon > 0$, there exists a refinement σ of σ_0 such that if X_i is any element of σ, then $axb \cdot X_i$ is just one point and $V_\epsilon(X_i) \supset I_i + I_{i+1}$.*

Proof. For convenience we will suppose the metric in M so chosen that all sets of the form $V_r(x)$, $x \in M$, are connected. Let $e = \epsilon/5$ and let us cover M

with a finite number of the sets $V_e(x)$, say $V_1 = V_e(x_1)$, $V_2 = V_e(x_2)$, $\cdots$, $V_m = V_e(x_m)$.

Now consider V_1. If $\bar{V}_1 \supset b$ we need go no further. Hence we suppose $\bar{V}_1 \cdot b = 0$ and let X_{i+1} be the first element of σ_0 such that $M_a(X_{i+1}) \supset \bar{V}_1$. Now if $\overline{V_{2e}(x_1)} \cdot X_{i+1} \neq 0$ we need go no further. If this is not so, we proceed as follows. Setting $\alpha = M_a(X_i) + X_i$, $\beta = M_b(X_{i+1}) + X_{i+1}$ and letting x' be a point of V_1 in the open interval (X_i, X_{i+1}) of σ_0, by (1.1) we can find a continuum $\alpha x'\beta$ in M which is a "simple arc" between the continua α and β, i.e., if α and β are shrunk to points we have an ordinary simple arc going through x'.

We shall treat first the case where $axb \cdot V_{2e}(x_1) = 0$. On $\alpha x'\beta$ proceeding from x' toward α and from x' toward β let α' and β', respectively, denote the first "points" of the set $\alpha + \beta + axb$.

(i) If $\alpha' \neq \alpha$ and $\beta' \neq \beta$, let c be the first and d the second of the points α' and β' on axb in the order a, b and let $E =$ arc cx' of $\alpha x'\beta$, $F =$ arc $x'd$ of $\alpha x'\beta$.

(ii) If $\alpha' \neq \alpha$ and $\beta' = \beta$, let $c = \alpha'$, $d = axb \cdot X_{i+1}$, $E =$ arc cx' of $\alpha x'\beta$, and $F = x'\beta$ of $\alpha x'\beta$.

(iii) If $\alpha' = \alpha$ and $\beta' \neq \beta$, let $c = axb \cdot X_i$, $d = \beta'$, $E = \alpha x'$ of $\alpha x'\beta$ and $F = x'd$ of $\alpha x'\beta$.

(iv) If $\alpha' = \alpha$ and $\beta' = \beta$, let $c = axb \cdot X_i$, $d = axb \cdot X_{i+1}$, $E = \alpha x'$ of $\alpha x'\beta$, $F = x'\beta$ of $\alpha x'\beta$.

In any case, let y be the last point of $\overline{V_{2e}(x_1)}$ on $\alpha' x'\beta'$ in the order α', β' and set

$$A = \alpha + ac \text{ (of } axb) + E + \bar{V}_1,$$

$$B = \beta + db \text{ (of } axb) + F - (x'y - y).$$

Let x be a point on axb between c and d. Since $cxd \cdot A = c$, $cxd \cdot B = d$, we may apply (1.2) and obtain a set X with $cxd \cdot X = x$ which separates M irreducibly between A and B into just two components.

Now in case $axb \cdot V_{2e}(x_1) \neq 0$, let d be the last point of $\overline{V_{2e}(x_1)}$ on axb in the order a, b and let x' be chosen as before. Either we can join x' to $ad - d$ by an arc $px' \subset M_a(X_{i+1})$ such that $px' \cdot axb = p$ or we can join x' to X_i by an arc px' such that $p \in X_i$ and $px' \cdot axb = 0$. If $p \in axb$, set $c = p$; if not, set $c = X_i \cdot axb$. Let x be between c and d on axb. Then if we set $A = \alpha + ac + px'$, $B = \beta + db$, we can proceed as before to find the set X.

Clearly X is on the "b side" of V_1, i.e., $M_a(X) \supset V_1$; and $X \cdot V_{2e}(x_1) \neq 0$ since $A \cdot V_{2e}(x_1) \neq 0 \neq B \cdot \overline{V_{2e}(x_1)}$. In exactly the same manner we construct a set Y on the "a side" of V_1 such that $Y \cdot V_{2e}(x_1) \neq 0$. Let us add X and Y to σ_0 and call σ_1 the resulting refinement of σ_0.

Similarly for V_2 we obtain a refinement σ_2 of σ_1 which retains the properties of σ_1 and in addition contains two elements W and Z each intersecting $V_{2e}(x_2)$ and such that $V_2 \subset M_b(W) \cdot M_a(Z)$. Proceeding in this manner to V_m we finally obtain a refinement σ_m of σ_0, which we call σ, which retains the essential properties of σ_0 and in addition has the property that for any $i \leq m$, there are two elements X_j and X_k of σ intersecting $V_{2e}(x_i)$ and such that $V_i \subset M_b(X_j) \cdot M_a(X_k)$. Since $e = \epsilon/5$ and the sets V_i cover M, clearly this is equivalent to our lemma.

2. Retractions into arcs.

(2.1) THEOREM. *In order that a locally connected continuum M be retractable into an arbitrarily given arc axb in M by a non-alternating interior transformation it is necessary and sufficient that M be identical with the cyclic chain $C(a, b)$ in M.*

Proof. We first show that the condition is sufficient. By lemma (2.2) we can find a subdivision σ_1 of M satisfying the conclusion of this lemma for $\epsilon = 1$. Let us set $\sigma_1 = \sigma_1'$ and suppose we have constructed σ_i and σ_i' for all $i < n$. Then let us choose points $x_1, x_2, \cdots, x_k$ on axb such that the diameter of each of the arcs $ax_1, x_1x_2, \cdots, x_kb$ is less than $1/n$. Then by lemma (1.3) we can find a refinement σ_n of σ_{n-1}' satisfying the conclusion of that lemma for $\epsilon = 1/n$; and by applying lemma (2.1) successively to the points $x_1, x_2, \cdots, x_k$ we can obtain a refinement σ_n' of σ_n such that each of the points $[x_i]$ belongs to exactly one element of σ_n' and such that σ_n' still satisfies the conclusion of lemma (2.2) for $\epsilon = 1/n$. To see this, consider x_i. Suppose x_i lies within the interval YZ of σ_n (if x_i belongs to an element of σ_n, we need make no construction for x_i). Then $axb \cdot YZ$ is an interval cx_id of axb; and if we set $M_a(Y) + Y = A$, $M_b(Z) + Z = B$ and apply lemma (2.1), we get a set X_i with $axb \cdot X_i = x_i$ which separates M irreducibly between A and B into just two components. If we add such a set X_i to σ_n for each x_i not on an element of σ_n, we obtain the required subdivision σ_n'.

Thus we have set up an infinite monotone sequence $\sigma_1', \sigma_2', \cdots$ such that for each n, σ_n' satisfies the conditions of lemma (1.3) for $\epsilon = 1/n$ and in addition such that each interval $I = X_iX_{i+1}$ of σ_n' intersects axb in an arc of diameter less than $1/n$.

We now define a transformation $f(p)$ of M into axb as follows. If p belongs to some element X_p of σ_n' for some n, let $f(p) = axb \cdot X_p$; if not, then let $Z_p = \prod_1^\infty I_n$, where I_n is the (minimum) interval of σ_n' containing p and let $f(p) = axb \cdot Z_p$.

Since for each n, $I_n \cdot axb$ is an interval of axb of diameter less than $1/n$, it follows that in either case $f(p)$ is a single point so that f is single-valued. Obviously $f(p) = p$ for $p \in axb$ so that f is retracting. Furthermore, since for any point p other than a or b and any $\epsilon > 0$ we can find an open interval $E = (X_j, X_k)$ of some σ_n' containing p whose intersection with axb is an open interval (x_jx_k) of axb of diameter less than ϵ and since $f(E) = (x_jx_k)$, it follows that f is continuous. Finally, since for any $x \in axb - (a + b)$, it follows by the construction of the subdivisions σ_n' that $f^{-1}(x)$ separates M irreducibly between a and b into just two components R_a and R_b which map under f onto the arcs $ax - x$ and $xb - x$, respectively, and since $a = f^{-1}(a)$, $b = f^{-1}(b)$, it is readily seen that f is non-alternating and interior.

To show the condition necessary, let $f(M) = axb$ be non-alternating, interior and retracting but suppose, contrary to our statement, that there is a component R of $M - C(a, b)$. The boundary of R is a single point x. If $x = a$ (or b), then since f is interior, $f(R) \supset axb - a$ so that R contains a point of $f^{-1}(b)$; but since R does not contain b, this contradicts the fact that f is non-alternating. If $a \neq x \neq b$, then since f is interior, $f(R)$ must contain either $ax - x$ or $xb - x$; but since R contains neither a nor b, again it follows that f alternates.

Remark. Attention may be called to the fact that the transformation f set up in the sufficiency part of the proof is so defined that $f^{-1}(a) = a, f^{-1}(b) = b$. Also, for any $x \in axb - (a + b), f^{-1}(x)$ separates M irreducibly between a and b into just two components. It results from this that *in case M is unicoherent, f is necessarily monotone.*

Now suppose axb is a simple arc and $g(x)$ is any interior transformation defined on axb. Let M be any locally connected continuum containing axb and such that M is identical with the cyclic chain $C(a, b)$. Let f be an interior transformation retracting M into axb as given by the above theorem. Then the transformation $\varphi = gf$ is interior; and since for $x \in axb$ we have $f(x) = x$ and thus $\varphi(x) = g(x)$, φ is an extension of $g(x)$ to M. Hence we have

(2.2) Extension theorem. *Any interior transformation on a simple arc axb can be extended (interiorly) to any locally connected continuum M containing axb which is of the form $M = C(a, b)$.*

Now let M be any locally connected continuum whatever and let axb be any arc in M. Since M can be retracted into any A-set in M by a monotone transformation, there exists a monotone transformation $g(x)$ which retracts M into $C(a, b)$. Applying (2.1), we get a non-alternating interior transformation $f(y)$ which retracts $C(a, b)$ into axb. The transformation $\varphi(x) = fg(x)$ is then non-alternating and it retracts M into axb. Since $\varphi(x) \equiv f(x)$ for $x \in C(a, b)$ and f is interior, it follows that when considered on $C(a, b)$ alone, φ is interior. Thus we have

(2.3) Theorem. *Any locally connected continuum M can be retracted into any arc axb in M by a non-alternating transformation $\varphi(x)$ which is interior when considered on the cyclic chain $C(a, b)$ alone.*

(2.31) *The simple arc is an absolute retract,* i.e., any separable metric space M can be retracted into any simple arc α in M.

For we may consider M imbedded in the Hilbert parallelotope Q_ω; and by (2.3), Q_ω may be retracted into α by a retraction f. Obviously f retracts M into α.

3. **Retractions into simple closed curves.** A locally connected continuum M is said to be unicoherent about a simple closed curve J in M provided that for every decomposition $M = H + K$ where H and K are closed sets such that $H \cdot J$ is an arc xry and $K \cdot J$ an arc xsy, the points x and y belong to the same component of $H \cdot K$.

(3.1) LEMMA. *If M is cyclic and not unicoherent about the simple closed curve $J \subset M$, there exist disjoint closed sets X and Y intersecting J in single points x and y, respectively, and disjoint regions R and S in M containing the open arcs xry and xsy, respectively, of J and such that $M - (X + Y) = R + S$ and $F(R) = X + Y = F(S)$.*

Proof. By hypothesis there exists a decomposition $M = H + K$ where H and K are closed sets such that $H \cdot J$ is an arc amb and $K \cdot J$ an arc anb and a and b lie in distinct components of $H \cdot K$. Let G be the component of $M - H$ containing $anb - (a + b)$, let U be the component of $M - \bar{G}$ containing $amb - (a + b)$ and let $W = M - U$. It follows from our hypothesis that $F(U) = A + B$ where A and B are disjoint and closed and $a \in A$, $b \in B$.

Let us decompose $\bar{U}$ upper semi-continuously into the sets A, B and individual points of U. Let U' be the hyperspace of this decomposition and $w(\bar{U}) = U'$ the associated continuous transformation. Clearly U' is a locally connected continuum. Furthermore, if $\alpha = w(A)$, $\beta = w(B)$, then since M is cyclic and neither A nor B cuts $\bar{U}$ (so that α and β are non-cut points of U'), it follows that U' is identical with the cyclic chain $C(\alpha, \beta)$ in U'. Hence if $\alpha\mu\beta$ denotes the arc $w(amb)$, by (2.1) there exists a non-alternating interior transformation $g(x)$ which retracts U' into $\alpha\mu\beta$ so that $\alpha = g^{-1}(\alpha)$, $\beta = g^{-1}(\beta)$ and for every $\mu \in \alpha\mu\beta - (\alpha + \beta)$, $g^{-1}(\mu)$ separates U' irreducibly between α and β into just two components.

Let v be a point of $\alpha\mu\beta - (\alpha + \beta)$ and let $V = w^{-1}g^{-1}(v)$. Since U is connected and V is a compact subset of U, there exists a (compact) continuum N with $V \subset N \subset U$. Since $gw(N)$ contains neither α nor β, we can find interior points ξ and η on $\alpha\mu\beta$ so that we have the order $\alpha, \xi, v, \eta, \beta$ and so that the open segment $Q = \xi v \eta - (\xi + \eta)$ of $\alpha\mu\beta$ contains $gw(N)$. Let

$$X = w^{-1}g^{-1}(\xi), \qquad Y = w^{-1}g^{-1}(\eta),$$

$$R = w^{-1}g^{-1}(Q), \qquad S = M - (R + X + Y).$$

These sets satisfy the conditions of our lemma as will now be shown.

Obviously $X \cdot J$ and $Y \cdot J$ are points x and y, respectively, and $R \cdot J$ and $S \cdot J$ are the open arcs xry and xsy, respectively, of J. Also $F(R) = X + Y = F(S)$, since each of the sets $g^{-1}(\xi)$ and $g^{-1}(\eta)$ separates U' irreducibly between α and β into just two components and w is topological on U. For the same reason, the closure of each component of R must intersect both X and Y. Hence each such component must intersect V and thus N. As $N \subset R$ it results that R is connected. Similarly, each component of S must intersect either A or B and hence W; and since W is connected, S is connected. This completes the proof.

(3.2) THEOREM. *In order that a locally connected continuum M be retractable into a given simple closed curve $J \subset M$ by a non-alternating interior transformation it is necessary and sufficient that M be cyclic and not unicoherent about J.*

We shall first show that the conditions are sufficient. Let X, Y, R and S be sets satisfying lemma (3.1). Let us decompose M into the sets X, Y and individual points of $M - (X + Y)$. Call M' the hyperspace of this decomposition and let $h(M) = M'$ be the associated transformation. Let $h(X) = a'$, $h(Y) = b'$, $h(J) = J' = h(xry) + h(xsy) = a'r'b' + a's'b'$, $h(R) = R'$, $h(S) = S'$. Then by (3.1) it follows that $\bar{R}'$ is identical with its cyclic chain $C(a', b')$ and $\bar{S}'$ is identical with its cyclic chain $C(a', b')$. Hence by (2.1) there exist non-alternating interior functions φ_r and φ_s retracting $\bar{R}'$ and $\bar{S}'$ into $a'r'b'$ and $a's'b'$, respectively, so that $\varphi_r^{-1}(a') = \varphi_s^{-1}(a') = a'$, $\varphi_r^{-1}(b') = \varphi_s^{-1}(b') = b'$. Thus if we let $\varphi(x)$ equal $\varphi_r(x)$ on $\bar{R}'$ and $\varphi_s(x)$ on $\bar{S}'$, then φ is a non-alternating interior function retracting M' into J'.

We now define

$$f(x) = h^{-1}[\varphi h(x)] \cdot J.$$

(*Note.* This equals $h^{-1}\varphi h(x)$ except when $x \in X + Y$.) Clearly $f(x)$ retracts M into J. Since φh is interior, if G is an open set in M, $\varphi h(G)$ is open in J'; hence $h^{-1}[\varphi h(G)]$ is open in M, so that the intersection of this set with J is open in J. Accordingly f is interior. To show that f is non-alternating, let $t \in J$. Then $f^{-1}(t) = h^{-1}\varphi^{-1}h(t)$. If $h(t)$ is either a' or b', say a', then $\varphi^{-1}h(t) = a'$ and $f^{-1}(t) = X$, which does not separate M. If $a' \neq h(t) \neq b'$, then $\varphi^{-1}h(t)$ cuts either $\bar{R}'$ or $\bar{S}'$, say $\bar{R}'$, irreducibly between a' and b'. Thus any component of $M - f^{-1}(t)$ must intersect either X or Y; and since $S + X + Y$ is connected, it follows that $M - f^{-1}(t)$ is connected. Thus as no set $f^{-1}(t)$ separates M, f is non-alternating.

To prove the conditions necessary, let $f(M) = J$ be non-alternating, interior and retracting, where M is a locally connected continuum and J is a simple closed curve in M. Let $C(J)$ be the cyclic element of M containing J. Then $M = C(J)$. For if not, there exists a component R of $M - C(J)$; and if u denotes the boundary of R and v is a point of $R - R \cdot f^{-1}f(u)$, $f(u) = x$, $f(v) = y$, then $f^{-1}(x)$ would separate the points y and v of $f^{-1}(y)$ in M contrary to the non-alternating property of f. Hence M is cyclic. To see that M is not unicoherent about J, we have only to take points a and b on J decomposing J into arcs axb and ayb and set $H = f^{-1}(axb)$, $K = f^{-1}(ayb)$. This gives $M = H + K$, $H \cdot J = axb$, $K \cdot J = ayb$; and since $f(H \cdot K) = a + b$, obviously a and b lie in different components of $H \cdot K$.

It will be noted that the transformation $f(x)$ set up in the sufficiency part of the proof just given is so constructed as to have the following additional properties:

(3.21) *There exist open intervals α and β of J such that if $x \in \alpha$, $y \in \beta$, $f^{-1}(x) + f^{-1}(y)$ separates M irreducibly into just two components*

(3.22) *If $r(M) = 1$, f is monotone (or if $r(M) = k$, then for each $x \in J$, $f^{-1}(x)$ has at most k components).*

The second of these is an immediate consequence of the first.

The same device as used above to obtain (2.2) yields

(3.3) EXTENSION THEOREM. *Any interior transformation on a simple closed curve J can be extended (interiorly) to any cyclic locally connected continuum containing J which is not unicoherent about J.*

Also the method of obtaining (2.3) from (2.1) when applied to (3.2) yields

(3.4) THEOREM. *If M is any locally connected continuum and J is a simple closed curve in M about which M is not unicoherent, M can be retracted into J by a non-alternating transformation which is interior when considered only on the cyclic element $C(J)$ of M containing J.*

Now if a locally connected continuum M is multicoherent, it can be written

$$M = H + K$$

where H and K are locally connected continua and $H \cdot K$ is not connected. Let $H \cdot K = A + B$ where A and B are closed and disjoint. Let axb be an arc in H such that $a \in A$, $b \in B$ and $axb \cdot K = a + b$, and let ayb be an arc in K. Then if $J = axb + ayb$, M is not unicoherent about J. Hence any multicoherent locally connected continuum contains a simple closed curve about which it is not unicoherent. (See Note 10, p. 280.) Accordingly we have

(3.5) THEOREM. *In order for M to be retractable into some simple closed curve by a non-alternating interior transformation it is necessary and sufficient that M be cyclic and multicoherent.*

We now readily obtain

(3.6) THEOREM. *In order that a simple closed curve J in a locally connected continuum M be a retract of M it is necessary and sufficient that M be not unicoherent about J.*

For the sufficiency is a special case of (3.4); and to prove the necessity we have only to let M be retractable into $J = axb + ayb$ by a mapping f and set

$$H = f^{-1}(axb), \qquad K = f^{-1}(ayb).$$

Then

$$M = H + K$$

and since

$$H \cdot K = f^{-1}(a) + f^{-1}(b),$$

clearly

$$H \cdot J = axb, \qquad K \cdot J = ayb$$

and a and b lie in distinct components of $H \cdot K$.

Thus we can assert that *if a locally connected continuum M can be retracted at all into a simple closed curve $J \subset M$, M can always be retracted into J by a non-alternating interior mapping provided M is cyclic.*

DEFINITION. *A set X is said to be an absolute neighborhood retract* (ANR) *provided that if M is any separable metric space and X' is a subset of M homeomorphic with X, there exists a neighborhood U of X' in M which can be retracted into X'.*

(3.7) THEOREM. *The simple closed curve is an absolute neighborhood retract.*

For let J be any simple closed curve in a separable metric space M. We may, without loss of generality, assume M to be a subset of the Q_ω in Hilbert space. Let J be divided into four simple arcs $J = a + b + h + k$ such that $a \cdot b = h \cdot k = 0$. Let Q_ω be broken up into a finite number of locally connected continua small enough so that if A_0, B_0, H_0, and K_0 denote, respectively, the sum of all these continua intersecting a, b, h, and k, then $A_0 \cdot B_0 = H_0 \cdot K_0 = 0$. Let N be $H_0 + K_0 + A_0 + B_0$. Then if h_0 and k_0 are subarcs of h and k intersecting A_0 in just one end point each and B_0 in just one end point each and a_0 and b_0 denote the complementary (closed) arcs on J containing a and b, respectively, and $A = A_0 + a_0$, $B = B_0 + b_0$, $H_1 = H_0 + A + B$, $K_1 = K_0 + A + B$, by applying (1.2) we can obtain sets X and Y in H_1 and K_1, respectively, which irreducibly separate H_1 and K_1 between A and B into just two components H_A, H_B and K_A, K_B, respectively, and so that $X \cdot J = X \cdot h_0 = x$, $Y \cdot J = Y \cdot k_0 = y$ where x and y are points. Then if we let

$$H = \overline{H_A + K_A}, \qquad K = \overline{H_B + K_B},$$

we have

$$N = H + K, \qquad H \cdot K = X + Y, \qquad H \cdot J = \text{one arc } xry, \qquad K \cdot J = \text{arc } xsy.$$

Hence H is not unicoherent about J. Accordingly N can be retracted into J by a retraction f. Since N contains a neighborhood of J in Q_ω it contains a neighborhood U of J in M. Clearly f retracts U into J.

(3.8) THEOREM. *Any continuous mapping f of a simple link E of a continuum M onto a simple closed curve (or* ANR*) S can be extended to a mapping of M into S.*

Proof. Decompose M upper semi-continuously into the sets $[f^{-1}(y)]$, $y \in S$, together with the individual points of $M - E$. Let $g(M) = M'$ be the associated transformation of M into the hyperspace M' of this decomposition.

Now since $S' = gf^{-1}(S)$ is homeomorphic with S, it is therefore an ANR. Hence there exists an open set U' in M' containing S' and a retraction $w(U') = S'$ of U' into S'. Let $U = g^{-1}(U')$ and let N be a continuum in U containing E and such that $M - N = N_1 + N_2 + \cdots + N_k$ where the sets N_i are disjoint open sets each having just one limit point x_i in N.

Let us define

$$\phi(x) = fg^{-1}wg(x) \qquad \text{for } x \in N,$$

$$\phi(x) = \phi(x_i) \qquad \text{for } x \in N_i,\ i = 1, 2, \cdots, k.$$

Then $\phi(x)$ is continuous and maps M into S. Also, since for $x \in E$, $wg(x) = g(x)$ so that

$$\phi(x) = fg^{-1}g(x) = ff^{-1}(y) = y = f(x),$$

ϕ is an extension of f.

4. **Interior non-alternating mappings onto the interval and circle.** The results of the preceding sections clearly give sufficient conditions for the mappability of a locally connected continuum M onto the unit interval or circle by a non-alternating interior mapping. For when M can be retracted into an arc α or a simple closed curve J by such a mapping f, all we need do is map α or J topologically onto the interval or circle, respectively, by a homeomorphism g and then gf will be the required mapping of M onto the interval or circle.

(4.1) THEOREM. *Let $f(x)$ be a non-alternating interior mapping of a locally connected continuum M onto the interval $(0, 1)$ and let $A = f^{-1}(0)$, $B = f^{-1}(1)$. Then for each y with $0 < y < 1$, $f^{-1}(y)$ separates M irreducibly between A and B into just two components. Neither A nor B separates M and if $a \in A$, $b \in B$, M is identical with the cyclic chain $C(a, b)$.*

Proof. By known properties of non-alternating transformations it follows that $f^{-1}(y)$ separates M into just two components R_a and R_b containing A and B, respectively. Furthermore R_a maps onto $(0, y) - y$ and R_b onto $(y, 1) - y$. Thus since f is interior, $f^{-1}(y)$ must be the boundary both of R_a and of R_b. Hence $f^{-1}(y)$ separates M irreducibly.

Since neither 0 nor 1 separates $(0, 1)$ and f is non-alternating, it follows, [see VIII, (2.31)] that neither A nor B separates M.

To show that $M = C(a, b)$ let us suppose on the contrary that there is a component R of $M - C(a, b)$. The boundary of R is a single point x and x cannot belong to A or to B since neither of these sets separates M. Thus $x \in f^{-1}(y)$ for some y with $0 < y < 1$. But this is impossible since $f^{-1}(y)$ would separate M into at least three components. Thus $M = C(a, b)$, and our proof is complete.

The above remarks together with (4.1) give

(4.2) THEOREM. *In order that a locally connected continuum M be mappable onto the interval $(0, 1)$ by a non-alternating interior transformation f so that $f(a) = 0$, $f(b) = 1$, $a, b \in M$, it is necessary and sufficient that M be identical with the cyclic chain $C(a, b)$.*

For the case of the circle we prove

(4.3) THEOREM. *In order that a locally connected continuum M be mappable onto a circle by a non-alternating interior mapping it is necessary and sufficient that M be cyclic and multicoherent.*

The sufficiency results as indicated above at the beginning of this section. To prove the necessity, let $f(M) = C$ be non-alternating and interior where C is a circle. Then M must be multicoherent because C is multicoherent and unicoherence is invariant under f. Also, M must be cyclic. For if $M = A + B$ where A and B are continua intersecting in a single point p, since $C = f(A) + f(B)$, $f(A) \cdot f(B)$ must contain a point x distinct from $y = f(p)$, because by interiority neither $f(A)$ nor $f(B)$ can reduce to y. But then $f^{-1}(y)$ separates the points of $f^{-1}(x)\cdot A$ from those of $f^{-1}(x)\cdot B$ in M, contrary to the non-alternating property of f.

(4.4) THEOREM (Roberts-Moore). *If A is any locally connected continuum and $a, b \in A$, there exists a monotone transformation $T(A) = B$ of A into a regular curve B such that $a = T^{-1}T(a)$, $b = T^{-1}T(b)$.*

Proof. Suppose first that a and b are non-cut points of A. Let $f_1(A) = C(a, b)$ be monotone and retracting. Let $\varphi[C(a, b)] = (0, 1) = I$ be non-alternating and interior and such that $\varphi^{-1}(0) = a$, $\varphi^{-1}(1) = b$. Let K be a non-dense perfect set on I containing 0 and 1, let G be the upper semi-continuous decomposition of I into the closures of the complementary intervals to K on I together with the individual points of K abutting on no such interval, and let $\phi(I) = J$ be the associated transformation. Then J is a simple arc which we may suppose is an interval (p, q) of the real axis. Let $w[C(a, b)] = \phi\varphi[C(a, b)] = J$. Let us factor w into the form

$$w[C(a, b)] = f_3f_2[C(a, b)]$$

where $f_2[C(a, b)] = B$ is monotone and $f_3(B) = J$ is light. Finally, let $T(A) = f_2f_1(A) = f_2[C(a, b)] = B$.

Since f_1 and f_2 are monotone, T is monotone. Since $\phi^{-1}(p) = 0$, $\phi^{-1}(q) = 1$, $\varphi^{-1}(0) = a$, $\varphi^{-1}(1) = b$, we have $w^{-1}(p) = \varphi^{-1}\phi^{-1}(p) = a$, $w^{-1}(q) = \varphi^{-1}\phi^{-1}(q) = b$. Whence $f_2^{-1}T(a) = a$, $f_2^{-1}T(b) = b$. Accordingly, $T^{-1}T(a) = f_1^{-1}f_2^{-1}T(a) = a$, $T^{-1}T(b) = f_1^{-1}f_2^{-1}T(b) = b$, since f_1 is monotone and retracting and a and b are non-cut points. Finally, B is a regular curve. For $f_3(B) = J$ is a light transformation mapping B into an interval J so that the set of points $e \in J$ such that $f_3^{-1}(e)$ is a finite set is dense in J. For if E is any point of J such that $\phi^{-1}(e)$ is the closure of an interval of I complementary to K, then since φ is interior, $\varphi^{-1}\phi^{-1}(e) = w^{-1}(e) = f_2^{-1}f_3^{-1}(e)$ has only a finite number of components; and hence $f_3^{-1}(e)$is a finite set. Thus if $x, y \in B$ and $f(x) \neq f(y)$, there is a point e between $f(x)$ and $f(y)$ on J such that $f_3^{-1}(e)$ is a finite set which therefore separates x and y in B; and since f_3 is light it results from [V, (4.4)] that B is a regular curve.

Now if a or b is a cut point, say a, we choose non-cut points $b_1, b_2, \cdots$ in the components of $A - a$ not containing b and map each of these components into a regular curve by the above method.

b. MAPPINGS ONTO THE CIRCLE

In the remainder of this chapter we shall consider the class S^X of all continuous mappings of a separable metric space X onto a circle S and our objective will be to show how this class of mappings may be used to yield intimate knowledge of the structure of the space X. The method of treatment given is due to S. Eilenberg.

For convenience we shall suppose S is the circle $|z| = 1$ in the complex plane and for $z_1, z_2 \in S$, $z_1 \cdot z_2$ will mean the ordinary complex product of z_1 and z_2.

We have already seen in Chapter II how the elements of S^X form a metric space by the distance definition

$$\rho(f_1, f_2) = \text{l.u.b.}\ \{\rho[f_1(x), f_2(x)]\}, \qquad x \in X.$$

Clearly also, if we use complex multiplication as the group operation, the elements of S^X form an abelian group, i.e.,

$$f_1 \cdot f_2 = f_3, \qquad \text{where } f_3(x) = f_1(x) \cdot f_2(x),\ x \in X.$$

We shall use the symbol S^X to denote either the metric space or the abelian group as the case may be, since no confusion is likely to result.

The symbol R (or R^1) will be used to denote the real number system and K (or K_1) to stand for the unit interval (0, 1). For convenience the metric on R or S will be denoted by $| x - y |$. In general, R^n and K_n denote the Euclidean n-space and the fundamental interval of n-space, respectively. Finally, for $z_1, z_2 \in S$, $| z_1 - z_2 | < 2$, $[z_1, z_2]$ will denote the signed angle less than 180° through which the radius Oz_1 must be rotated in order to coincide with Oz_2. Clearly we have

(i) $$e^{i[z_1, z_2]} = z_2/z_1.$$

5. **Equivalence to 1.** A continuous mapping $f(x)$ of X into S, $f \in S^X$, is said to be *equivalent to* 1 on a set $Y \subset X$, written $f \sim 1$ on Y, provided there exists a continuous mapping $\phi(x)$ of Y into R $[\phi(x) \in R^Y]$ such that

$$f(x) \equiv e^{i\phi(x)}, \qquad x \in Y.$$

In other words, $f \sim 1$ on Y provided it can be factored into the form

$$f(x) \equiv g\phi(x), \qquad x \in Y,$$

where $\phi(x)$ is a continuous real-valued function and $g(y)$ is the exponential function e^{iy}, $y \in R$. It will be noted here that $\phi(x)$ is a value of $(1/i) \log f(x)$, i.e.,

(ii) $$\phi(x) = (1/i) \log f(x), \qquad x \in Y, \text{ holds.}$$

To illustrate, let $X = S$. Then if f is defined by

$$f(x + iy) = | x | + iy, \qquad x + iy \in S,$$

$f \sim 1$. For we have only to define

$$\phi(x + iy) = \text{arc tan } (y/|x|), \quad -\pi/2 \leq \text{arc tan } (y/|x|) \leq \pi/2.$$

ϕ is continuous and

$$e^{i\phi} = \cos \phi + i \sin \phi = | x | + iy.$$

On the other hand, it readily follows that if $f(S) = S$ is any homeomorphism,

f cannot be ~ 1; for if we had $f(x) = e^{i\phi(x)}$, $\phi(x)$ would have to be a homeomorphism; and this is impossible since R contains no topological image of S.

(5.1) *If X is connected, ϕ_1 and ϕ_2 are continuous real functions defined on X and*

$$e^{i\phi_1(x)} \equiv e^{i\phi_2(x)}, \qquad x \in X,$$

there exists an integer k such that

$$\phi_1(x) - \phi_2(x) \equiv 2k\pi, \qquad x \in X.$$

For since

$$f(x) = e^{i(\phi_1(x)-\phi_2(x))} \equiv 0 \text{ on } X,$$

it results from (ii) that

$$i[\phi_1(x) - \phi_2(x)] = \log 1.$$

Thus $\phi_1(x) - \phi_2(x)$ is of the form $2k\pi$ and since X is connected, k is a fixed integer.

(5.2) THEOREM. *Let the sets X_1 and X_2 be closed (or open) in their sum $X_1 + X_2$, let $X_1 \cdot X_2$ be connected (possibly empty) and suppose $f(x)$ is a mapping of $X_1 + X_2$ into S. Then if $f \sim 1$ on X_1 and on X_2, $f \sim 1$ on $X_1 + X_2$.*

Proof. By hypothesis, there exist real functions $\phi_1(x)$ and $\phi_2(x)$ defined on X_1 and X_2, respectively, such that

$$f(x) = e^{i\phi_1(x)} \qquad \text{on } X_1$$

$$= e^{i\phi_2(x)} \qquad \text{on } X_2.$$

Hence by (5.1) there exists an integer k such that

$$\phi_1(x) - \phi_2(x) = 2k\pi \qquad \text{for } x \in X_1 \cdot X_2.$$

Define

$$\phi(x) = \phi_1(x), \qquad x \in X_1$$

$$\phi(x) = \phi_2(x) + 2k\pi, \qquad x \in X_2.$$

Then $\phi(x)$ is continuous since X_1 and X_2 are closed in $X_1 + X_2$. Thus since $f(x) = e^{i\phi(x)}$, $x \in X_1 + X_2$, we have $f(x) \sim 1$ on $X_1 + X_2$.

(5.3) THEOREM. *If the mapping $f(x)$ of X into S is ~ 1 on a subset Y of X it is also ~ 1 on an open subset U of X containing Y.*

Proof. By hypothesis there exists a continuous real function $\phi(y)$ defined on Y such that

$$f(y) = e^{i\phi(y)}, \qquad y \in Y.$$

For each $y \in Y$ there exists an $\epsilon_y > 0$ such that

$$(*) \qquad \begin{aligned} &\delta[f(V_{\epsilon_y}(y)] < 1, \\ &\delta[\phi\{Y \cdot V_{2\epsilon_y}(y)\}] < \pi. \end{aligned}$$

Let us define

$$\phi(x) = \phi(y) + [f(y), f(x)] \qquad \text{for } x \in V_{\epsilon_y}(y).$$

Then

$$e^{i\phi(x)} = e^{i\phi(y)} e^{i[f(y), f(x)]} = f(y)\frac{f(x)}{f(y)} = f(x), \qquad \text{by (i).}$$

Hence it remains only to show that $\phi(x)$ is single-valued on the set $U = \sum_{y \in Y} V_{\epsilon_y}(y)$. To this end, let $x \in V_{\epsilon_{y_1}}(y_1) \cdot V_{\epsilon_{y_2}}(y_2)$. Set

$$g_1 = \phi(y_1) + [f(y_1), f(x)], \qquad g_2 = \phi(y_2) + [f(y_2), f(x)].$$

Then

$$e^{ig_1} = e^{ig_2}.$$

Whence

$$g_1 = g_2 \pm 2k\pi.$$

But by relations (*),

$$\begin{aligned} |g_1 - g_2| &\leqq |\phi(y_1) - \phi(y_2)| + |[f(y_1), f(x)]| + |[f(y_2), f(x)]| \\ &< \pi + \pi/2 + \pi/2 = 2\pi. \end{aligned}$$

Accordingly, $g_1 = g_2$ and hence $\phi(x)$ is single-valued and continuous on U.

(5.40) LEMMA. *If f is any mapping of X into S, the property of a set $Y \subset X$ to have f not ~ 1 on Y is inducible.*

For let $Y_1 \supset Y_2 \supset \cdots$ be a sequence of compact sets in X with f not ~ 1 on Y_n, $(n = 1, 2, \cdots)$ and let $Y = \prod Y_n$. Then if $f \sim 1$ on Y, by (5.3) $f \sim 1$ on some open set U in X containing Y. But since $U \supset Y_n$ for some n, this gives $f \sim 1$ on Y_n.

(5.4) THEOREM. *If f is any mapping of a compact space X into S which is not ~ 1 on X, there exists a compact subset Y of X such that f is not ~ 1 on Y but is ~ 1 on any closed proper subset of Y. Further, any such set Y is a continuum which is not separated by any one of its proper subcontinua.*

The first conclusion results at once from (5.40). To prove the second assertion, suppose there exists a separation $Y - Z = Y_1 + Y_2$ where Z is a continuum in Y. Then if $X_i = Y_i + Z$ $(i = 1, 2)$, $Y = X_1 + X_2$, X_1 and X_2 are closed proper subsets of Y and $X_1 \cdot X_2 = Z$. Hence $f \sim 1$ on X_1 and on X_2; and by (5.2) this gives $f \sim 1$ on Y, contrary to the definition of Y.

(5.41) CYCLIC EXTENSIBILITY THEOREM. *If X is compact, any mapping of X ino S which is ~ 1 on each E_0-set (or on each simple link) in X is ~ 1 on X.*

(5.411) COROLLARY. *If X is compact and all the E_0-sets in X reduce to single points—in particular if X is a simple arc or a dendrite—every mapping of X into S is ~ 1.*

(5.5) THEOREM. *Suppose that for each component Y of X and any open set $U \supset Y$ there exists a set U' which is both open and closed in X such that $Y \subset U' \subset U$. Then any mapping f of X into S which is ~ 1 on each component of X is ~ 1 on X.*

For by (5.3) and our hypothesis each component Y of X can be covered by a set $U(Y)$ both open and closed in X and such that $f \sim 1$ on $U(Y)$. By the Lindelöf Theorem, a countable collection $[U(Y_i)]$ of these sets covers X. If we set

$$V_n = U(Y_n) - U(Y_n)\cdot\sum_1^{n-1} U(Y_i), \qquad n = 1, 2, \cdots,$$

we have $X = \sum_1^\infty V_n$ and the sets V_n are disjoint and each of them is both open and closed in X.

Since for each n, $f \sim 1$ on V_n, there exists a real function $\phi_n(x)$ defined and continuous on V_n such that $f(x) = e^{i\phi_n(x)}$, $x \in V_n$. If we define $\phi(x) = \phi_n(x)$ $x \in V_n$ $(n = 1, 2, \cdots)$ clearly $\phi(x)$ is continuous on X and $f(x) = e^{i\phi(x)}$, $x \in X$. Hence $f \sim 1$ on X.

(5.51) COROLLARY. *If X is $\left\{\begin{array}{l}\text{compact}\\ \text{locally connected}\\ \text{0-dimensional}\\ \text{a subset of } R\end{array}\right\}$, any mapping of X into S which is ~ 1 on each component of X is ~ 1 on X.*

Our next result is concerned with the equivalence to 1 of a mapping on the cartesian product $X \times Y$ of two metric spaces X and Y. It will be recalled that $X \times Y$ consists of all ordered pairs (x, y) where $x \in X$ and $y \in Y$ and the metric in $X \times Y$ is given by

$$\rho[(x_1, y_1), (x_2, y_2)] = \sqrt{\rho(x_1, x_2)^2 + \rho(y_1, y_2)^2}.$$

(5.6) THEOREM. *Suppose Y is a continuum and f is a mapping of $X \times Y$ into S. If for each $x \in X$, $f \sim 1$ on $(x) \times Y$ and there exists a $y_0 \in Y$ such that $f \sim 1$ on $X \times (y_0)$, then $f \sim 1$ on $X \times Y$.*

Proof. By hypothesis, for each x there exists a real function $\phi_x(y)$ on Y such that

$$f(x, y) = e^{i\phi_x(y)} \qquad \text{for } (x, y) \in X \times Y$$

and there exists a real function $\phi_0(x)$ on X such that

$$f(x, y_0) = e^{i\phi_0(x)} \qquad \text{for } x \in X.$$

Since $e^{i\phi_0(x)} = e^{i\phi_x(y_0)}$, we may suppose without loss of generality that $\phi_x(y)$ is so chosen that $\phi_x(y_0) = \phi_0(x)$.

Let us define

$$\phi(x, y) = \phi_x(y), \qquad (x, y) \in X \times Y.$$

Then

$$(*)\qquad \begin{aligned} f(x) &= e^{i\phi(x,y)}, & (x, y) \in X \times Y,\\ \phi(x, y_0) &= \phi_0(x), & x \in X.\end{aligned}$$

It remains only to prove that $\phi(x, y)$ is continuous on $X \times Y$. Clearly for this purpose we may suppose X consists of a single sequence $[x_n]$ converging to a point x_0. In this case X, and hence $X \times Y$, is compact. Furthermore the components of $X \times Y$ are the sets $[x_n] \times Y$, $(n = 0, 1, 2, \cdots)$. Thus since $f \sim 1$ on each component of $X \times Y$, it results by (5.51) that $f \sim 1$ on $X \times Y$. Hence there exists a continuous real function $\psi(x, y)$ defined on $X \times Y$ such that

$$f(x, y) = e^{i\psi(x,y)}, \qquad (x, y) \in X \times Y.$$

Now to prove the continuity of $\phi(x, y)$ it suffices to establish the continuity of $\psi(x, y) - \phi(x, y)$. To this end let $(\bar{x}, \bar{y}) \in X \times Y$ and let (x_n, y_n) be a sequence converging to $(\bar{x}, \bar{y})$. (The only case presenting any difficulty is that in which $\bar{x} = x_0$.) Since Y is connected and $e^{i\psi(x,y)} = e^{i\phi(x,y)} = f(x, y)$, by (5.1) we have

$$\psi(x_n, y_n) - \phi(x_n, y_n) = \psi(x_n, y_0) - \phi(x_n, y_0).$$

Hence

$$\begin{aligned}\lim_{n\to\infty} [\psi(x_n, y_n) - \phi(x_n, y_n)] &= \lim_{n\to\infty} [\psi(x_n, y_0) - \phi(x_n, y_0)]\\ &= \psi(\bar{x}, y_0) - \phi(\bar{x}, y_0) = \psi(\bar{x}, \bar{y}) - \phi(\bar{x}, \bar{y}),\end{aligned}$$

because

$$\phi(x_n, y_0) = \phi_0(x_n) \to \phi_0(\bar{x}) = \phi(\bar{x}, y_0).$$

(5.61) COROLLARY. *If X and Y are continua and $f(x, y)$ is a mapping of $X \times Y$ into S such that for some $(x_0, y_0) \in X \times Y$ we have $f \sim 1$ on $(x_0) \times Y$ and on $X \times (y_0)$, then $f \sim 1$ on $X \times Y$.*

For since $X \times Y$ is compact, uniform continuity of f on $X \times Y$ gives a $\delta > 0$ such that $\rho[(x_1, y), (x_2, y)] = \rho(x_1, x_2) < \delta$ implies $|f(x_1, y) - f(x_2, y)| < 2$. Now suppose that $f \sim 1$ on $(x_1) \times Y$ and that x_2 is any point in X with $\rho(x_1, x_2) < \delta$. There exists a real function $\phi(x_1, y)$ on $(x_1) \times Y$ such that $f(x_1, y) = e^{i\phi(x_1,y)}$, $y \in Y$. If we define $\phi(x_2, y) = \phi(x_1, y) + [f(x_1, y), f(x_2, y)]$, $(x_2, y) \in (x_2) \times Y$, we have

$$e^{i\phi(x_2,y)} = e^{i\phi(x_1,y)} \cdot e^{i[f(x_1,y),f(x_2,y)]} = f(x_1, y) \cdot \frac{f(x_2, y)}{f(x_1, y)} = f(x_1, y),$$

by (i) of the introduction of this part of Chapter XI. Hence $f \sim 1$ on $(x_2) \times Y$. Accordingly the set X' of all $x \in X$ such that $f \sim 1$ on $(x) \times Y$ is both open and closed in X and since $x_0 \in X'$, we have $X' \equiv X$. Our corollary then follows by (5.6). This theorem together with (5.411) yields at once

(5.62) *Any mapping of the n-dimensional interval K_n into S is ~ 1.*

(5.7) THEOREM. *Any mapping f of X into a proper subset of S is ~ 1.*

For let $a \in S - f(x)$, $\alpha = \text{amp}(-a)$. Then if for each $x \in X$, $A(x)$ is the value of amp $[-f(x)/a]$ on the open interval $(-\pi, \pi)$, $A(x)$ is continuous and hence so is

$$\phi(x) = A(x) + \alpha.$$

Now

$$e^{i\phi(x)} = e^{iA(x)} \cdot e^{i\alpha} = \left[-\frac{f(x)}{a} \right] (-a) = f(x), \qquad x \in X.$$

(5.71) COROLLARY. *Any mapping $f(X) = c$, $c \in S$, is ~ 1 on X.*

6. **Homotopy, exponential equivalence, essentiality.** Two mappings f_1 and f_2 of X into any metric space Σ are said to be *homotopic* provided there exists a continuous mapping $g(x, t)$ of $X \times K$ (i.e., $x \in X$, $0 \leqq t \leqq 1$) into Σ such that $g(x, 0) \equiv f_1(x)$ and $g(x, 1) \equiv f_2(x)$ for $x \in X$. A mapping of X into S homotopic to the transformation $f_0(x) \equiv 1$ of X into S is said to be *homotopic to* 1. It results at once that two mappings f_1 and f_2 of X into S are homotopic if and only if f_1/f_2 is homotopic to 1.

Two mappings f_1 and f_2 of X into S will be said to be *exponentially equivalent* or simply *equivalent* on a set $Y \subset X$ provided their ratio f_1/f_2 is ~ 1 on Y. Since $f \sim 1$ on Y implies $1/f \sim 1$ on Y, clearly this is a symmetric relation. Indeed, both homotopy and equivalence are easily seen to be reflexive, symmetric and transitive and thus equable.

A transformation f of X into S is said to be *inessential* or *essential* according as it does or does not belong to the same component of the space S^X as the mapping $f_0(X) = 1$, $x \in X$.

(6.1) THEOREM. *Any continuous mapping f of the interval K into S is inessential.*

For the set of transformations

$$f_t(x) = \frac{f(tx)}{t + (1 - t)f(0)} \, | t + (1 - t)f(0) |$$

is a connected subset of S^X containing $f_0(x) \equiv 1$ for $t = 0$ and $f(x)$ for $t = 1$.

(6.2) THEOREM. *In order that a mapping f of X into S be inessential it is necessary and sufficient that there exist a bounded real continuous function $\phi(x)$ on X such that $f(x) \equiv e^{i\phi(x)}$, $x \in X$.*

To prove the sufficiency, we have only to set $f_t(x) = e^{it\phi(x)}$, $0 \leqq t \leqq 1$, to obtain a connected subset $[f_t(x)]$ of S^X containing $f_0(x) \equiv 1$ for $t = 0$ and $f(x)$ for $t = 1$.

To prove the necessity, we note first that the function $f_0(x) \equiv 1$ satisfies the condition since $e^{i\phi(x)} \equiv f_0(x) \equiv 1$ for $\phi(x) \equiv 0$. Now let $f_1(x)$ be any mapping in S^X satisfying the condition, i.e., there exists a bounded real function $\phi_1(x)$ on X such that $f_1(x) = e^{i\phi_1(x)}$, $x \in X$; and let f_2 be any mapping in S^X with $\rho(f_1, f_2) < 2$. If we set

$$\phi_2(x) = \phi_1(x) + [f_1(x), f_2(x)],$$

$\phi_2(x)$ is bounded and continuous and by (i),

$$e^{i\phi_2(x)} = e^{i\phi_1(x)} \cdot e^{i[f_1(x), f_2(x)]} = f_1(x) \frac{f_2(x)}{f_1(x)} = f_2(x).$$

Thus f_2 satisfies the condition. Hence the set Q of all elements of S^X satisfying the condition is both open and closed in S^X and $Q \supset f_0(x) \equiv 1$; accordingly, Q contains all inessential mappings in S^X.

(6.11) COROLLARY. *If X is compact, a mapping f of X into S is inessential if and only if $f \sim 1$.*

(6.12) COROLLARY. *Any mapping of the n-dimensional interval K_n ($n = 1, 2, \cdots$) into S is inessential.*

(6.2) THEOREM. *In order that a mapping $f(x)$ of X into S be ~ 1 it is necessary and sufficient that $f(x)$ be homotopic to* 1.

To prove the necessity, suppose $f(x) = e^{i\phi(x)}$ where $\phi(x)$ is a real continuous, function on X. Then if we set

$$g(x, t) = e^{it\phi(x)} \qquad (x, t) \in X \times K$$

$g(x, t)$ is continuous and $g(x, 0) \equiv 1$, $g(x, 1) \equiv f(x)$, so that f is homotopic to 1.

On the other hand, suppose $f(x)$ is homotopic to 1, so that there exists a continuous mapping $g(x, t)$ of $X \times K$ into S with $g(x, 0) \equiv 1$, $g(x, 1) \equiv f(x)$ for $x \in X$. Since $g(x, 0) \equiv e^{i0} \equiv 1$ on $X \times (0)$, $g \sim 1$ on $X \times (0)$. Further, by (6.12), $g \sim 1$ on $(x) \times K$ for each $x \in X$. Hence, by (5.6), $g \sim 1$ on $X \times K$; and in particular $f(x) \equiv g(x, 1) \sim 1$ on X.

(6.21) COROLLARY. *In order that two mappings f_1 and f_2 of X into S be equivalent it is necessary and sufficient that they be homotopic.*

(6.22) COROLLARY. *If X is compact the properties*: (i) "*inessential*" (ii) "$\sim$1," (iii) "*homotopic to* 1," *are equivalent.*

7. **Property** (b) **and unicoherence.** A space X will be said to have *property* (b) provided every mapping of X into S is $\sim$1. By §6, property (b) is equivalent to the property that every mapping of X into S be homotopic to 1; and in case X is compact, this is the same as the property that every mapping of X into S be inessential, or what amounts to the same thing, that the space S^X be be connected. Also any dendrite has property (b) as does also the n-dimensional interval K_n.

We note further that for continua X, property (b) is cyclicly extensible, or E_0-extensible, in the sense that *if each E_0-set in X has property* (b), *so also has X.*

Also, property (b) is "simple link reducible," i.e., *if X has property* (b) *so also has each simple link in X.* This results at once from (3.8).

We begin with

(7.1) *Property* (b) *is a topological property.*

For let X have property (b) and let $X' = h(X)$ where h is any homeomorphism. Then if f is any mapping of X' into S, the mapping fh of X into S is ~ 1. Hence there exists a real function $\psi(x)$ on X such that $fh(x) = e^{i\psi(x)}$, $x \in X$. Now if we set $\phi(x') = \psi h^{-1}(x')$, $x' \in X'$, ϕ is continuous on X' and $e^{i\phi(x')} = e^{i\psi h^{-1}(x')} = fhh^{-1}(x') = f(x')$, $x' \in X'$. Hence $f \sim 1$ on X', so that X' has property (b).

(7.2) THEOREM. *Suppose $X = \sum_1^\infty X_n$, where the sets $X_1 \subset X_2 \subset X_3 \subset \cdots$ are connected, and suppose that for each sequence $x_i \to x$ in X there exists an n such that $\sum_1^\infty x_i \subset X_n$. Then any mapping f of X into S which is ~ 1 on each X_n is ~ 1 on X.*

Proof. For each n there exists a continuous real function $\phi_n(x)$ on X_n such that

$$f(x) = e^{i\phi_n(x)}, \qquad x \in X_n .$$

Let $x_1 \in X_1$. Since for any n, $e^{i\phi_n(x_1)} = e^{i\phi_1(x_1)}$, there is no loss in generality in supposing the functions $\phi_n(x)$ chosen for $n > 1$ so that

$$\phi_n(x_1) = \phi_1(x_1).$$

Now since X_n is connected and $\phi_n(x_1) = \phi_{n+1}(x_1)$, it follows by (5.1) that

$$\phi_n(x) \equiv \phi_{n+1}(x), \qquad x \in X_n .$$

Hence if we define

$$\phi(x) = \phi_n(x), \qquad x \in X_n ,\ n = 1, 2, \cdots ,$$

$\phi(x)$ is single-valued, $f(x) = e^{i\phi(x)}$ for all $x \in X$. To prove the continuity of $\phi(x)$, let x_i be a sequence in X converging to a point $x_0 \in X$. By hypothesis there exists an n such that $\sum_0^\infty x_i \subset X_n$. Hence $\lim_{i\to\infty} \phi(x_i) = \lim_{i\to\infty} \phi_n(x_i) = \phi_n(x_0) = \phi(x_0)$.

(7.21) COROLLARY. *The Euclidean n-space R^n ($n = 1, 2, \cdots$) has property* (b).

For we have only to let X_k be the n-dimensional interval in R^n having center at the origin and one vertex at $(k, k, \cdots, k)$ and apply (7.2).

It will be recalled that a connected space X is unicoherent provided that for every decomposition of X into two closed and connected sets X_1 and X_2, the set $X_1 \cdot X_2$ is connected.

(7.3) THEOREM. *Every connected space X having property* (b) *is unicoherent.*

Proof. For suppose X is not unicoherent. Then there exist two closed and connected sets X_1 and X_2 and two closed disjoint sets P_1 and P_2 such that

$$X = X_1 + X_2, \qquad X_1 \cdot X_2 = P_1 + P_2 .$$

Define

$$\psi(x) = \pi \frac{\rho(x, P_1)}{\rho(x, P_1) + \rho(x, P_2)}, \qquad x \in X;$$

$$f(x) = e^{i\psi(x)}, \qquad x \in X_1;$$

$$f(x) = e^{-i\psi(x)}, \qquad x \in X_2.$$

Then $\psi(x) = 0$ for $x \in P_1$ and $\psi(x) = \pi$ for $x \in P_2$, so that f is single-valued and continuous.

Since S has property (b), $f \sim 1$ on X. Hence there exists a real continuous function $\phi(x)$ on X such that $f(x) = e^{i\phi(x)}$, $x \in X$. Since X_1 and X_2 are connected, by (5.1) we have

$$\psi(x) = \phi(x) + 2k_1 \qquad \text{on } X_1,$$

$$-\psi(x) = \phi(x) + 2k_2 \qquad \text{on } X_2;$$

whence,

$$\psi(x) = \pi(k_1 - k_2) \qquad \text{on } X_1 \cdot X_2.$$

This is absurd, since $\psi(x)$ takes each of the values 0 and π on $X_1 \cdot X_2$.

(7.31) COROLLARY. *R^n and K_n are unicoherent.*

(7.4) THEOREM. *In order that a locally connected continuum have property* (b) *it is necessary and sufficient that it be unicoherent.*

Proof. The necessity results from (7.3). To establish the sufficiency, let X be a unicoherent locally connected continuum and let f be any mapping of X into S.

Let $f(x) = f_2 f_1(x)$ be the monotone-light factorization of f and let $f_1(X) = X'$. Then since f_1 is monotone, X' must be a unicoherent locally connected continuum; and since $f_2(X') = S$ is light, X' must be at most 1-dimensional. Hence X' must be a dendrite. Thus X' has property (b) and $f_2 \sim 1$ on X'. But clearly $f_2 \sim 1$ on X' gives $f \sim 1$ on X; for if $f_2(x') = e^{i\psi(x')}$, $x' \in X'$, and we define

$$\phi(x) = \psi f_1(x), \qquad x \in X,$$

we have $e^{i\phi(x)} = f_2 f_1(x) = f(x)$, $x \in X$. Accordingly, $f \sim 1$ on X and X has property (b).

(7.41) COROLLARY. *Property* (b) *is invariant under monotone or under interior transformations on locally connected continua.*

From (5.6) we have

(7.5) *If each of two continua X and Y has property* (b), *so also has their cartesian product $X \times Y$.*

This together with (7.4) gives

(7.51) COROLLARY. *If X and Y are unicoherent locally connected continua, so also is their product $X \times Y$.*

Similarly, from (5.2) we get

(7.6) *If X and Y are unicoherent locally connected continua and $X \cdot Y$ is connected, $X + Y$ is unicoherent.*

(7.7) THEOREM. *Property* (b) *is invariant under retraction.*

For let X have property (b), let $r(X) = Y$ be a retraction, and let f be any mapping of Y onto S. Since fr is a mapping of X onto S, we have $fr \sim 1$ on X. But this gives $fr \sim 1$ on Y or $f \sim 1$ on Y because on Y, fr is identical with f.

(7.71) COROLLARY. *Every retract of a locally connected unicoherent continuum is unicoherent*

(7.72) *Every retract of a connected space having property* (b) [*in particular, of R^n or K_n*] *is unicoherent.*

8. The groups S^X, $P(X)$, $B(X)$, and so on. It has already been noted that if complex multiplication is used as a group operation, the elements of S^X form an abelian group which we also denote by S^X. Let $P(X)$ be the set of all $f \in S^X$ such that $f \sim 1$ on X. Then since $f_1 \sim 1, f_2 \sim 1$ clearly gives $f_1 \cdot f_2 \sim 1$ and $f \sim 1$ gives $1/f \sim 1$, $P(X)$ is a subgroup of S^X. Let $B(X)$ denote the factor group S^X/P^X.

Similarly, if $X = X_1 + X_2$, where X_1 and X_2 are closed sets, the set $P(X_1, X_2)$ of all $f \in S^X$ such that $f \sim 1$ on X_1 and on X_2 is a subgroup of S^X and, further, $P(X)$ is a subgroup of $P(X_1, X_2)$.

Let $b(X)$ and $p(X_1, X_2)$ denote the ranks of the groups $B(X)$ and $P(X_1, X_2)/P(X)$, respectively, i.e., the maximum number of linearly independent elements in these groups if this number is finite, and otherwise ∞. A set of mappings $f_1, f_2, \cdots, f_n$ in S^X is said to be linearly independent provided that

$$f_1^{p_1} \cdot f_2^{p_2} \cdots f_n^{p_n} \sim 1, \qquad p_i \text{ integers},$$

implies $p_1 = p_2 = \cdots = p_n = 0$, i.e., provided the elements of $B(X)$ containing $f_1, f_2, \cdots, f_n$ are linearly independent.

Finally, let us define

$$\rho(X) = \text{l.u.b. } p[X_1, X_2],$$

for all decompositions of X into closed sets X_1 and X_2.

(8.1) THEOREM. *If f is any mapping of X into S such that $f^p \sim 1$ on X for an integer $p \neq 0$, then $f \sim 1$ on X.*

Proof. Let $\phi(x)$ be a real function on X such that $[f(X)]^p = e^{i\phi(x)}$, $x \in X$. Define

$$\psi_j(x) = \frac{\phi(x) + 2j\pi}{p}, \qquad x \in X, j = 0, 1, \cdots, p - 1.$$

Let X_j be the set of all $x \in X$ such that $f(x) = e^{i\psi_j(x)}$. Since the sets X_j $(j = 0, \cdots, p-1)$ are closed and disjoint and $f \sim 1$ on X_j, clearly our theorem follows as soon as we show that $X = X_0 + X_1 + \cdots + X_{p-1}$.

To this end let $x \in X$ and let y be any definite real number such that $f(x) = e^{iy}$. Then

$$f(x) = e^{i(y+2n\pi)} \qquad \text{for } n = 0, \pm 1, \pm 2, \cdots.$$

Hence

$$e^{ip(y+2n\pi)} = e^{i\phi(x)}.$$

Thus there exists an integer k_n such that

$$p(y + 2n\pi) = \phi(x) + 2k_n\pi,$$

or

$$\text{(i)} \qquad y + 2n\pi = \frac{\phi(x) + 2k_n\pi}{p}.$$

Whence,

$$\text{(ii)} \qquad y - \frac{\phi(x)}{p} + 2n\pi = 2\frac{k_n}{p}\pi, \qquad n = 0, \pm 1, \pm 2, \cdots.$$

Now if n is chosen so that the left member of (ii) is greater than or equal to 0 and less than 2π, we have $0 \leqq k_n \leqq p - 1$. Hence, by (i), for this n, $y + 2n\pi = \psi_{k_n}(x)$ or $x \in X_{k_n}$ and our theorem is proved.

(8.11) COROLLARY. *$P(X)$ and $P(X_1, X_2)$ are division subgroups of S^X.*

(8.111) *$B(X)$ and $P(X_1, X_2)/P(X)$ have no elements of finite period other than the neutral element. Hence, if either of them has a finite generating system, it is a free group.*

(8.2) THEOREM. *For any mapping f of S into S there exists an integer p such that $f \sim z^p$ on S.*

Proof. Define

$$g(x) = e^{ix} \qquad \text{for } 0 \leqq x \leqq 2\pi,$$

$$f^*(x) = fg(x) \qquad \text{for } 0 \leqq x \leqq 2\pi.$$

Then since f^* is a mapping of the interval $(0, 2\pi)$ into S, we have $f \sim 1$ on $(0, 2\pi)$. Hence there exists a real function $\phi^*(x)$ on $(0, 2\pi)$ such that

$$f^*(x) = fg(x) = e^{i\phi^*(x)}.$$

Now since $g(0) = g(2\pi)$ and hence $f^*(0) = f^*(2\pi)$, there exists an integer p such that

$$\phi^*(2\pi) - \phi^*(0) = 2p\pi.$$

Thus if we define

$$\phi(z) = \phi^*(x) - px, \qquad z = e^{ix}, 0 \leqq x \leqq 2\pi,$$

$\phi(z)$ is continuous and is single-valued because

$$\phi(2\pi) = \phi^*(2\pi) - 2p\pi = \phi^*(0) - p\cdot 0 = \phi(0).$$

Further,

$$e^{i\phi(z)} = \frac{e^{i\phi^*(x)}}{e^{ipx}} = \frac{fg(x)}{e^{ipx}} = \frac{f(e^{ix})}{(e^{ix})^p} = \frac{f(z)}{z^p}.$$

Whence,

$$\frac{f(z)}{z^p} \sim 1 \quad \text{or} \quad f(z) \sim z^p \quad \text{on } S.$$

(8.21) *The group $B(S)$ is isomorphic with the additive group of all integers. In other words, $B(S)$ is a free group of rank* 1.

This results from (8.2) together with the fact established in (8.1) and the introduction to §5 that z^p is never ~ 1 for $p \neq 0$.

Since for any integer p, the mappings z^p and z^{-p} are topologically equivalent (though not homotopic to each other for $p \neq 0$) it results that any mapping of S into S is topologically equivalent to z^p for $p \geqq 0$. Furthermore, since clearly any mapping of one simple closed curve J into another (or the same) simple closed curve C is topologically equivalent to a mapping of S into S, we have

(8.22) *Any mapping of one simple closed curve J into another one C is homotopic to a mapping of J into C which is topologically equivalent to the mapping $f(z) = z^p$ of S into S for some integer $p \geqq 0$.*

In particular, then, we have shown that *any mapping of one simple closed curve into another is homotopic to an interior mapping of the one curve into the other.*

Definition. *Let X and Y be metric spaces and let $g(x)$ be any continuous transformation of X into Y. To any $f \in S^Y$ we associate the mapping fg of X into S. Since for $f_1, f_2 \in S^Y$ we have*

$$(f_1\cdot f_2)g(x) \equiv f_1g(x)\cdot f_2g(x), \qquad x \in X,$$

it follows that this correspondence is a homomorphism of S^Y into S^X, i.e., onto a subgroup of S^X. Further, since clearly $f \sim 1$ on Y gives $fg \sim 1$ on X so that $P(Y)$ maps into $P(X)$ under this correspondence, this homomorphism generates a homomorphism h of $B(Y)$ into $B(X)$. We shall call h the homomorphism of $B(Y)$ into $B(X)$ generated by $g(x)$.

Since h will be an isomorphism if and only if its kernel reduces to the neutral element $P(Y)$ of $B(Y)$, we have

(8.3) Theorem. *The homomorphism h of $B(Y)$ into a subgroup of $B(X)$ generated by a continuous mapping $g(x)$ of X into Y is an isomorphism if and* *only if the condition*

(†) $$fg \sim 1 \text{ on } X \text{ implies } f \sim 1 \text{ on } Y$$

is satisfied for every mapping f of Y into S.

(8.4) THEOREM. *If the condition* (†) *in* (8.3) *is satisfied by a mapping g of X onto Y, we have*

$$b(X) \geqq b(Y), \qquad \rho(X) \geqq \rho(Y).$$

Proof. The first inequality results at once from the fact that, by (8.3), $B(Y)$ is isomorphic with a subgroup of $B(X)$. To prove the second, let $\rho(Y)$ be greater than or equal to n. There exists, then, a decomposition of Y into closed sets, $Y = Y_1 + Y_2$ and n linearly independent mappings $f_1, f_2, \cdots, f_n$ of Y into S such that $f_i \sim 1$ on Y_1 and on Y_2 for $i = 1, \cdots, n$, because $p(Y_1, Y_2) =$ rank of $P(Y_1, Y_2)/P(Y) \geqq n$. Now if we set

$$X_1 = g^{-1}(Y_1), \qquad X_2 = g^{-1}(Y_2)$$

we have the decomposition

$$X = X_1 + X_2$$

and $f_i g \sim 1$ on X_1 and on X_2, for $i = 1, 2, \cdots, n$. Hence $f_i g \in P(X_1, X_2)$ $(i = 1, 2, \cdots, n)$. Furthermore, since condition (†) is satisfied, the mappings $f_i g$ $(i = 1, 2, \cdots, n)$ are linearly independent, since $\prod_1^n (f_i g)^{p_i} = (\prod_1^n f_i^{p_i})g \sim 1$ gives $\prod_1^n f_i^{p_i} \sim 1$. Hence, $\rho(X) \geqq p(X_1, X_2) \geqq n$. Accordingly, $\rho(X) \geqq \rho(Y)$.

(8.5) THEOREM. *If X is compact and* $g(X) = Y$ *is either monotone or interior,*

$$b(X) \geqq b(Y), \qquad \rho(X) \geqq \rho(Y).$$

Proof. By (8.4) we have only to show that the condition (†) is satisfied for every mapping f of Y into S. To this end, suppose $fg \sim 1$ on X. Then there exists a real function $\psi(x)$ on X such that

$$fg(x) = e^{i\psi(x)}, \qquad x \in X.$$

Now if g is monotone we have only to define $\phi(y) = \psi g^{-1}(y)$, $y \in Y$. Then since $g^{-1}(y)$ is connected for each $y \in Y$ it follows that ψ can take only one value on $g^{-1}(y)$, so that ϕ is single-valued and continuous. Finally, we have

$$e^{i\phi(y)} = e^{i\psi g^{-1}(y)} = fgg^{-1}(y) = f(y), \qquad y \in Y.$$

On the other hand, if g is interior, we define

$$\phi(y) = \text{g.l.b. } [\psi(x)] \qquad \text{for } x \in g^{-1}(y).$$

Then since $\phi(y) = \psi(x_0)$ for some $x_0 \in g^{-1}(y)$,

$$e^{i\phi(y)} = e^{i\psi(x_0)} = fg(x_0) = f(y), \qquad y \in Y.$$

Furthermore, by interiority of q, $y_i \to y$ in Y implies $g^{-1}(y_i) \to g^{-1}(y)$ and then uniform continuity of g on X gives

$$\phi(y_i) = \underset{x \in g^{-1}(y_i)}{\text{g.l.b.}} [\psi(x)] \rightarrow \underset{x \in g^{-1}(y)}{\text{g.l.b.}} [\psi(x)] = \phi(y).$$

Hence $\phi(y)$ is continuous on Y and our proof is complete.

(8.6) THEOREM. *If $X = X_1 + X_2$ where X_1 and X_2 are closed and $X_1 \cdot X_2$ contains at most one point, we have*

$$B(X) = B(X_1) \cdot B(X_2),$$

i.e., $B(X)$ is the direct product of $B(X_1)$ and $B(X_2)$.

Proof. Since for any mapping $f(x)$ of X_1, X_2 or X into S we have by (5.71)

$$f(x) \sim \frac{f(x)}{f(a)} = g(x) \qquad \text{on } X_1, X_2 \text{ or } X$$

(where $a = X_1 \cdot X_2$ if $X_1 \cdot X_2 \neq 0$ and $f(a) = 1$ if $X_1 \cdot X_2 = 0$), and since $g(a) = 1$ it follows that every element of $B(X_1)$, $B(X_2)$ or $B(X)$ contains a mapping sending a into 1 if $X_1 \cdot X_2 \neq 0$.

If we extend the mappings of S^{X_1} and S^{X_2} to X by defining $f(x) = f(a)$ for $x \in X_2$, $f \in S^{X_1}$, and $f(x) = f(a)$ for $x \in X_1$, $f \in S^{X_2}$, so that S^{X_1} and S^{X_2} become subgroups of S^X, then since the extended f is ~ 1 on X if and only if the original was ~ 1 on X_1 or X_2, respectively, it follows that $B(X_1)$ and $B(X_2)$ become subgroups of $B(X)$.

Now to prove our theorem we have only to show that every element F of $B(X)$ is expressible in the form $F = F_1 \cdot F_2$, $F_i \in B(X_i)$, $(i = 1, 2)$, and that the intersection of $B(X_1)$ and $B(X_2)$ reduces to the neutral element $P(X)$ of $B(X)$. To this end let $F \in B(X)$ and let $f \in F$ be a mapping of X into S such that $f(a) = 1$ if $X_1 \cdot X_2 \neq 0$. Let us define

$$\begin{aligned} f_1(x) &= f(x) & x &\in X_1, \\ &= 1 & x &\in X_2, \\ f_2(x) &= 1 & x &\in X_1, \\ &= f(x) & x &\in X_2. \end{aligned}$$

Then since $f(x) = f_1(x) \cdot f_2(x)$, $x \in X$, it follows that if $f_1 \in F_1 \in B(X_1)$, $f_2 \in F_2 \in B(X_2)$ we have

$$F = F_1 \cdot F_2.$$

On the other hand, let $F \in B(X_1) \cdot B(X_2)$. Then $F \in B(X_i)$ gives a mapping $f_i(x)$ of X_i into S such that $f_i \in F$ and $f_i(x) \equiv 1$ on X_j for $i, j = 1, 2, j \neq i$. Since $f_1, f_2 \in F$ we have $f_1/f_2 \sim 1$ on X. Whence $f_1/f_2 \sim 1$ on X_1. Thus on X_1, $f_1 = f_1/f_2 \sim 1$ and since $f_1(x) \equiv 1$ on X_2, and $X_1 \cdot X_2$ is connected or vacuous, this gives $f_1 \sim 1$ on X. Accordingly, F must be the neutral element of $B(X)$, and our theorem is proved.

COROLLARY. *If $X = X_1 + X_2$ where X_1 and X_2 are closed and $X_1 \cdot X_2$ contains at most one point,*

$$b(X) = b(X_1) + b(X_2).$$

(8.7) THEOREM. *The number $b(X)$ is cyclicly additive for continua X, i.e.,*

$$b(X) = \sum b(E_\alpha) \tag{*}$$

where (E_α) is the collection of simple links of X.

Proof. Since for any set N, the condition $b(N) = 0$ is equivalent to the condition that N have property (b) [see §7], it results from the extensibility and reducibility of property (b) noted at the beginning of §7 that if either side of (*) is 0 the other side also is 0 and (*) holds.

Now suppose that for all integers $k \geqq 0$, $k \leqq n - 1$, if either side of (*) has the value k the other side also has the value k. It then follows that this statement also holds for $k = n$.

For suppose either side of (*) has the value n. Suppose first that there are at least two distinct simple links E_i and E_j such that $b(E_i) \neq 0 \neq b(E_j)$. Then there exist continua X_1 and X_2 containing E_i and E_j, respectively, and such that $X = X_1 + X_2$ and $X_1 \cdot X_2$ reduces to a single point a. Now if the right-hand member of (*) has the value n, then

$$\sum b(E_\alpha) = \sum b(E'_\alpha) + \sum b(E''_\alpha), \tag{‡}$$

where the E'_α are true simple links of X on X_1 and the E''_α are true simple links of X on X_2. Since each term on the right of (‡) is less than n we have

$$b(X_1) = \sum b(E'_\alpha)$$
$$b(X_2) = \sum b(E''_\alpha)$$

whence by (8.6)

$$b(X) = b(X_1) + b(X_2) = \sum b(E_\alpha).$$

On the other hand if $b(X) = n$, then by (8.6)

$$b(X) = b(X_1) + b(X_2) = n.$$

Furthermore, neither $b(X_1)$ nor $b(X_2)$ can vanish. For $b(X_1) = 0$ would imply $b(E_i) = 0$ and $b(X_2) = 0$ would give $b(E_j) = 0$. Hence we have $b(X_1) < n$, $b(X_2) < n$; and thus

$$b(X_1) = \sum b(E'_\alpha),$$
$$b(X_2) = \sum b(E''_\alpha),$$

or

$$n = b(X) = b(X_1) + b(X_2) = \sum b(E'_\alpha) + \sum b(E''_\alpha) = \sum b(E_\alpha).$$

There remains only the case in which there is only a single simple link E of

X with $b(E) \neq 0$. In this case the right side of (*) reduces to $b(E)$ and we have to prove that if either $b(X)$ or $b(E)$ has the value n so also has the other. Suppose $b(X) \geqq k$. Let $f_1, f_2, \cdots, f_k$ be a set of k linearly independent mappings of X into S. Then $f_1, f_2, \cdots, f_k$ are linearly independent on E. For let $p_1, p_2, \cdots, p_k$ be any integers not all 0. Then since $g(x) = \prod_1^k f_i^{p_i}(x)$ is not ~ 1 on X it follows by (5.41) that $g(x)$ is not ~ 1 on E. Hence the assumption $b(X) \geqq k$ for any k implies $b(E) \geqq k$. On the other hand, suppose $b(E) \geqq k$ and let $f_1, f_2 \cdots, f_k$ be linearly independent mappings of E into S. By (3.8) there exist extensions $\phi_1, \phi_2, \cdots, \phi_k$ of $f_1, f_2, \cdots, f_k$, respectively, to X, where $\phi_i \in S^X$. Then the mappings $\phi_1, \phi_2, \cdots, \phi_k$ of X into S are linearly independent. For if $\prod \phi_i^{p_i} \sim 1$ on X, not all $p_i = 0$, we would have $\prod \phi_i^{p_i} = \prod f_i^{p_i} \sim 1$ on E. Thus $b(X) \geqq k$. Accordingly $b(X) = b(E)$. Hence in any case we have shown that if either side of (*) has the value n so also has the other. Thus by the principle of induction, (*) holds in the sense that if either side is finite the other is finite and the two are equal.

9. $\rho(X)$ **and** $r(X)$ **for locally connected continua.** For any two closed sets X_1 and X_2 let us understand by $r(X_1, X_2)$ the number defined by the relation

$$r(X_1, X_2) + 1 = \text{number of components of } X_1 \cdot X_2 .$$

It will be recalled that for continua X, $r(X)$ is by definition equal to the l.u.b. $r(X_1, X_2)$ for all decompositions of X into continua X_1 and X_2.

For continua X in general it is readily seen that the numbers $r(X)$ and $\rho(X)$ may be different. For example, let N be the continuum consisting of the circle S together with a spiral approaching S asymptotically. Since N is unicoherent, we have $r(N) = 0$, whereas since a radial projection $p(x)$ of N into S is not ~ 1 on N but is ~ 1 on any set $p^{-1}(S')$ where S' is any proper subset of S, we have $\rho(N) = 1$.

In this section, however, results will be established leading to the conclusion that for locally connected continua X, the numbers $\rho(X)$ and $r(X)$ are identical.

(9.1) *If the sets X_1 and X_2 are closed (or open) in $X = X_1 + X_2$ and $X_1 \cdot X_2 \neq 0$, we have*

$$p(X_1, X_2) \leqq r(X_1, X_2).$$

Proof. We may suppose $r(X_1, X_2)$ finite and equal to n, so that

$$X_1 \cdot X_2 = P_0 + P_1 + \cdots + P_n, \qquad n \geqq 0,$$

where the sets P_t are closed (or open), non-empty, disjoint, and connected. Now to prove $p(X_1, X_2) \leqq n$, suppose

$$f_1, f_2, \cdots, f_m \in P(X_1, X_2), \qquad m > n.$$

Then

$$f_j(x) = e^{i\phi_j(x)} \qquad \text{on } X_1$$

$$= e^{i\psi_j(x)} \qquad \text{on } X_2,$$

where ϕ_j and ψ_j are real functions which we may suppose so chosen that $\phi_j(x_0) = \psi_j(x_0)$ for a fixed $x_0 \in P_0$ $(j = 1, 2, \cdots, m)$. Since P_t is connected, on P_t we have integers k_{tj} so that

$$\phi_j(x) - \psi_j(x) = 2k_{tj}\pi \qquad t = 1, 2, \cdots, n; j = 1, 2, \cdots, m.$$

Since $m > n$ there exist integers $p_1, p_2, \cdots, p_m$ not all 0 and such that

$$\sum_{j=1}^{m} p_j k_{tj} = 0, \qquad t = 1, 2, \cdots, n.$$

Define

$$\phi^*(x) = \sum_1^m p_j \phi_j(x) \qquad x \in X_1,$$

$$= \sum_1^m p_j \psi_j(x) \qquad x \in X_2.$$

Then $\phi^*(x)$ is single-valued and continuous on X because $\sum_1^m p_j[\phi_j(x) - \psi_j(x)] = 2\pi \sum_1^m p_j k_{tj} = 0$ for $x \in X_1 \cdot X_2$. Further,

$$e^{i\phi^*(x)} = e^{ip_1\phi_1(x)} \cdot e^{ip_2\phi_2(x)} \cdots e^{ip_m\phi_m(x)}, \qquad x \in X_1,$$

$$= e^{ip_1\psi_1(x)} \cdot e^{ip_2\psi_2(x)} \cdots e^{ip_m\psi_m(x)}, \qquad x \in X_2,$$

$$= f_1^{p_1} \cdot f_2^{p_2} \cdots f_m^{p_m} \qquad \text{in any case.}$$

Hence $\prod_1^m f_j^{p_j} \sim 1$ on X, and the mappings $f_1, f_2, \cdots, f_m$ are linearly dependent for any $m > n$. Accordingly $p(X_1, X_2) \leqq n$, which proves (9.1).

(9.2) THEOREM. *If the sets* X_1 *and* X_2 *are closed in* $X = X_1 + X_2$, *connected and* $X_1 \cdot X_2 \neq 0$,

$$p(X_1, X_2) = r(X_1, X_2).$$

Proof. In view of (9.1) we have only to prove

$$p(X_1, X_2) \geqq r(X_1, X_2).$$

To this end, suppose

$$X_1 \cdot X_2 = P_0 + P_1 + \cdots + P_n, \qquad n \geqq 0,$$

where the sets P_j are closed, disjoint, and different from 0. Define

$$\psi_j(x) = \frac{\rho(x, X_1 \cdot X_2 - P_j)}{\rho(x, P_j) + \rho(x, X_1 \cdot X_2 - P_j)}, \qquad j = 1, 2, \cdots, n,$$

$$f_j(x) = e^{i\psi_j(x)} \qquad x \in X_1;$$

$$= 1 \qquad x \in X_2.$$

Then by defintion $f_j \sim 1$ on X_1 and on X_2 $(j = 1, 2, \cdots, n)$, so that $f_j \in P(X_1, X_2)$. We shall show that the mappings $f_1, f_2, \cdots, f_n$ are linearly independent. To this end, suppose there is a relation

$$\prod_1^n f_j^{p_j}(x) = e^{i\phi(x)}, \qquad x \in X,$$

where $\phi(x)$ is real and continuous on X and the p_j are integers. By the definition of the f_j we get

$$e^{ip_1\psi_1} \cdot e^{ip_2\psi_2} \cdots e^{ip_n\psi_n} = e^{i\Sigma p_j\psi_j} = e^{i\phi(x)}, \qquad x \in X_1 .$$

Whence, by connectedness of X_1 and X_2,

$$\sum_1^n p_j\psi_j(x) = \phi(x) - 2k_1\pi \qquad \text{on } X_1$$

$$2k_2\pi = \phi(x) \qquad \text{on } X_2,$$

where k_1 and k_2 are integers. Whence

$$\sum_1^n p_j\psi_j(x) = 2(k_2 - k_1)\pi \qquad \text{on } X_1 \cdot X_2 .$$

Now let x_0 be a fixed point of P_0. By definition

$$\psi_j(x_0) = 0 \qquad \text{for } j = 1, 2, \cdots, n.$$

Hence

$$\sum_1^n p_j\psi_j(x) = 0 \qquad \text{on } X_1 \cdot X_2 .$$

Now take any $j \geqq 1$ and take a point $x_j \in P_j$. Then

$$\psi_j(x_j) = 2\pi \quad \text{and} \quad \psi_k(x_j) = 0 \qquad \text{for } k \neq j.$$

Whence

$$\sum_1^n p_j\psi_j(x_j) = p_j 2\pi = 0,$$

or

$$p_j = 0, \qquad j = 1, 2, \cdots, n.$$

Thus the f_j ($j = 1, 2, \cdots, n$) are linearly independent and $p(X_1, X_2) \geqq n$. Since the assumption $r(X_1, X_2) \geqq n$ entails $p(X_1, X_2) \geqq n$ we have

$$p(X_1, X_2) \geqq r(X_1, X_2)$$

and our proof is complete.

(9.3) LEMMA. *If X is a locally connected continuum and $f_1, f_2, \cdots, f_n$ are mappings of X into S, then for any closed set Y in X such that $f_i \sim 1$ on Y for $i = 1, 2, \cdots, n$, there exists a locally connected subcontinuum Y' of X containing Y and such that $f_i \sim 1$ on Y' for $i = 1, 2, \cdots, n$.*

Proof. By (5.3) it follows that if we decompose X into a finite number of sufficiently small locally connected subcontinua and let X' be the sum of all of these subcontinua which intersect Y, we will have $f_i \sim 1$ on X' for $i = 1, 2, \cdots, n$. Now by joining components of X' by arcs in X successively for a finite number of steps we can imbed X' in a locally connected continuum Y' in X such that any point of $Y' - X'$ is a cut point of Y' and hence so that every true cyclic element of Y' is in X'. But then $f_i \sim 1$ on each cyclic element of Y' for $i = 1, 2, \cdots, n$ and hence by (5.7), $f_i \sim 1$ on Y' for $i = 1, 2, \cdots, n$.

(9.4) THEOREM. *For any locally connected continuum X, we have*

$$\rho(X) = r(X).$$

Proof. By (9.2), for any decomposition $X = X_1 + X_2$ of X into continua X_1 and X_2 we have

$$p(X_1, X_2) = r(X_1, X_2).$$

Whence,

$$\rho(X) \geqq \text{l.u.b. } [r(X_1, X_2)] = r(X)$$

because $\rho(X) = \text{l.u.b. } [p(X_1, X_2)]$ for all decompositions of X into *closed sets* X_1 and X_2.

Now to prove the reverse inequality, suppose $\rho(X) \geqq n$. Then there exists a decomposition $X = X_1 + X_2$ of X into closed sets X_1 and X_2 and a set of linearly independent mappings $f_1, f_2, \cdots, f_n$ of X into S such that $f_i \sim 1$ on X_1 and on X_2 for $i = 1, 2, \cdots, n$. By (9.3) there exist locally connected continua X'_j in X ($j = 1, 2$) with $X_j \subset X'_j$ and such that $f_i \sim 1$ on X_j for $i = 1, 2, \cdots, n$, $j = 1, 2$. Thus $X = X'_1 + X'_2$ is a decomposition of X into (locally connected) continua and $p(X'_1, X'_2) \geqq n$. By (9.2) this gives

$$r(X) \geqq r(X'_1, X'_2) = p(X'_1, X'_2) \geqq n.$$

Whence $r(X) \geqq \rho(X)$ and our proof is complete.

REFERENCES

Part a. Whyburn [34, 37, 38]. For (4.3) see Moore [1] and Roberts [1]. In connection with retractions and particularly §3, see Borsuk [2, 3] and Wilson [2]. For other existence results in connection with continuous decompositions, see Knaster [3] and Roberts [4].

Part b. Eilenberg [2, 3]. Kuratowski [6]. Results (5.40), (5.4), (5.41) and (8.7) are published here for the first time.

CHAPTER XII

PERIODICITY. FIXED POINTS

1. Preliminaries. Invariant sets. We shall be concerned with a continuous mapping $f(X) \subset X$ of a set X into itself. In general conditions sufficient to make f a homeomorphism of X onto itself will be imposed, though for the present this is not assumed.

Let us set

$$\begin{aligned} f(x) &= f^1(x), \\ ff(x) &= f^2(x), \\ fff(x) &= f^3(x), \\ \cdots\cdots & \cdots\cdots, \\ \underbrace{ff \cdots ff}_{n \text{ iterations}}(x) &= f^n(x), \end{aligned}$$

and speak of the mappings $f^n(x)$, $n = 1, 2, \cdots$, as the *powers* of $f(x)$. Clearly

$$f^m f^n(x) \equiv f^n f^m(x) \equiv f^{m+n}(x), \qquad m, n > 0.$$

The mapping $f(x)$ is said to be *periodic* provided there exists an integer n such that

$$f^n(x) \equiv x, \qquad x \in X.$$

The least such integer n is called the *period* of $f(x)$. Similarly, $f(x)$ is said to be *pointwise periodic* at a point $x \in X$ provided there exists an integer n_x such that $f^{n_x}(x) = x$; and the least such integer n_x is called the period of f at x or simply the period of x. If $f(x)$ is pointwise periodic at each point of a set $Y \subset X$, f is said to be pointwise periodic on Y.

Obviously if $f(x)$ is periodic on X, it is pointwise periodic on X. The converse is not true, however. For if X is the dendrite formed by an infinite sequence of intervals a_ip, $i = 1, 2, \cdots$, with a common end point p and with lengths converging to 0 with $1/i$ and f maps a_1p topologically onto a_2p so that $f(p) = p$ and a_2p into a_1p, maps the next 3 intervals a_3p, a_4p, a_5p cyclically into each other, the next 4 intervals cyclically into each other, and so on indefinitely, clearly f is pointwise periodic but not periodic on X.

Now if a point $p \in X$ has the period n or if merely $f^n(p) = p$, we have $f^{mn}(p) = p$ for any integer $m > 0$, for

$$f^{mn}(p) = \underbrace{f^n f^n f^n \cdots f^n}_{m \text{ times}}(p) = p$$

since $f^n(p) = p$.

(1.1) *If $f(X) \subset X$ is a pointwise periodic mapping of X into itself, f is a* (1 - 1) *mapping of X onto itself.*

Proof. Clearly $f(X) = X$, because if $x \in X$ is of period n, we have

$$x = f[f^{n-1}(x)] \subset f(X).$$

To show th at f is (1-1) suppose

$$f(p) = f(q) = x, \qquad p, q, x \in X.$$

Let m and n be the periods of p and q, respectively, and let $k = mn$. Then by the above remark,

$$p = f^k(p) = f^{k-1}f(p) = f^{k-1}(x),$$
$$q = f^k(q) = f^{k-1}f(q) = f^{k-1}(x);$$

whence $p = q$.

(1.11) COROLLARY. *Any pointwise periodic continuous mapping of a compact set X into itself is topological.*

Without compactness on X this result may fail to hold, i.e., even though the mapping $f(X) = X$ is (1-1) and continuous, f^{-1} may fail to be continuous. However, if f is actually *periodic* on S, say of period n, then since $f^n(x) \equiv x$ on X we have $f^{n-1}(x) \equiv f^{-1}(x)$. Thus $f^{-1}(x)$ is continuous. Whence

(1.12) *Any periodic continuous mapping on an arbitrary set X is topological.*

(1.2) THEOREM. *If X is any set and $f(X) \subset X$ is any mapping whatever, in order that f be pointwise periodic it is necessary and sufficient that for any subset Y of X, $f(Y) \subset Y$ imply $f(Y) = Y$.*

Proof. First let f be pointwise periodic, let Y be any subset of X and suppose $f(Y) \subset Y$. Then if $y \in Y$ and n is the period of y,

$$f(y) \subset Y, f^2(y) \subset Y, \cdots, f^{n-1}(y) \subset Y.$$

Now $f[f^{n-1}(y)] = y$. Hence $y \in f(Y)$ so that $f(Y) = Y$.

On the other hand, suppose $f(Y) \subset Y$ implies $f(Y) = Y$ for any subset Y of X. Let $p \in X$ and let

$$Y = p + f(p) + f^2(p) + f^3(p) + \cdots.$$

Then $f(Y) \subset Y$. Hence $f(Y) = Y$. Accordingly, for some n we have

$$f^n(p) = p.$$

(1.21) COROLLARY. *If $f(X) \subset X$ is pointwise periodic, then for any subset Y of X, either $f(Y) \supset Y$ or $f(Y) \subset Y$ implies $f(Y) = Y$.*

For if $f(Y) \supset Y$ we have $f(X - Y) \subset X - Y$ by (1.1). Whence $f(X - Y) = X - Y$ so that $f(Y) = Y$.

DEFINITION. *If $f(X) \subset X$ is continuous, a subset Y of X is said to be invariant provided $f(Y) = Y$. An invariant point, i.e., a point p such that $f(p) = p$, is said to be fixed or to be a fixed point.*

Clearly, then, in terms of periodicity a fixed point is simply a point of period 1. On the other hand, in terms of fixed points, a mapping $f(X) \subset X$ is pointwise periodic at $x \in X$ provided x is fixed under some power of f.

Invariant sets are readily constructed for any mapping $f(X) \subset X$. For if Y is any subset of X, the set

$$O(Y) = \sum_{-\infty}^{\infty} f^n(Y),$$

where $f^0(Y) = Y$ and $f^{-n}(Y) = f^{-1}f^{-(n-1)}(Y)$, is invariant. If Y is a single point p and f is topological $O(p)$ is a countable set

$$\cdots p_{-3}, p_{-2}, p_{-1}, p_0, p_1, p_2, \cdots$$

where $p_n = f^n(p)$, $-\infty < n < \infty$. Clearly f is pointwise periodic at p if and only if $O(p)$ is a finite set of points (see 6.16 below).

2. Cyclic element invariance.

(2.1) THEOREM. *If M is a semi-locally-connected continuum and $f(M) \subset M$ is continuous, there exists a cyclic element E of M such that if $x \in E$, $f(x)$ is not contained in any component of $M - E$ bounded by x.*

Proof. We show first that the property of being a true A-set in M such that for $x \in A$, $f(x)$ is not in any component of $M - A$ bounded by x is inducible.

Let $A_1 \supset A_2 \supset A_3 \supset \cdots$ be a sequence of sets having this property and let $A = \prod A_i$. Then A is a true A-set. Now let $p \in A$ and suppose, contrary to what we wish to show, that $f(p) \subset Q$ where Q is a component of $M - A$ with $F(Q) = p$. Since Q is open in M and $Q + p$ is s.l.c. at p, there exists a neighborhood V of p such that $N = Q \cdot (M - V)$ is connected and $N \supset f(V)$. Choose an integer k such that $A_k \cdot N = 0$. Let R be the component of $M - A_k$ containing N and let $x = F(R)$. Then since $N \subset R \subset Q$, we have $x \in V$. Whence $f(x) \subset f(V) \subset N \subset R$, contrary to the fact that A_k has our property. Thus this property is inducible.

Since M obviously has this property, it therefore contains a set E having the property irreducibly. We shall show that E is a cyclic element of M. If E is degenerate, this is true and, further, E is then a fixed point. If E is nondegenerate we have only to show that it has no cut point. Suppose, on the contrary that E has a cut point p. Let R be the component of $E - p$ lying in the component R_1 of $M - p$ containing $f(p)$ and let $A = R + p$. Then if $x \in R$ and S is any component of $M - A$ having x as its boundary, since S is a component of $M - E$ bounded by x we have $f(x) \cdot S = 0$; and if $x = p$ the same holds because $f(p) \in R_1$. Hence A is an A-set having our property contrary to the fact that A is a proper subset of E.

(2.11) COROLLARY. *If E is a single point p, p is a fixed point under f.*

(2.2) THEOREM. *If M is a s.l.c. continuum, $f(M) \subset M$ is continuous and E*

is a cyclic element of M such that $f(E)$ is nondegenerate, for no $x \epsilon E$ is $f(x)$ contained in a component of $M - E$ bounded by x and no cut point or end point of M is a fixed point, then $E \cdot f(E)$ is nondegenerate.

For suppose $E \cdot f(E)$ contains at most one point. Then there exists a point $x \epsilon E$ such that $f(E) \subset x$ plus components of $M - E$ bounded by x. Since $f(x) \subset f(E)$ but $f(x)$ is in no component of $M - E$ bounded by x, this gives $f(x) = x$, contrary to hypothesis.

(2.3) THEOREM. *If M is a s.l.c. continuum, $f(M) \subset M$ is continuous and has no fixed point and A is an A-set in M such that for no $x \epsilon A$ does $f(x)$ belong to a component of $M - A$ bounded by x and $g(x)$ is the monotone retraction of M onto A, then $gf(A) \subset A$ has no fixed point.*

For suppose $gf(x) = x, x \epsilon A$. Since $f(x)$ does not lie in R, where R is the sum of all components of $M - A$ bounded by x, and $g^{-1}(x) = x + R$ we must have $f(x) = x$, because $f(x) \subset g^{-1}(x)$ since $g[f(x)] = x$.

If every continuous mapping of a set X into itself has at least one fixed point, X is said to have the *fixed point property*.

(2.4) THEOREM (Borsuk). *The fixed point property is cyclicly extensible in s.l.c. continua.*

For let M be s.l.c. and let $f(M) \subset M$ be continuous. If f has no fixed point, by theorems (2.1) and (2.3) there exists a true cyclic element E of M and a continuous transformation $gf(E) \subset E$ which has no fixed point. Thus if every cyclic element of M has the fixed point property, so also does M.

(2.5) THEOREM (Ayres). *If M is a s.l.c. continuum and $h(M) \subset M$ is a homeomorphism, there exists a cyclic element E of M such that $h(E) \subset E$.*

Proof. If any cut point or end point p of M is fixed we have only to take $E = p$. If not, then by (2.1) and (2.2), there exists a true cyclic element E of M such that $E \cdot h(E)$ is nondegenerate. But since $h(E)$ is cyclic and $h(E) \subset M$, this gives $h(E) \subset E$.

(2.51) COROLLARY. *If $h(M) = M$, there exists an invariant cyclic element E, i.e., a cyclic element E such that $h(E) = E$.*

3. The fixed point property. It will be recalled that a set X has the fixed point property provided every continuous mapping of X into itself has at least one fixed point.

(3.1) THEOREM. *The fixed point property is a topological property. Furthermore it is invariant under retraction.*

Proof. Let X have the fixed point property. To prove the first conclusion, let

$$h(X) = X'$$

be any homeomorphism and let

$$f(X') \subset X'$$

be any continuous mapping of X' into itself. Then since $\phi(x) = h^{-1}fh(x)$, $x \in X$, maps X continuously into itself there exists a point $p \in X$ satisfying

$$\phi(p) = h^{-1}fh(p) = p.$$

Clearly this gives $f[h(p)] = h(p)$ so that the point $h(p)$ is fixed under f. Hence X' has the fixed point property.

To establish the second conclusion, let $r(X) = Y$ be any retraction and let $f(Y) \subset Y$ be any continuous mapping of Y into itself. Then $fr(x)$ is a continuous mapping of X into itself and hence has a fixed point p. Since $Y \supset fr(p) = p$, we have $r(p) = p$ so that $f(p) = p$ and p is fixed under f. Thus Y has the fixed point property.

(3.2) THEOREM. *The fixed point property is cyclicly extensible and reducible in semi-locally-connected continua.*

The extensibility was established in (2.4) and the reducibility results from (3.1) in view of the fact that each cyclic element of a semi-locally-connected continuum X is a retract of X.

(3.21) COROLLARY. *Any dendrite has the fixed point property.*

(3.3) BROUWER FIXED POINT THEOREM. *The Euclidean n-cell has the fixed point property.*

Since the fixed point property is topological, it suffices to prove this property for a convex n-dimensional simplex, i.e., the smallest convex set $|p_0, p_1, \cdots, p_n| = S$ in R^m containing the $n + 1$ independent points (vertices) $p_0, p_1, \cdots, p_n$. We first establish two lemmas:

LEMMA 1. *If K is any simplicial subdivision of the n-simplex $S = |p_0, p_1, \cdots, p_n|$ and $v(e)$ is a mapping of the vertices of K into the vertices of S such that for any vertex e of K, $v(e)$ is a vertex of the carrier side of S, i.e., of the side of least dimension which contains e, then there exist an odd number of n-simplexes in K whose vertices map in (1-1) fashion onto the vertices of S.*

Proof. Let us call such an n-simplex in K an R-simplex. Further, let us call an $(n - 1)$-simplex in K an R-side provided its vertices map onto the points $p_0, p_1, \cdots, p_{n-1}$ under v.

Our proof will be by induction on the dimensionality n of S. The lemma being obvious for $n = 0$, let us suppose it established for dimensionality $n - 1$. Let

$$\rho = \text{number of } R\text{-simplexes in } K,$$

$$\alpha(T) = \text{number of } R\text{-sides on an } n\text{-simplex } T \text{ of } K,$$

$$\alpha = \text{number of } R\text{-sides of } K \text{ lying on the boundary of } S.$$

Now if T is an R-simplex of K, clearly $\alpha(T) = 1$; while if T is not an R-simplex, we have $\alpha(T) = 0$ or 2 according as one of the points $p_0, p_1, \cdots, p_{n-1}$ is or is not missing from the set of images of the vertices of T. Hence

$$\rho \equiv \sum \alpha(T), \text{ mod } 2, \qquad T \text{ in } K.$$

On the other hand, an R-side appears exactly twice or exactly once in $\sum \alpha(T)$ according as it is interior to S or on the boundary of S. Accordingly

$$\sum \alpha(T) \equiv \alpha, \text{ mod } 2.$$

Whence

$$\alpha \equiv \rho, \text{ mod } 2.$$

Now obviously any R-side on the boundary of S must be on the side $S^{n-1} = |p_0, p_1, \cdots, p_{n-1}|$ of S, since otherwise one of the points $p_0, p_1, \cdots, p_{n-1}$ would fail to be the image of a vertex of this side. Thus α is the number of $(n-1)$-simplexes in the subdivision of S^{n-1} whose vertices map in (1-1) fashion onto the vertices of S^{n-1} under v. By our induction hypothesis, α is odd and hence ρ is odd.

LEMMA 2. *If $A_0, A_1, \cdots, A_n$ are closed subsets of $S = |p_0, p_1, \cdots, p_n|$ such that for each set of integers $i_0, i_1, \cdots, i_k$ $(0 \leqq i_j \leqq n)$ the side $|p_{i_0}, p_{i_1}, \cdots, p_{i_k}|$ of S is contained in $A_{i_0} + A_{i_1} + \cdots + A_{i_k}$, then $A_0 \cdot A_1 \cdots A_n \neq 0$.*

Proof. Clearly this is equivalent to showing that for every $\epsilon > 0$ there exists a subset of S of diameter less than ϵ intersecting every one of the sets A_i. To this end, let K be a simplicial subdivision of S of norm less than ϵ, i.e., each simplex of K is of diameter less than ϵ. For any vertex e of K, let $|p_{i_0}, p_{i_1}, \cdots, p_{i_k}| = S^k$ be the carrier side of S for e, and let A_{i_j} be one of the sets $A_{i_0}, A_{i_1}, \cdots, A_{i_k}$ containing e and define $v(e) = p_{i_j}$—note that such an A_{i_j} must exist because $S^k \subset A_{i_0} + A_{i_1} + \cdots + A_{i_k}$. By Lemma 1 there exists at least one simplex T of K whose vertices map under v onto the vertices of S in (1-1) fashion. Now in order for a vertex e of T to map into p_i it is necessary that $e \subset A_i$ by definition of $v(e)$. Accordingly T has a vertex in each of the sets $A_0, A_1, \cdots, A_n$ and since $\delta(T) < \epsilon$, our lemma follows.

Proof of the Theorem. Let $S = |p_0, p_1, \cdots, p_n|$ be the simplex in R^n with vertices $p_0 = (0, 0, 0, \cdots, 0)$, $p_1 = (1, 0, 0, \cdots, 0)$, $\cdots$, $p_n = (0, 0, 0, \cdots, 0, 1)$. Let

$$f(x_1, x_2, \cdots, x_n) = (x_1', x_2', \cdots, x_n')$$

be any continuous mapping of S into itself. Let A_0 be the set of all $x = (x_1, x_2, \cdots, x_n) \in S$ such that

$$\text{(i)} \qquad \sum_1^n x_i' \geqq \sum_1^n x_i$$

and for each i, $1 \leqq i \leqq n$, let A_i be the set of all $x \in S$ such that

$$\text{(ii)} \qquad x_i' \leqq x_i .$$

Now let $S^k = | p_{i_0}, p_{i_1}, \cdots, p_{i_k} |$ be any side of S. We shall show that

(iii) $$S^k \subset A_{i_0} + A_{i_1} + \cdots + A_{i_k}.$$

To that end let $p = (x_1, x_2, \cdots, x_n) \in S^k$.

We distinguish two cases:

(a) If all i_j are different from 0, then

$$x_{i_0} + x_{i_1} + \cdots + x_{i_k} = 1,$$

$$x_i = 0, \quad i \neq i_0, i_1, \cdots, i_k.$$

Since

$$\sum x_i' \leqq 1$$

and $x_i' \geqq 0 = x_i$ for $i \neq i_0, i_1, \cdots, i_k$, we must have

$$x_{i_j}' \leqq x_{i_j} \qquad \text{for some } i_j.$$

This gives

$$p = (x_1, x_2, \cdots, x_n) \in A_{i_j}.$$

(b) If some i_j, say i_0, equals 0, then

$$x_i = 0 \qquad \text{for } i \neq i_1, i_2, \cdots, i_k.$$

Hence if

$$x_{i_j}' > x_{i_j} \qquad \text{for } j = 1, 2, \cdots, k,$$

we have

$$\sum_1^n x_i' \geqq \sum_{j=1}^k x_{i_j}' > \sum_{j=1}^k x_{i_j} = \sum_1^n x_i,$$

and this gives $p \in A_0 = A_{i_0}$.

Hence in any case the condition in Lemma 2 is satisfied. Accordingly, there exists a point

$$p = (x_1, x_2, \cdots, x_n) \in A_0 \cdot A_1 \cdots A_n.$$

Thus for this point conditions (i) and (ii) must hold simultaneously; and obviously this requires

$$x_i' \equiv x_i, \qquad i = 1, 2, \cdots, n,$$

so that $f(p) = p$.

Although we make no attempt at a comprehensive treatment of the subject of fixed point properties, mention should be made of several additional noteworthy fixed point theorems which are beyond our scope. In particular, a theorem of very general nature has been proved by Lefschetz [1] which has as a consequence that any connected Euclidean polyhedron with vanishing Betti numbers p^r for $r > 0$ has the fixed point property. Also Kelley [1] proved that if X is any continuum and $f(X) \subset X$ is any continuous mapping of X into itself, there exists either a fixed point or a subcontinuum π of some simple link of X such that $f(\pi) \supset \pi$. Hence, in particular, if there are no true simple links in X, X has the fixed point property. Hamilton [1] showed that if every subcontinuum of a continuum X is both decomposable and unicoherent, X has the fixed point property. Smith and Eilenberg have shown that any periodic mapping of prime period p of space $X \subset R^m$ into itself such that every continuous image in X of a sphere of dimension less than or equal to $pm - m - 1$ is deformable in X to a single point has a fixed point. [See Smith [1]; Eilenberg [7]. Eilenberg extends Smith's result by using homology instead of homotopy assumptions.]

The Lefschetz Theorem, of course, contains the Brouwer Theorem as a special case. It may be remarked that the Brouwer result, the Lefschetz Theorem, and the theorems of Smith and Eilenberg seem to impose *homogeneity* conditions on the space X and use these effectively to follow the action of the transformation and locate the fixed points. On the other hand, the results presented above and the theorems of Kelley and Hamilton seem to make use of the *heterogeneity* of the structure of a continuum X in order to limit the action of the mapping and show that some sort of recurrence must take place. The same may be said of the results to be presented in the next section.

4. Almost periodicity. A mapping $f(X) \subset X$ is said to be *almost periodic* provided that for any $\epsilon > 0$ there exists an integer n such that

$$\rho[x, f^n(x)] < \epsilon, \qquad x \in X.$$

Also, f is said to be *pointwise almost periodic* (abbreviated p.a.p.) at a point $p \in X$ provided there exists, for any $\epsilon > 0$, an integer n_p such that

$$\rho[p, f^{n_p}(p)] < \epsilon,$$

and f is said to be *pointwise almost periodic* on a subset Y of X provided this condition holds at each $p \in Y$. Obviously almost periodicity implies pointwise almost periodicity, but not conversely; periodicity implies almost periodicity but not conversely (e.g., a rotation of a circle through an irrational multiple of 2π); and pointwise periodicity implies pointwise almost periodicity, but not conversely.

That an almost periodic mapping $f(X) \subset X$ is (1-1) follows at once. For suppose $f(p) = f(q) = y$, $p \neq q$, and let $\epsilon = (1/3)\rho(p, q)$ and determine n so that

$$\rho[x, f^n(x)] < \epsilon, \qquad x \in X.$$

Then $\rho[p, f^n(p)] < \epsilon$, $\rho[q, f^n(q)] < \epsilon$ and clearly this is impossible since

$$f^n(p) = f^{n-1}f(p) = f^{n-1}(y) = f^{n-1}f(q) = f^n(q).$$

However, this argument will not apply if only pointwise almost periodicity is assumed. Also it no longer follows that an almost periodic mapping of X into itself is necessarily a mapping *onto* X unless X is compact. Since most of the theorems to be proved are concerned with pointwise almost periodicity, it is therefore frequently necessary to *assume* that the mappings are homeomorphisms of the set X onto itself.

(4.11) Lemma. *If $f(X) \subset X$ is a homeomorphism and K is a compact subset of X such that $f(K) \subset K$, f cannot be p.a.p. at any point of $K - f(K)$.*

For $f(K) \subset K$ gives $f^2(K) \subset f(K), f^3(K) \subset f^2(K) \subset f(K), \cdots, f^n(K) \subset f^{n-1}(K) \subset f(K)$, so that if $x \in K - f(K), f^n(x) \subset f(K)$ for all $n = 1, 2, \cdots$ and $\rho[x, f^n(x)] \geqq \rho[x, f(K)]$ for all $n = 1, 2, \cdots$.

(4.12) Theorem. *If X is compact, in order that a homeomorphism $f(X) \subset X$ be pointwise almost periodic it is necessary and sufficient that for any compact subset K of X, $f(K) \subset K$ imply $f(K) = K$.*

The necessity results at once from (4.11). To prove the sufficiency of the condition, let us suppose it satisfied and suppose there exists a point $p \in X$ and a neighborhood U of p such that $U \cdot f^n(p) = 0$ for $n = 1, 2, \cdots$. Let

$$P = \sum_1^\infty f^n(p) \qquad \text{and} \qquad K = \bar{P} + p.$$

Then

$$f(\bar{P}) \subset \bar{P} \qquad \text{since } f(P) \subset P,$$

and hence

$$f(K) \subset \bar{P} = K - p,$$

contrary to our condition.

(4.21) Lemma. *If X is a continuum, $X = H + K$ is a division of X into continua H and K such that $H \cdot K = p \in X$ and $f(X) \subset X$ is a homeomorphism such that $H \cdot f(H) \neq 0 \neq K \cdot f(K)$ and f is p.a.p. at p, then p is fixed under f.*

Proof. Suppose $f(p) \neq p$. Then $f(p)$ belongs either to $H - p$ or to $K - p$, say to $K - p$. Since $f(H)$ is connected and intersects both H and K we have $f(H - p) \supset p$. Thus $p \in K - f(K)$. Also, $f(K) \subset K$ since $f(K) \cdot K \neq 0$. But then by (4.11), f could not be p.a.p. at p.

(4.22) Lemma. *Let X be a semi-locally-connected continuum and let f be a pointwise almost periodic homeomorphism on X. If a node N of X is invariant and $N \neq X$, there exists a fixed cut point of X.*

Proof. If N is nondegenerate and p is the point $N \cdot \overline{X - N}$, then since $f(N) = N$ we must have $f(p) = p$ because p is the only cut point of X on N. On the other hand, if N is an end point, say q, then there exists a cut point p of X such that if R is the component of $X - p$ containing q, $H = R + p$ and $K = X - R$, then $f(K) \cdot K \neq 0$. Since $f(H) \cdot H \supset q$, it follows by (4.21) that p is fixed under f.

In the next three theorems [(4.3), (4.4), and (4.5)] it is assumed that X is a semi-locally-connected continuum and that $f(X) = X$ is a pointwise almost periodic homeomorphism.

(4.3) THEOREM. *If the cyclic elements C_1 and C_2 are invariant, every cyclic element of the chain $C(C_1, C_2)$ is invariant.*

Proof. Let a and b be C_1 and C_2 or non-cut points of X on C_1 and C_2, respectively, according as C_1 and/or C_2 are degenerate or nondegenerate. Then $C(a, b) \equiv C(C_1, C_2)$. By (4.21) every point of $E(a, b)$ is fixed. Thus if E is any true cyclic element of X in $C(a, b)$ other than C_1 and C_2, since $E \cdot E(a, b)$ consists of two fixed points, we must have $E = f(E)$, because $f(E)$ is a cyclic element of X. This completes the proof.

(4.31) COROLLARY. *Any point separating a point of C_1 and a point of C_2 is fixed.*

(4.4) THEOREM. *The sum I_f of all invariant cyclic elements of X is a non-empty A-set.*

By the preceding theorem, I_f is a connected sum of cyclic elements and hence is an H-set. To prove I_f is closed, we have only to note that any point x of $\bar{I}_f - I_f$ would be a limit point of cut points of I_f and hence must be fixed since these latter points are fixed. Finally, by (2.51), I_f is not empty.

(4.5) THEOREM. *If C_1 and C_2 are distinct cyclic elements of X and $f(C_1) = C_2$, the chain $Y = C(C_1, C_2)$ has one and only one cyclic element which is invariant.*

Proof. We first show that there is a cyclic element of Y which belongs to I_f. If not, there would exist a component R of $X - I_f$ with boundary point p such that $Y \subset R + p = A$. Since $f(I_f) = I_f$ and $f(C_1) = C_2 \subset A$, we have $f(R) = R$, $f(p) = p$ so that $f(A) = A$. Hence by (4.4) there exists a cyclic element E of A which is invariant. Since E is also a cyclic element of X, this gives $E \subset A \cdot I_f = p$. Hence p must be an end point of A. But then by (4.22) some other cyclic element of A is invariant, and this is impossible since it could not belong to I_f.

Now to show that only one cyclic element of Y can be invariant, we note that if this were not so a cut point p of Y would be fixed. Furthermore, if R_1 and R_2 were the components of $X - p$ containing $C_1 - p$ and $C_2 - p$, respectively, since $f(C_1) = C_2$ we would have $f(R_1) = R_2$; and since $R_1 + R_2 + p \supset Y$ and $R_1 \cdot R_2 = 0$, no cyclic element of Y other than p would be invariant.

DEFINITION. *If X is a semi-locally-connected continuum, a homeomorphism $f(X) \subset X$ is said to be elementwise periodic on a cyclic element E of X provided there exists an integer n such that $f^n(E) = E$.*

(4.6) THEOREM. *Let L be the set of all end points of a s.l.c. continuum X. If the homeomorphism $f(X) = X$ is p.a.p. on the set $X - L$, it is elementwise* *periodic on all cyclic elements E of X that lie in X − L, i.e., on all cut points and all true cyclic elements of X.*

Proof. First let E be any cut point p of X. By (2.51) there exists an invariant cyclic element C of X under f. If $C = p$, obviously p is of period 1. If $C \neq p$, let $X = H + K$ be a division of X into continua H and K such that $H \cdot K = p$ and $H \supset C$. Since $K \cdot (X - L) \neq 0$, there exists an integer n such that $f^n(K) \cdot K \neq 0$. Thus since $f^n(C) = C$ so that $f^n(H) \cdot H \neq 0$, it results from (4.21) that $f^n(p) = p$ and again p is of finite period.

Now let E be a true cyclic element of X. If E contains two cut points p and q of X, by what was just shown there exist integers m and n such that $f^m(p) = p$, $f^n(q) = q$ and hence if $k = mn$,

$$f^k(p) = p, \qquad f^k(q) = q.$$

This gives $f^k(E) \cdot E \supset p + q$ so that

$$f^k(E) = E.$$

If E contains not more than one cut point p of X let $x \in E - p$. Since $E - p$ is open in X there exists an integer n such that $f^n(x) \in E - p$. Hence $E \cdot f^n(E)$ contains the non-cut point $f^n(x)$ so that

$$f^n(E) = E.$$

This completes the proof. (See Note 7, p. 279.)

(4.61) Corollary. *If $f(X) \subset X$ is p.a.p. at all points of $X - L$ where L is the set of all end points of the s.l.c. continuum X, then f is p.a.p. also at the points of L.*

For let x be an end point of X, let $\epsilon > 0$ be given and choose cut points p and q of X such that if R is the component of $X - p$ containing x, then $q \in R$ and $\delta(R) < \epsilon$. Then if k is an integer such that

$$f^k(p) = p, \qquad f^k(q) = q,$$

we must have

$$f^k(R) = R$$

since $R \cdot f^k(R) \supset q$. Hence

$$\rho[x, f(x)] \leqq \delta(R) < \epsilon.$$

Note. For any mapping $f(X) \subset X$ we continue to denote by I_f the sum of all invariant cyclic elements of X under f.

(4.7) Theorem. *If L is the set of all end points of a s.l.c. continuum X and $f(X) \subset X$ is a p.a.p. homeomorphism, there exists a sequence of integers $n_1, n_2, \cdots$ such that*

(a) $I_{f^{n_1}} \subset I_{f^{n_2}} \subset \cdots,$

(b) $\sum_{i=1}^{\infty} I_{f^{n_i}} \supset X - L,$

(c) *for any* $\epsilon > 0$ *there exists an* i *such that each component of* $X - I_{f^{n_i}}$ *is of diameter less than* ϵ.

Proof. By [IV, (7.1)] there exists a monotone increasing sequence of A-sets $H_1, H_2, \cdots$ in X such that (1) $\sum_1^\infty H_n = X - L$, (2) each H_n has only a finite number of nodes and (3) $\delta_n \to 0$, where δ_n is the least upper bound of the diameter of the components of $X - H_n$.

Now by the preceding theorem if N is any node of H_1, N has a definite finite period, i.e., a least integer k such that $f^k(N) = N$. Let n_1 be the lowest common multiple of the periods of the nodes of H_1. Then for any node N of H_1 we have

$$f^{n_1}(N) = N;$$

and if E is any cyclic element in H_1, then since E belongs to some cyclic chain $C(N_1, N_2)$, where N_1 and N_2 are nodes of H_1, we have by (4.3),

$$f^{n_1}(E) = E.$$

Accordingly, $H_1 \subset I_{f^{n_1}}$. Just as in the case of H_1 we can find an integer n_2 which is a multiple of n_1 and such that

$$H_2 \subset I_{f^{n_2}}.$$

Similarly, we can find n_3, a multiple of n_2, such that

$$H_3 \subset I_{f^{n_3}}$$

and so on indefinitely. We have then

$$I_{f^{n_1}} \subset I_{f^{n_2}} \subset I_{f^{n_3}} \subset \cdots \text{ and } H_i \subset I_{f^{n_i}}.$$

Whence we get (a), (b), and (c) using (1), (2), and (3).

(4.71) COROLLARY. *If* X *is a dendrite, any p.a.p. homeomorphism* f *of* X *into itself is almost periodic and is pointwise periodic at all cut points of* X.

The latter statement follows from (4.7). To see that f is almost periodic, let $\epsilon > 0$ be given. Then, applying (4.7), let $n = n_k$ be an integer such that every component of $X - I_{f^n}$ is of diameter less than $\epsilon/2$. Then

$$\rho[x, f^n(x)] < \epsilon, \qquad x \in X.$$

For if $x \in I_{f^n}$, $x = f^n(x)$; and if x is in a component R of $X - I_{f^n}$ and p is the point $\bar{R} - R$, then since $f^n(p) = p$ we have $p = \overline{f^n(R)} - f^n(R)$; and since $f^n(R)$ is also a component of $X - I_{f^n}$, this gives

$$\rho[x, f^n(x)] \leqq \delta[R + f^n(R)] \leqq \delta(R) + \delta[f^n(R)] \leqq \epsilon/2 + \epsilon/2 = \epsilon.$$

5. **Regular almost periodicity.** A continuous transformation $f(X) = X$ will be said to be *regularly almost periodic* provided that for any $\epsilon > 0$ an integer n exists such that

$$\rho[x, f^{mn}(x)] < \epsilon \qquad \text{for all } x \in X \text{ and all } m = 1, 2, \cdots.$$

Similarly, f is *pointwise regularly almost periodic* if for each $x \in X$ and $\epsilon > 0$ an n_x exists so that

$$\rho[x, f^{mn_x}(x)] < \epsilon \qquad \text{for all } m = 1, 2, \cdots.$$

It results at once that pointwise regular almost periodicity implies that the mapping is (1-1). In most of the theorems in this section it will be assumed that all mappings are homeomorphisms.

It is clear also that the properties "pointwise regular almost periodicity" and "regular almost periodicity" are intermediate in strength between, respectively, "pointwise almost periodicity" and "pointwise periodicity" on the one hand and "almost periodicity" and "periodicity" on the other hand.

(5.1) Theorem. *Regular almost periodicity is cyclicly extensible in s.l.c. continua X.*

Proof. Let $\epsilon > 0$ be given. Let A be an A-set in X having only a finite number of nodes $N_1, N_2, \cdots, N_a$ and such that any component of $X - A$ is of diameter less than $\epsilon/3$. Let $N_{a+1}, N_{a+2}, \cdots, N_t$ be the remaining true cyclic elements, if any, in A which are of diameter greater than or equal to $\epsilon/3$. For each $i \leqq t$ there exists an integer n_i such that for any m,

(i) $\rho[x, f^{mn_i}(x)] < \epsilon/3$, $x \in N_i$, and $f^{mn_i}(N_i) = N_i$. Thus if $N = \max (n_i)$ and $n = m \cdot N!$, where m is any integer greater than 0,

(ii) $\rho[x, f^n(x)] < \epsilon/3$, $x \in \sum_1^t N_i$, and $f^n(N_i) = N_i$. It follows from (4.3) that for any cyclic element E in A,

$$\text{(iii)} \qquad f^n(E) = E.$$

Thus

$$\text{(iv)} \qquad \rho[x, f^n(x)] < \epsilon/3 \qquad \text{for any } x \in A.$$

For if $x \in \sum N_i$, this follows by (ii); and if x is not on any N_i, any cyclic element E in A containing x is of diameter less than $\epsilon/3$, and by (iii) this gives $x + f^n(x) \subset E$ so that

$$\rho[x, f^n(x)] \leqq \delta(E) < \epsilon.$$

Now let p be any point of $X - A$, let R be the component of $X - A$ containing p and let x be the boundary of R. Since by (4.3), $f^n(A) = A$, $f^n(R)$ is a component of $X - A$ bounded by $f^n(x)$. Therefore

$$\rho[p, f(p)] \leqq \delta[R + f(R)] \leqq \delta(R) + \rho[x, f^n(x)] + \delta[f(R)] < \epsilon/3 + \epsilon/3 + \epsilon/3 = \epsilon$$

because $x \in A$ so that (iv) applies. Hence

$$\rho[x, f^n(x)] < \epsilon \qquad \text{for any } x \in X.$$

Since n was any multiple of $N!$, this gives regular almost periodicity on X.

(5.11) COROLLARY. *Any pointwise periodic transformation on a dendrite is regularly almost periodic.*

Since the A-set A can be chosen so that none of its nodes are end points of X (and thus so that A contains no end points of X) the above proof makes no essential use of regular almost periodicity on the end points of X. Thus we have

(5.12) *If the homeomorphism $f(X) = X$ on the semi-locally-connected continuum X is regularly almost periodic on all cyclic elements of X except end points, f is regularly almost periodic on X.*

If X is a dendrite, then by (4.61) and (4.71) any homeomorphism $f(X) = X$ which is pointwise almost periodic on the set of all cut points of X is pointwise periodic on this set and hence is pointwise regularly almost periodic on this set. Hence by (5.12) we have

(5.13) *If the homeomorphism $f(X) = X$ on the dendrite X is pointwise almost periodic on the set of all cut points of X, it is regularly almost periodic on X.*

DEFINITION. *A continuous mapping $f(X) \subset X$ is said to have equicontinuous powers provided that for any $p \in X$ and any $\epsilon > 0$ a $\delta > 0$ exists such that if $\rho(p, x) < \delta$, then for any integer $n > 0$,*

$$\rho[f^n(p), f^n(x)] < \epsilon.$$

It results at once by standard methods of proof that if X is compact, any mapping on X with equicontinuous powers has *equi-uniformly continuous powers*, i.e., for any $\epsilon > 0$ a $\delta > 0$ exists such that $\rho(x_1, x_2) < \delta$ implies

$$\rho[f^n(x_1), f^n(x_2)] < \epsilon \qquad \text{for all } n > 0.$$

(5.2) THEOREM. *If X is compact and $f(X) \subset X$ is pointwise regularly almost periodic, in order that f have equicontinuous powers it is necessary and sufficient that it be regularly almost periodic.*

Proof. To prove the necessity, let $\epsilon > 0$ be given. For any $x \in X$, let n_x be an integer such that $\rho[x, f^{mn_x}(x)] < \epsilon/3$ for all $m = 1, 2, \cdots$. By equicontinuity of powers (uniform), for each x there is a neighborhood U_x of x of diameter less than $\epsilon/3$ such that $\delta[f^m(U_x)] < \epsilon/3$ for all $m = 1, 2, \cdots$. Let us cover X by a finite collection $U_{x_1}, U_{x_2}, \cdots, U_{x_k}$ of these sets U_x. Then for each $p \in U_{x_i}$,

$$\rho[p, f^{mn_{x_i}}(p)] \leqq \rho(p, x_i) + \rho[x_i, f^{mn_{x_i}}(x_i)] + \rho[f^{mn_{x_i}}(x_i), f^{mn_{x_i}}(p)]$$
$$< \epsilon/3 + \epsilon/3 + \epsilon/3 = \epsilon.$$

Thus if $N = \max (n_{x_i})$ and $n = N!$, for any $p \in X$ we have $\rho[p, f^{mn}(p)] < \epsilon$, for all $m = 1, 2, \cdots$, because $p \in U_{x_i}$ for some i and n is a multiple of n_{x_i}.

To prove the sufficiency, let $f(X) \subset X$ be regularly almost periodic and let $\epsilon > 0$ be given. Choose n such that

(i) $$\rho[x, f^{mn}(x)] < \epsilon/3, \qquad x \in X, \; m = 1, 2, \cdots.$$

Using uniform continuity of $f, f^2, \cdots, f^n$, determine δ such that if $\rho(x, y) < \delta$,

(ii) $$\rho[f^i(x), f^i(y)] < \epsilon/3, \qquad i = 1, 2, \cdots, n.$$

Now let x and y satisfy $\rho(x, y) < \delta$ and let k be any positive integer. Write

$$k = mn + i, \qquad 0 \leqq i < n.$$

Then, letting $x^i = f^i(x)$, $y^i = f^i(y)$, we have

$$\rho[f^k(x), f^k(y)] = \rho[f^{mn}f^i(x), f^{mn}f^i(y)] = \rho[f^{mn}(x^i), f^{mn}(y^i)] \leqq \rho[x^i, f^{mn}(x^i)] + \rho[x^i, y^i] + \rho[y^i, f^{mn}(y^i)] < \epsilon/3 + \epsilon/3 + \epsilon/3 = \epsilon$$

by (i) and (ii).

6. **Orbit decompositions.** Let $f(X) \subset X$ be any continuous mapping—not necessarily topological. For any $x \in X$ the set A_x of all $y \in X$ such that for some two integers $m, n \geqq 0$

$$f^n(y) = f^m(x)$$

will be called the *orbit* of x. We have the following elementary properties of the orbits A_x.

(6.11) *$y \in A_x$ implies $A_y = A_x$.*

For obviously $x \in A_y$. Thus if $z \in A_y$,

$$f^k(z) = f^t(y), \qquad f^n(y) = f^m(x);$$

and since obviously $f^{n+i}(y) = f^{m+i}(x)$, $i > 0$, there is no loss in supposing $n > t$,

$$f^{k+n-t}(z) = f^n(y) = f^m(x),$$

so that $z \in A_x$. Hence $A_y \subset A_x$.

By symmetry, $A_x \subset A_y$. Whence $A_y = A_x$.

Thus the orbits A_x are disjoint when different so that we have a decomposition of X into disjoint sets A_x which will be called the *orbit decomposition.* Clearly this same decomposition could be obtained by the equivalence relation: x eq. y provided $f^m(x) = f^n(y)$ for some $m, n \geqq 0$. This obviously is reflexive and symmetric and by (6.11) is transitive.

(6.12) $f(A_x) = A_x$, $f^{-1}(A_x) = A_x$.

For if $y \in A_x$, for an $n > 0$ we have $f^n(y) = f^m(x) = f^{n-1}f(y)$, so that $f(y) \in A_x$. Also if $z \in f^{-1}(y)$, $y \in A_x$, we have

$$f^m(x) = f^n(y) = f^{n+1}f^{-1}(y) = f^{n+1}(z)$$

so that $z \subset A_x$.

(6.13) *Any subset A of X containing x and satisfying relations* (6.12) *must contain A_x.*

Definition. *If $A \subset X$ satisfies $f(A) = A = f^{-1}(A)$, A will be said to be completely invariant. Thus for any $x \in X$, A_x is the smallest completely invariant* *subset of X containing x. Further, if we define*

$$A_x^k = \sum_{n=1}^{\infty} f^{-n}f^{n+k}(x), \qquad -\infty < k < \infty,$$
$$= \sum_{n=1}^{\infty} A_x^{k,n},$$

where $A_x^{k,n}$ is the set of all $y \in X$ such that $f^n(y) \subset f^{n+k}(x)$, obviously,

$$A_x = \sum_{-\infty}^{\infty} A_x^k.$$

Since the sets $A_x^{k,n}$ are closed,

(6.14) *A_x is an F_σ, $A_x = \sum_{k=-\infty}^{\infty} \sum_{n=1}^{\infty} A_x^{k,n}$.*

(6.15) *If f is interior, any compact A_x must be a finite set.*

Proof. We can consider that f acts only on A_x, since by (6.12), $f^{-1}(A_x) = f(A_x) = A_x$. We first prove (using only the completeness of A_x)

(*) *For any $y \in A_x$ there exists an open set U_y in A_x containing y and an integer n_y such that f^{n_y} is constant on U_y.*

To prove (*), we note that since the sets $A_x^{k,n}$ in the above definition are closed and $A_x = \sum \sum A_x^{k,n}$, there exist a k and an n such that $A_x^{k,n}$ contains an open set V. Let $z \in V$. Then

$$f^n(V) = f^n(z) = f^{n+k}(x) = x^{n+k}.$$

Since f (and hence f^n) is interior, x^{n+k} is an open subset of A_x. Then for all $r > n + k$, $f^{n+k+r}(x) = f^r(x^{n+k}) = x^{n+k+r}$ must be isolated points of A_x.

Now $f^m(y) = f^r(x)$ for some positive integers m and r, $r > n + k$. This gives

$$f^{m+n+k}(y) = f^{n+k+r}(x) = w.$$

Accordingly, if we set

$$U_y = (f^{m+n+k})^{-1}(w), \qquad n_y = m + n + k,$$

we have for $y \in U_y$,

$$f^{n_y}(U_y) = w$$

and U_y is open since w is isolated and hence open. This proves (*).

To prove (6.15), let us select a finite number $U_{y_1}, \cdots, U_{y_k}$ of the sets $[U_y]$, $y \in A_x$, covering A_x. Let n_y be the largest of the integers $n_{y_1}, n_{y_2}, \cdots, n_{y_k}$ and set $n = n_y!$. Then $f^n(U_{y_i})$ is a constant w_i for each i, $0 \leqq i \leqq k$. Then since

$$f^n(A_x) = A_x = \sum_1^k U_{y_i}$$

$$A_x = w_1 + w_2 + \cdots + w_k.$$

Thus A_x is a finite set.

(6.16) *If A_x is finite and of power m, f is periodic of period m at x.*

For set $x = x_0$. Let x_{-1} be a point of $f^{-1}(x)$ distinct from x. [Note: if $f^{-1}(x) \equiv x$, we must have $f(x) = x = A_x$ by (6.13).] Choose $x_{-2} \epsilon f^{-1}(x_{-1})$, $x_{-3} \epsilon f^{-1}(x_{-2})$, $\cdots$. Since the points $x_0, x_{-1}, x_{-2}, \cdots$ cannot all be distinct, there exists a least integer n such that $x_{-n-1} = x_{-j}$ for some $j < n$. Since

$$f(x_{-j}) = x_{-n} = x_{-j+1}$$

and $n - 1 < n$, we must have $j = 0$. Whence

$$\begin{aligned} f(x_0) &= x_{-n}, \\ f(x_{-n}) &= x_{-n+1}, \\ &\cdots\cdots\cdots\cdots, \\ f(x_{-1}) &= x_0. \end{aligned}$$

Thus f is periodic of period $n + 1$ at x.

This proves f pointwise periodic on A_x. Hence f is (1-1) on A_x. Since then if $B = x + f(x) + \cdots + f^n(x)$, $f(B) = B$, $f^{-1}(B) = B$ by (6.13) we have $B = A_x$. Hence $m = n + 1$.

Thus we have

(6.2) THEOREM. *In order that the mapping $f(X) = X$ be pointwise periodic it is sufficient that the orbits A_x be compact and f be interior. If X is compact, this condition is also necessary (and in this case f is topological).*

(6.3) THEOREM. *If the continuous mapping $f(X) = X$ has equicontinuous powers, any compact orbit A_x must be a finite set and hence f must be periodic on this set.*

Proof. Since A_x is compact, it results from (6.14) that there exists, as in (6.15), an $\epsilon > 0$ such that for some integer k, $f^k[A_x \cdot V_\epsilon(x)]$ is a single point y. Now if A_x is an infinite set, there must exist an infinite sequence of distinct points

$$x = x_0, x_{-1}, x_{-2}, \cdots$$

such that for each $n > 0$, $f(x_{-n}) = x_{-n+1}$. For let y_{-1} be any point different from x in $f^{-1}(x)$, let y_{-2} be a point different from x in $f^{-1}(y_{-1})$ if there is such a point, and so on. If this process can be continued indefinitely, we have only to set $x_{-n} = y_{-n}$ and we have the required sequence. On the other hand, if for some n, $f^{-1}(y_{-n})$ reduces to x, then since A_x is infinite there must be an $m < n$ such that $f^{-1}(y_{-m})$ contains a point x_{-m-1} distinct from y_{-m-1}. In this case we set $x_{-i} = y_{-i}$ for $0 < i \leqq m$, let x_{-m-2} be an arbitrary point of $f^{-1}(x_{-m-1})$, x_{-m-3} an arbitrary point of $f^{-1}(x_{-m-2})$ and so on indefinitely and we obtain a sequence satisfying our requirements.

Now by (uniform) equicontinuity of the powers of f on A_x, there exists a $\delta > 0$ such that for $p, q \epsilon A_x$, $\rho(p, q) < \delta$ implies $\rho[f^n(p), f^n(q)] < \epsilon$ for all $n > 0$. We let s be either 0 or the integer such that $y = x_{-s}$ according as $y \cdot \sum x_{-n}$ is vacuous or not.

Since the sequence (x_{-n}) is infinite, and A_x is compact, there exist integers m and n satisfying

$$n - m > k + s$$

and

$$\rho(x_{-m}, x_{-n}) < \delta.$$

This gives

$$\rho[f^m(x_{-m}), f^m(x_{-n})] = \rho(x, x_{m-n}) < \epsilon.$$

But since thus $x_{m-n} \in V_\epsilon(x)$, this gives $f^k(x_{m-n}) = y$ whereas

$$f^k(x_{m-n}) = x_{m-n+k} \neq x_{-s}$$

since $m - n + k < -s$, and either $y = x_{-s}$ or $y \cdot \sum x_{-i} = 0$. Thus our supposition that A_x is infinite leads to a contradiction. That f is periodic on A_x results from (6.16).

(6.31) COROLLARY. *If $f(X) = X$ has equicontinuous powers and the orbits A_x are compact, f is pointwise periodic.*

That compactness of the orbits alone is not sufficient to give pointwise periodicity is shown by the following example:

Let x and y be distinct points and let (x_n) and (y_n) be disjoint sequences of distinct points converging to x and y, respectively. Define:

$$f(x) = f(x_1) = y, \qquad f(y) = f(y_1) = x,$$

$$f(x_n) = y_{n-1}, \qquad f(y_n) = x_{n-1}, \qquad n > 1;$$

$$X = x + y + \sum_1^\infty (x_n + y_n).$$

Then $A_x = X$, but f is not even (1-1).

It is interesting to note that by putting the sequences (x_n) and (y_n) on disjoint lemniscates C_x and C_y connected by an interval xy, a mapping f exhibiting the same phenomena can be defined on the curve $C_x + C_y + xy$.

Assuming now that the orbits A_x are compact, we next study conditions under which the orbit decomposition is continuous. To see that, even for pointwise periodic mappings, some conditions are necessary, let I be the interval (0, 0) to (2, 0) and for each $n > 0$, $a_0^n = (0, 1/n)$, $a_1^n = (2/n, 1/n)$, $\cdots$, $a_n^n = (2n/n, 1/n)$, $a_{n+1}^n = ((2n-1)/n, 1/n)$, $a_{n+2}^n = ((2n-3)/n, 1/n)$, $\cdots$, $a_{2n}^n = (1/n, 1/n)$. Then if we define $f(x) = x$ for $x \in I$, $f(a_i^n) = a_{i+1}^n$ for $i < 2n$, and $f(a_{2n}^n) = a_0^n$, f is pointwise periodic on the set

$$X = I + \sum_{n=1}^\infty \sum_{i=0}^{2n} a_i^n$$

but clearly the orbit decomposition is not upper semi-continuous since every point of I is identical with its own orbit.

That the orbit decomposition may be semi-continuous but not continuous is readily seen if we add to the example $X = x + y + \sum_1^\infty (x_n + y_n)$ given after the statement of (6.31) two additional sequences, $a_n \to x$, and $b_n \to y$ and define

$$f(a_n) = b_n, \qquad f(b_n) = a_n, \qquad\qquad n = 1, 2, \cdots.$$

Then for each n, $A_{a_n} = a_n + b_n$ whereas $A_{a_n} \to x + y$ and $x + y$ is a proper subset of $A_x = X$.

If we assume the transformation pointwise periodic, however, the situation is greatly simplified. We proceed to study this important case.

We have at once

(6.41) *If $f(X) = X$ is pointwise periodic, the orbit decomposition is always lower semi-continuous [i.e., given any A_x and any $\epsilon > 0$ a $\delta > 0$ exists such that $A_y \cdot V_\delta(A_x) \neq 0$ implies $V_\epsilon(A_y) \supset A_x$].*

For if x is of period n, this results at once from the continuity of the functions $f, f^2, \cdots, f^n$.

As a consequence, if for each $x \in X$, $p(x)$ is the period of x, so that $p(x)$ is the *period function* of $f(x)$,

(6.411) *The period function $p(x)$ is lower semi-continuous.*

(6.4) THEOREM. *If $f(X) = X$ is a pointwise periodic homeomorphism, in order that the orbit decomposition be continuous it is necessary and sufficient that f have equicontinuous powers.*

Proof. To prove the sufficiency, by (6.41) we have only to show that the orbit decomposition is upper semi-continuous. If $x \in X$ and $\epsilon > 0$, by equicontinuity of powers at x, there exists a $\delta > 0$ such that $p \in V_\delta(x)$ gives $f^n(p) \subset V_\epsilon[f^n(x)] \subset V_\epsilon(A_x)$ for all n. Hence $A_p \subset V_\epsilon(A_x)$ and since A_x is a finite set, upper semi-continuity follows.

To prove the necessity, let $y \in X$, let n be the period of y and let $\epsilon > 0$ be given. Let U_y be a neighborhood of y such that

$$f^i(U_y) \cdot f^j(U_y) = 0 \qquad \text{for } 0 \leqq i, j < n, i \neq j, \text{ where } f^0(U_y) = U_y,$$

$$\delta[f^i(U_y)] < \epsilon \qquad \text{for } 0 \leqq i \leqq n,$$

$$f^n(U_y) \cdot f^i(U_y) = 0 \qquad \text{for } 0 < i < n,$$

Let V be a neighborhood of A_y such that any orbit intersecting V lies wholly in $U = \sum_0^{n-1} f^i(U_y)$. Then if $z \in V \cdot U_y$ and k is any integer, let

$$k = mn + i, \qquad\qquad 0 \leqq i < n.$$

Then since $f^n(U_y) \cdot U_y \neq 0$, $f^n(z) \subset U_y$. Whence

$$f^{2n}(z) \subset f^n(U_y)$$

and since $f^n(U_y) \cdot U_y \neq 0$ and $f^{2n}(z) \subset U$, we have

$$f^{2n}(z) \subset U_y.$$

Whence

$$f^{3n}(z) \subset f^n(U_y)$$

and again

$$f^{3n}(z) \subset U_y\,,$$

and so on so that

$$f^{mn}(z) \subset U_y\,, \qquad m = 1, 2, \cdots.$$

This gives

$$f^k(z) = f^i f^{mn}(z) \subset f^i(U_y)$$

and since

$$f^k(y) = f^i f^{mn}(y) = f^i(y) \subset f^i(U_y)$$

and

$$\delta[f^i(U_y)] < \epsilon,$$

we have

$$\rho[f^k(z), f^k(y)] < \epsilon \qquad \text{for any } z \in U_y \cdot V \text{ and any } k.$$

Thus f has equicontinuous powers.

Note. It is immediate, of course, that any *periodic* mapping on a set S has equicontinuous powers and hence yields a continuous orbit decomposition of X. However, pointwise periodicity together with equicontinuous powers is still insufficient to imply periodicity even on compact spaces X. For let X consist of a pair of distinct points a and b together with two disjoint sequences of disjoint sets $A_n \to a$ and $B_n \to b$ where for each n, A_n and B_n each consists of exactly n points:

$$A_n = a_n^1 + a_n^2 + \cdots + a_n^n\,; \qquad B_n = b_n^1 + b_n^2 + \cdots + b_n^n\,.$$

Then if we define

$$f(a) = b, \qquad f(b) = a$$

and for any n,

$$f(a_n^i) = b_n^i\,, \quad f(b_n^i) = a_n^{i+1} \qquad \text{for } 1 \leqq i \leqq n-1,$$

$$f(a_n^n) = b_n^n\,, \quad f(b_n^n) = a_n^1\,,$$

clearly f has equicontinuous powers and is pointwise periodic with orbits $a + b$, $A_1 + B_1, \cdots$ but is not periodic.

As a result of (5.2) we have

(6.42) *If X is compact and $f(X) = X$ is pointwise periodic, the properties "regular almost periodicity," "to have equicontinuous powers" and "to give a continuous orbit decomposition" are equivalent for f.*

In this connection the following result is of interest.

(6.43) *If X is locally compact, $f(X) = X$ is topological and A_x is a finite orbit such that for any convergent sequence of orbits (A_{x_i}) with limit L, $L \cdot A_x \neq 0$ implies $L \subset A_x$, then the orbit decomposition is continuous at A_x.*

For by the remark given in proof of (6.41), we have only to prove upper semi-continuity. If we suppose this is not true, then there exists a neighborhood U_0 of A_x and a convergent sequence of orbits (A_{x_i}) with limit set L such that $L \cdot A_x \neq 0$ but $A_{x_i} \cdot (X - U) \neq 0$ for $i = 1, 2, \cdots$. Let $U \subset U_0$ be any neighborhood of A_x such that $\bar{U}$ is compact and let V be a neighborhood of A_x with $\bar{V} \subset U$. For each n we suppose A_{x_n} ordered:

$$\cdots, p_{-2}, p_{-1}, p_0, p_1, p_2, \cdots$$

so that $f(p_i) = p_{i+1}$, $-\infty < i < \infty$. For each n let y_n be the first point of A_{x_n} which is not in U and let us suppose, as we may without loss, that $x_n = f^{-1}(y_n)$. Then $x_n \subset U$ and since $\lim\sup A_{x_n} = L \subset V$, almost all x_n must be in V, since $\bar{U} - V$ is compact and does not intersect L. Now if we take a convergent subsequence of (x_n), which we may suppose is the whole sequence, we have $x_n \to \bar{x} \in L$ since $\bar{V}$ is compact. By continuity of f, this gives $y_n \to f(\bar{x}) \in A_{\bar{x}} = A_x$ contrary to $A_x \subset U$ and $y_n \cdot U = 0$.

Definition. *If $f(X) = X$ is pointwise periodic, a closed invariant set L in X is said to be a component orbit provided the components of L can be so ordered into a finite sequence*

$$L_1, L_2, \cdots, L_n$$

that $f(L_i) = L_{i+1}$, $i < n$, and $f(L_n) = L_1$. Clearly the period of any point of L is a multiple of n. Further, any point orbit A_x (i.e., ordinary orbit) is a component orbit.

(6.5) Theorem. *Let $f(X) \subset X$ be a pointwise periodic homeomorphism on a compact space X. Then if $G_1, G_2, \cdots$ is a convergent sequence of point or component orbits in X, the limit set L is a component orbit.*

Proof. Obviously $f(L) = L$ because $f(G_n) = G_n$ $(n = 1, 2, \cdots)$. Let K_0 be a component of L and let $f(K_0) = K_1$, $f(K_1) = K_2, \cdots, f(K_{k-1}) = K_k$, $f(K_k) = K_0$, so that $K_0, K_1, \cdots, K_k$ is a component orbit, and let

$$K = \sum_0^k K_i.$$

We have to prove

$$K = L.$$

If this is not so, let V be an open set containing K and such that $F(V) \cdot L = 0$, $(X - V) \cdot L \neq 0$.

Choose points $x_i \in G_i$ such that $x_i \to x \in K_0$. Since almost all x_i are in V, clearly we may suppose all are in V. For each i let y_i be the first point in the sequence

$$f(x_i), f^2(x_i), \cdots$$

belonging to $X - V$. Again we may suppose y_i exists for all i, because almost all G_i must intersect $X - V$. A subsequence of $[y_i]$, which we may suppose is the whole sequence, converges to a point $y \in L \cdot (X - V)$. The points $f^{-1}(y_i)$ are in V and by continuity of f^{-1},

$$f^{-1}(y_i) \to f^{-1}(y) \in V \cdot L.$$

Since $f^{-1}(y) \in V \cdot L - K$, there exists an integer N_1 such that

$$f^{-1}(y_i) \neq x_i, \qquad i > N_1.$$

In the same way, the points $f^{-2}(y_i)$, $i > N_1$, are in V and

$$f^{-2}(y_i) \to f^{-2}(y) \in V \cdot L - K.$$

Further, there exists an integer $N_2 > N_1$ such that

$$f^{-2}(y_i) \neq x_i, \qquad i > N_2.$$

Again $f^{-3}(y_i)$, $i > N_2$, are in V and

$$f^{-3}(y_i) \to f^{-3}(y) \in V \cdot L - K.$$

Again we can find $N_3 > N_2$ such that

$$f^{-3}(y_i) \neq x_i, \qquad i > N_3,$$

and so on indefinitely. But this gives

$$f^{-n}(y) \in V \qquad \text{for all } n = 1, 2, \cdots,$$

contrary to pointwise periodicity at y which would require that $f^{-n}(y) = y \in X - V$ for some n.

(6.51) COROLLARY. *If L contains a fixed point or an invariant connected set, L is connected.*

(6.6) THEOREM. *Let X be compact and suppose $f(X) \subset X$ is pointwise periodic. Any non-empty disconnected invariant closed set L in X has the property that for every separation*

$$L = L_1 + L_2$$

there exists an integer N such that $f^N(L_i) = L_i$ $(i = 1, 2)$.

Proof. Suppose this is not so. Then for each n there exists a point $x_n \in L_1$ such that

$$f^{n!}(x_n) \in L_2.$$

Choose a subsequence (x_{n_i}) of (x_n) converging to a point $x \in L_1$. Let k be the period of x so that

$$f^k(x) = x$$

and define $\phi(p) = f^k(p)$, $p \in X$. Then ϕ is pointwise periodic on X. For each i, let G_i be the ϕ-orbit of x_{n_i}, i.e.,

$$G_i = x_{n_i} + \phi(x_{n_i}) + \cdots + \phi^p(x_{n_i}), \qquad \phi^{p+1}(x_{n_i}) = x_{n_i}.$$

Select a convergent subsequence $[G_{i_j}]$ of $[G_i]$ with limit set G. Since $x \in G$ and $\phi(x) = x$, it follows by the preceding theorem that G is connected. Hence $G \subset L_1$. But since for each j

$$f^{n_{i_j}!}(x_{n_{i_j}}) \in L_2$$

and for $n_{i_j} > k$

$$f^{n_{i_j}!}(x_{n_{i_j}}) = \phi^{n_{i_j}!/k}(x),$$

it follows that $G_{i_j} \cdot L_2 \neq 0$ for $i > k$. Whence $G \cdot L_2 \neq 0$, contrary to $G \subset L_1$.

We proceed to apply (6.5) to reduce, in a certain sense, every pointwise periodic mapping on a compact space to a regular almost periodic mapping.

Let X be compact and suppose $f(X) = X$ is pointwise periodic and continuous (hence topological). Since clearly the property of being a closed subset of X containing all sets L such that L has a nondegenerate component and is the limit of some sequence of orbits is inducible, there exists a smallest subset A of X having this property. Incidentally, since the period function $p(x)$ is lower semi-continuous and integer-valued [see (6.411)], the set $X - A$ is dense in X.

Let

$$\phi(X) = X'$$

be the monotone mapping associated with the decomposition of X into the components of A together with single points of $X - A$. Let

$$\phi(A) = A'.$$

Define

$$g(x') = \phi f \phi^{-1}(x'), \qquad x' \in X'.$$

Then if

$$n_{x'} = \min [p(x)], \qquad x \in \phi^{-1}(x'),$$

$$f^{n_{x'}}\phi^{-1}(x') = \phi^{-1}(x')$$

so that

$$(*) \qquad g^{n_{x'}}(x') = \overbrace{\overbrace{\phi f \phi^{-1}} \cdots \overbrace{\phi f \phi^{-1}}\overbrace{\phi f \phi^{-1}}}^{n_{x'} \text{ groups}}(x')$$
$$= \phi f^{n_{x'}} \phi^{-1}(x')$$
$$= \phi\phi^{-1}(x') = x'.$$

Thus g is pointwise periodic on X'.

The orbit decomposition $[A_{x'}]$ *of* X' *given by* g *is upper semi-continuous.*

To prove this, by (6.5) it suffices to show that the limit of any convergent sequence of the sets $A_{x'}$ is totally disconnected. Suppose, on the contrary, that the limit L of such a sequence $(A_{x_i'})$ has a nondegenerate component C. Since $\phi(A) = A'$ is totally disconnected, C contains a nondegenerate subcontinuum N not intersecting A' and we may suppose without loss of generality that $A' \cdot A_{x_i} = 0$ $(i = 1, 2, \cdots)$. Since ϕ is topological on $X - A$, a subsequence of the sets $\phi^{-1}(A_{x_i'})$ converges to a limit set L which contains $\phi^{-1}(N)$. Now by (*)

$$\phi^{-1}(x_i') = \phi^{-1}\phi f^{n_{x'}}\phi^{-1}(x_i') = f^{n_{x'}}\phi^{-1}(x_i').$$

Thus $\phi^{-1}(A_{x_i'})$ is the f-orbit of $\phi^{-1}(x_i')$. This is impossible since L has a nondegenerate component and yet is not contained in A.

COROLLARIES. 1. *The orbit decomposition of* X' *given by* g *is continuous.*

2. g *is regularly almost periodic.*

3. g *has equicontinuous powers.*

Thus we have proved the

(6.7) THEOREM. *Any pointwise periodic homeomorphism* $f(X) \subset X$ *on a compact set* X *is monotonically equivalent to a pointwise periodic and regularly almost periodic mapping* $g(X') = X'$, *i.e., there is a monotone mapping* $\phi(X) = X'$ *such that*

$$g(x') = \phi f \phi^{-1}(x'), \qquad x' \in X'.$$

Further, ϕ *is topological on an open dense subset of* X *and for any* $x' \in X'$, *the period of* x' *under* g *is the minimum of the periods of the points of* $\phi^{-1}(x')$ *under* f.

7. Applications. The manifold cases. Although we have seen that in general pointwise periodicity does not imply periodicity, in the case where the set X acted on is a connected Euclidean manifold of any dimensionality, such implication is known to hold by virtue of a theorem of Montgomery. We proceed to establish the 2-dimensional case of this theorem, the general case being beyond our scope. Montgomery's proof of the general theorem leans heavily on a deep theorem of Newman concerning periodic mappings. Our proof of the 2-dimensional case avoids Newman's theorem and may be of interest in that it uses instead properties of interior mappings.

(7.1) THEOREM. *Any pointwise periodic mapping* $f(M) = M$ *on a 2-dimensional manifold* M *is periodic.*

Proof. We consider the orbit decomposition $[A_x]$ of M given by f. Let Q_0 denote the set of all points $x \in M$ such that the orbit decomposition is upper semi-continuous (and hence continuous) on the orbit A_x of x, i.e., if U is any open set containing A_x there exists an open set V with $A_x \subset V \subset U$ such that any orbit intersecting V lies wholly in U. Let $L_0 = M - Q_0$, $L = \overline{L}_0$. Since

Q_0 contains all points x where the period function $p(x)$ is continuous, it results that L contains no open set in M and hence is 1-dimensional. We shall show that $L = 0$.

Let R be any component of $M - L$. If $x \in R$ is of period p we have $f^p(R) = R$ so that the set

$$N = R + f(R) + \cdots + f^{p-1}(R)$$

is a 2-dimensional manifold which is invariant under f. Hence N contains the whole orbit of each of its points. Obviously the orbit decomposition is continuous on N. Let N' be the hyperspace and $g(N) = N'$ the associated interior transformation. Then g is finite-to-one and since $g(R) = N'$, N' is connected. Therefore by [X, (5.21)] there exists a finite degree (= maximum multiplicity) n of g. But this means that every point of R is of period less than or equal to n.

Now let q be any point of L where the period function on L alone is continuous and let E be an open 2-cell neighborhood of p such that the period of any point of $L \cdot E$ is k. It follows from the above that infinitely many components of $M - L$ must intersect E and indeed the period function cannot be bounded on E. For otherwise $p(x)$ would be bounded, say $p(x) \leqq P$, on E; and continuity of the mappings $f, f^2, \cdots, f^n$ would give continuity of the orbit decomposition at the orbit of any point of $L \cdot E$ so that no such point could be in L. Now let R be any one such component of $M - L$, let n be the period of f on R as given above and let

$$N = R + f(R) + \cdots + f^{n-1}(R).$$

On the set $W = N + E$, define

$$\phi(x) = f^k(x), \qquad x \in N,$$

$$\phi(x) = x, \qquad x \in (E - E \cdot N).$$

Then clearly ϕ is a periodic homeomorphism of period $m \leqq n$. Hence the orbit decomposition of W given by ϕ is continuous and if $g(W) = W'$ is the associated interior transformation, g is of degree m. But since every point of $E - E \cdot N$ is fixed under ϕ and this set contains interior points of M as shown above, it follows by [X, (5.2)] that $m = 1$. But this gives $n \leqq k$; and since R was an arbitrary component of $M - L$ intersecting E, we have $p(x) \leqq k$ on E contrary to what was shown above.

Accordingly $L = 0$. Thus $M - L$ is connected and, as shown above, f is periodic on $M - L = M$.

(7.11) Corollary. *If each cyclic element of a locally connected continuum X is a 2-dimensional manifold, any pointwise periodic transformation on S is regularly almost periodic.*

In case X is a 1-dimensional compact manifold, i.e., a simple arc or a simple closed curve, not only does pointwise periodicity imply periodicity but we easily obtain a complete analysis of all possible periodic mappings. We proceed to give these:

(7.2) THEOREM. *If f is a pointwise periodic mapping of an arc ab into itself it must be periodic of period n; furthermore, either $n = 1$ (all points fixed) or $n = 2$ and there is exactly one interior point p of ab which is fixed and f is equivalent to a reflection about p.*

Proof. If either a or b is fixed, they must both be fixed since end points map into end points; and in this case every point is fixed, by (4.3), so that $n = 1$. If neither a nor b is fixed, we have $f(a) = b$, $f(b) = a$, $f^2(a) = a$, $f^2(b) = b$ and hence $f^2(x) \equiv x$ for $x \in ab$. Whence $n = 2$. Also if p is a fixed point, we have $f(ap) = bp$ and no $q \neq p$ is fixed.

(7.3) THEOREM. *Any pointwise periodic mapping f of the circle S: $|z| = 1$ into itself is equivalent either to the reflection $w = \bar{z}$ of period 2 with two fixed points or to the rotation $w = ze^{2\pi i/n}$ $(n = 1, 2, \cdots)$ (all points fixed for $n = 1$, no fixed points for $n > 1$).*

Proof. If all points of S are fixed obviously f is the identity $w = z$. Suppose there is a fixed point p and a point x with $f(x) \neq x$. Let px, $pf(x)$ and $xf(x)$ be the arcs on S each containing only two of the points p, x, and $f(x)$. Then since $f(px) = pf(x)$, we have $f[xf(x)] = xf(x)$. Hence by (1) there is exactly one fixed point q on $xf(x)$ and also $f^2(x) = x$; and since $f(qxp) = qf(x)p$, f is equivalent to the reflection $w = \bar{z}$ on $|z| = 1$.

Finally, if there are no fixed points it results that if $p = p_0 \in S$, $p_1 = f(p), \cdots, p_{n-1} = f^{n-1}(p)$, $p_0 = f^n(p)$, then

$$f(p_0p_1) = p_1p_2\,,\ f(p_1p_2) = p_2p_3\,,\ \cdots\,,\ f(p_{n-1}p_0) = p_0p_1$$

so that f is equivalent to the mapping $w = ze^{2\pi i/n}$, $n > 1$, on $|z| = 1$.

Using these simple results for the case of the simple arc and simple closed curve Kerékjártó and Eilenberg have obtained an equally complete analysis of the possible periodic mappings on the sphere, i.e., every periodic mapping on the sphere is topologically equivalent either to a rotation (2 fixed points), a reflection (a simple closed curve of fixed points), or to a rotation followed by a reflection (no fixed points). Although only elementary methods of proof are required to establish this result, it seems superfluous to reproduce the proof here and hence it is not given. Similar results are known for other 2-dimensional manifolds though in not quite so simple a form.

For higher-dimensional manifolds, far reaching results have been obtained by P. A. Smith, though here a great deal is yet to be done. Smith has proved, for example, that if X is the 3-sphere, any periodic mapping on X has as its set of fixed points either the null set, a pair of points, a simple closed curve or a 2-sphere. Thus if S_n $(n = 0, 1, 2, \cdots)$ denotes the n-sphere, for $n \leqq 3$ it can be said that any periodic mapping on S_n has as its set of fixed points an S_i where $i \leqq n$.

References

§2. Kelley [1], Ayres [7], Borsuk [1].

§3. Borsuk [1], Brouwer [2]. The proof given for (3.3) is a slightly modified version of that given by Knaster, Kuratowski, and Mazurkiewicz. (See Knaster [2].)

§4. Ayres [8]. The definitions and main theorems are due to Ayres. The treãtment given is a modification and extension to s.l.c. continua of that of Ayres, the preliminaries through (4.22) and also the theorem (4.7) being given here for the first time. For treatments of other types of almost periodicity see Montgomery [1] and reference therein to Cameron.

§5. Whyburn [39]. The property of equicontinuous powers was introduced under the name "regular" by Kerékjártó [2] and has been used also by Montgomery [1] and Hall [2] and Kelley.

§6. Whyburn [39], Hall [2] and Kelley, Hall [4] and Schweigert, and Schweigert [2].

§7. Montgomery [2], Kerékjártó [3, 4], Eilenberg [8], Smith [2]. See also Newman [2], Nielsen [1], Brouwer [4]. For closely related concepts treated from the dynamical viewpoint see Birkhoff [1].

BIBLIOGRAPHY

Adkisson, V. W.

1. (With MacLane, S.) *Extending maps of plane Peano continua*, Duke Mathematical Journal, vol. 6 (1940), pp. 216–228.

Aitchison, B.

1. *Concerning regular accessibility*, Fundamenta Mathematicae, vol. 20 (1933), pp. 119–228.

Alexander, J. W.

1. *A proof of Jordan's theorem about a simple closed curve*, Annals of Mathematics, (2), vol. 21 (1919), pp. 180–181. *A proof and extension of the Jordan-Brouwer separation theorem*, Transactions of the American Mathematical Society, vol. 23 (1922), pp. 333–349.

Alexandroff, P.

1. *Über die Metrization der im kleinen kompakten topologischen Räume*, Mathematische Annalen, vol. 92 (1924), pp. 294–301.
2. *Sur les ensembles de la première classe et les ensembles abstraits*, Comptes Rendus de l'Académie des Sciences, Paris, vol. 178 (1924), pp. 185–187.
3. *Darstellung der Grundzüge der Urysohnschen Dimensionstheorie*, Mathematische Annalen, vol. 98 (1928), pp. 31–63.
4. *Une généralisation nouvelle du theoreme de Phragmén-Brouwer*, Comptes Rendus de l'Académie des Sciences, Paris, vol. 184 (1927), pp. 575–577.
6. (With Hopf, H.) *Topologie*, I, Berlin, 1935, Springer.
7. (With Urysohn, P.) *Mémoire sur les espaces topologiques compacts*, Verhandelingen der Akademie van Wetenschappen, Amsterdam, vol. 14 (1929), pp. 1–96.

Arnold, H. A.

1. *Homology in set-intersections, with an application to r-regular convergence, and on r-regular convergence of sets*, Bulletin of the American Mathematical Society, abstracts 47-7-325 and 47-7-326.

Aronszajn, N.

1. *Ein Urbildproblem*, Fundamenta Mathematicae, vol. 17 (1931), pp. 92–121.

Ayres, W. L.

1. *On continua which are disconnected by the omission of any point and some related problems*, Monatshefte für Mathematik und Physik, vol. 36 (1929), pp. 135–148.
2. *Continuous curves homeomorphic with the boundary of a plane domain*, Fundamenta Mathematicae, vol. 14 (1929), pp. 92–95.
3. *Concerning continuous curves in metric space*, American Journal of Mathematics, vol. 51 (1929), pp. 577–594.
4. *Continuous curves which are cyclically connected*, Bulletin International de l'Académie Polonaise des Sciences et des Lettres, Classe des Sciences Mathématiques et Naturelles, A (1928), pp. 127–142.
5. *Concerning the arc-curves and basic sets of a continuous curve*, Transactions of the American Mathematical Society, vol. 30 (1928), pp. 567–578.
6. *Concerning the arc-curves and basic sets of a continuous curve. Second paper*, ibid., vol. 31 (1929), pp. 595–612.
7. *Some generalizations of the Sherrer fixed-point theorem*, Fundamenta Mathematicae, vol. 16 (1930), pp. 332–336.
8. *On transformations having periodic properties*, ibid., vol. 33 (1939), pp. 95–103 (reprint).

Birkhoff, G. D.

1. *Dynamical Systems*, American Mathematical Society Colloquium Publications, vol. 9, 1927.

Borsuk, K.

1. *Einige Sätze über stetige Streckenbilder*, Fundamenta Mathematicae, vol. 18 (1932), pp. 198–214.
2. *Sur les rétractes*, ibid., vol. 17 (1931), pp. 152–170.
3. *Quelques théorèmes sur les ensembles unicoherent*, ibid., vol. 17 (1931), pp. 171–209.

Brouwer, L. E. J.

1. *Over de structur der perfecte punktnerzamelingen*, Akademie Versammlungen, Amsterdam, vol. 18 (1910), pp. 833–842 and vol. 19 (1911), pp. 1416–1426. See also Proceedings, Akademie van Wetenschappen, Amsterdam, vol. 14 (1911), p. 138.
2. *Beweis des Jordanschen Kurvensatzes*, Mathematische Annalen, vol. 69 (1910), pp. 169–175.
3. *Über Abbildungen von Mannigfaltigkeiten*, ibid., vol. 71 (1912), pp. 97–115.
4. *Aufzählung der periodischen Transformationen des Torus*, Proceedings, Akademie van Wetenschappen, Amsterdam, vol. 21 (1919), pp. 1352–1356.

Cairns, S. S.

1. *The cellular division and approximation of regular spreads*, Proceedings of the National Academy of Sciences, vol. 16 (1930), pp. 488–490.

Carathéodory, C.

1. *Über der Begrenzung einfach zusammenhängender Gebiete*, Mathematische Annalen, vol. 73 (1913), pp. 323–370.

Claytor, S.

1. *Topological immersion of peanian continua in a spherical surface*, Annals of Mathematics, (2), vol. 35 (1934), pp. 809–835.

Cleveland, C. M.

1. *Concerning points of a continuous curve that are not accessible from each other*, Proceedings of the National Academy of Sciences, vol. 13 (1927), pp. 175–276.

Eilenberg, S.

1. *Sur la multicohérence des surfaces closes*, Odbitka ze Sprawozdania z posiedzen Towarzystwa Naukawego Warszawskiego, vol. 30 (1937), pp. 109–111.
2. *Transformations continues en circonférence et la topologie du plan*, Fundamenta Mathematicae, vol. 26 (1936), pp. 61–112.
3. *Sur les espaces multicohérents* I, ibid., vol. 27 (1936), pp. 151–190.
4. *Sur les transformations d'espaces métriques en circonférence*, ibid., vol. 24 (1935), pp. 160–176.
5. *Sur les transformations continues d'espaces métriques compacts*, ibid., vol. 22 (1934), pp. 292–296.
6. *Sur quelques propriétés des transformations localement homéomorphes*, ibid., vol. 24 (1935), pp. 33–42.
7. *On a theorem of P. A. Smith concerning fixed points for periodic transformations*, Duke Mathematical Journal, vol. 6 (1940), pp. 428–437.
8. *Sur les transformations périodiques de la surface de sphere*, Fundamenta Mathematicae, vol. 22 (1934), pp. 28–41.

Forder, H. G.

1. *The Foundations of Euclidean Geometry*, Cambridge University Press, 1927.

Fréchet, M.

1. *Les Espaces Abstracts*, Gauthier-Villars, Paris, 1926.

Gehman, H. M.

1. *Concerning certain types of non-cut points, with application to continuous curves*, Proceedings of the National Academy of Sciences, vol. 14 (1928), pp. 431–433.
2. *Concerning acyclic continuous curves*, Transactions of the American Mathematical Society, vol. 29 (1927), pp. 553–568.

Hahn, H.

1. *Über die Komponenten offenen Mengen*, Fundamenta Mathematicae, vol. 2 (1921), pp. 189–192.
2. *Mengentheoretische Characterisierung der stetige Kurven*, Sitzungsberichte, Akademie der Wissenschaften, Vienna, vol. 123 (1914), p. 2433.

Hall, D. W.

1. *On a decomposition of true cyclic elements*, Transactions of the American Mathematical Society, vol. 47 (1940), pp. 305–321.
2. (With Kelley, J. L.) *Periodic types of transformations*, Duke Mathematical Journal, vol. 8 (1941), pp. 625–630.
3. (With Schweigert, G. E.) *Non n-alternating transformations*, ibid., vol. 3 (1937), pp. 623–626.
4. (With Schweigert, G. E.) *Properties of invariant sets under pointwise periodic homeomorphisms*, ibid., vol. 4 (1938), pp. 719–724.

Hamilton, O. H.

1. *Fixed points under transformations of continua which are not connected im kleinen*, Transactions of the American Mathematical Society, vol. 44 (1938), pp. 18–24.

Harrold, O. G.

1. *A note on strongly irreducible maps of an interval*, Duke Mathematical Journal, vol. 6 (1940), pp. 750–752.

Harry, C. H.

1. *Concerning the geometry of acyclic sets*, American Journal of Mathematics, vol. 56 (1934), pp. 233–253.

Hausdorff, F.

1. *Mengenlehre*, de Gruyter, Leipzig, 1927.
2. *Die Mengen G_δ in vollständigen Räumen*, Fundamenta Mathematicae, vol. 6 (1924), pp. 147–148.

Hildebrandt, T. H.

1. *The Borel theorem and its generalizations*, Bulletin of the American Mathematical Society, vol. 32 (1926), pp. 423–474.

Hopf, H.

1. See Alexandroff [6].

Hurewicz, W.

1. *Grundrisse der Mengerschen Dimensionstheorie*, Mathematische Annalen, vol. 98 (1928), pp. 63–88.

Janiszewski, S.

1. *Sur les continus irréductibles entre deux points*, Journal de l'École Polytechnique, (2), vol. 16 (1912), pp. 76–170.

Jones, F. B.

1. *Almost cyclic elements and simple links of a continuous curve*, Bulletin of the American Mathematical Society, vol. 46 (1940), pp. 775–783.

Jordan, C.

1. *Cour d'Analyse*, Paris, 1893, 2d edition, p. 92.

Kaluzsay, A.

1. See Kerékjártó [4], p. 103.

Kampen, E. R. v.

1. *On some characterizations of 2-dimensional manifolds*, Duke Mathematical Journal, vol. 1 (1935), pp. 74–93.

Kelley, J. L.

1. *A decomposition of compact continua and related theorems on fixed sets under continuous transformations*, Proceedings of the National Academy of Sciences, vol. 26 (1940), pp. 192–194.
2. See Hall [2].

Kerékjártó, B. v.

1. *Involutions et surfaces continues*, Szeged, Acta Litterarum ac Scientiarum, vol. 3 (1927), pp. 49–67.
2. Ibid., vol. 6 (1932–1934), p. 235.
3. *Über die periodischen Transformationen der Kreisscheibe und der Kugelfläche*, Mathematische Annalen, vol. 80 (1919), pp. 36–38.
4. *Topologie*, Berlin, 1923.

Kline, J. R.

1. *A condition that every subcontinua of a continuous curve be a continuous curve*, Fundamenta Mathematicae, vol. 10 (1927), pp. 298–301.

Knaster, B.

1. *Über rationale Kurven ohne Bögen*, Monatshefte für Mathematik und Physik, vol. 42 (1935), pp. 37–44.
2. (With Kuratowski, C., and Mazurkiewicz, S.) *Ein Beweis des Fixpunktsatzes für n-dimensionale Simplexe*, Fundamenta Mathematicae, vol. 14 (1929), pp. 132–137.
3. *Un continu irréductible à décomposition continue en tranches*, ibid., vol. 15 (1935), pp. 568–577.

Kuratowski, C.

1. *Topologie* I, Warsaw, 1933.
2. *Une définition topologique de la ligne de Jordan*, Fundamenta Mathematicae, vol. 1 (1920), pp. 40–43.
3. *Sur les espaces complets*, ibid., vol. 15 (1930), pp. 301–309.
4. *Sur une propriété des continus Péaniens plans*, ibid., vol. 14 (1930), pp. 180–184.
5. *Sur les décompositions semi-continues d'espaces métriques compacts*, ibid., vol. 11 (1928), pp. 169–185.
6. *Sur les espaces des transformations continues en certains groupes abeliens*, ibid., vol. 31 (1938), pp. 231–246.
7. *Sur le problème des courbes gauches en topologie*, ibid., vol. 15 (1930), pp. 271–283.
8. *Sur la structure des frontières communes à deux régions*, ibid., vol. 12 (1928), pp. 20–42.
9. *Sur les familles monotones d'ensembles fermés et leurs applications à la théorie des espaces connexes*, ibid., vol. 30 (1937), pp. 17–33.
10. (With Whyburn, G. T.) *Sur les éléments cycliques et leurs applications*, ibid., vol. 16 (1930), pp. 305–331.
11. See Knaster [2].

Lefschetz, S.

1. *Topology*, American Mathematical Society Colloquium Publications, vol. 12, 1930.

Lubben, R. G.

1. *Concerning limiting sets in abstract spaces*, Transactions of the American Mathematical Society, vol. 30 (1928), pp. 668–685.
2. *Separation theorems with applications to questions concerning approachability and plane continua*, ibid., vol. 31 (1929), pp. 503–522.

MacLane, S.

1. *A structural characterization of plane combinatorial graphs*, Duke Mathematical Journal, vol. 3 (1937), pp. 460–472.
2. See Adkisson [1].

Mazurkiewicz, S.

1. *Sur les lignes de Jordan*, Fundamenta Mathematicae, vol. 1 (1920), pp. 166–209.
2. See Knaster [2].

Menger, K.

1. *Kurventheorie*, Teubner, Berlin-Leipzig, 1932.

2. *Zur Begründung einer axiomatischen Theorie der Dimension*, Monatshefte für Mathematik und Physik, vol. 36 (1929), pp. 193–218.

3. *Über reguläre Baumkurven*, Mathematische Annalen, vol. 96 (1926), pp. 572–582.

4. *Grundzüge einer Theorie der Kurven*, ibid., vol. 95 (1925), pp. 272–306.

5. *Dimensionstheorie*, Teubner, Berlin-Leipzig, 1928.

Montgomery, D.

1. *Almost periodic transformation groups*, Transactions of the American Mathematical Society, vol. 42 (1937), pp. 322–332.

2. *Pointwise periodic homeomorphisms*, American Journal of Mathematics, vol. 59 (1937), pp. 118–120.

Moore, R. L.

1. *Foundations of Point Set Theory*, American Mathematical Society Colloquium Publications, vol. 13, 1932.

2. *Concerning the sum of a countable number of mutually exclusive continua in the plane*, Fundamenta Mathematicae, vol. 6 (1924), pp. 189–202.

3. *Continuous sets that have no continuous sets of condensation*, Bulletin of the American Mathematical Society, vol. 25 (1919), pp. 174–176.

4. *Report on continuous curves from the viewpoint of analysis situs*, ibid., vol. 29 (1923), pp. 289–302.

5. *Concerning connectedness im kleinen and a related property*, Fundamenta Mathematicae, vol. 3 (1922), pp. 232–237.

6. *On the foundations of plane analysis situs*, Transactions of the American Mathematical Society, vol. 17 (1916), pp. 131–164.

7. *Concerning the cut points of continuous curves and of other closed and connected point sets*, Proceedings of the National Academy of Sciences, vol. 9 (1923), pp. 101–106.

8. *Concerning simple continuous curves*, Transactions of the American Mathematical Society, vol. 21 (1920), pp. 333–347.

9. *Concerning upper semi-continuous collections*, Monatshefte für Mathematik und Physik, vol. 36 (1929), pp. 81–88.

10. *Sets of metrical hypotheses for geometry*, Transactions of the American Mathematical Society, vol. 9 (1908), pp. 487–512.

11. *Concerning the separation of point sets by curves*, Proceedings of the National Academy of Sciences, vol. 11 (1925), pp. 469–476.

12. *A characterization of Jordan regions by properties having no reference to their boundaries*, ibid., vol. 4 (1918), pp. 364–370.

13. *Concerning upper semi-continuous collections of continua*, Transactions of the American Mathematical Society, vol. 27 (1925), pp. 416–428.

Morrey, C. B.

1. *The topology of (path) surfaces*, American Journal of Mathematics, vol. 57 (1935), pp. 17–50.

Mullikin, A. M.

1. *Certain theorems relating to plane connected point sets*, Transactions of the American Mathematical Society, vol. 24 (1922), pp. 144–162.

Newman, M. H. A.

1. *Elements of the Topology of Plane Sets of Points*, Cambridge University Press, 1939.

2. *A theorem on periodic transformations of spaces*, Quarterly Journal of Mathematics, vol. 2 (1931), pp. 1–8.

Nielsen, J.

1. *Über topologische Abbildungen geschlossener Flächen*, Hamburger Abhandlungen, vol. 3 (1924), pp. 246–260.

Nöbeling, G.

1. *Eine Verschärfung des n-Beinsatzes*, Fundamenta Mathematicae, vol. 18 (1931), pp. 23–38.

2. *Reguläre Kurven als Bilder der Kreislinie*, ibid., vol. 20 (1933), pp. 30–46.

Puckett, W. T.

1. *Regular transformations*, Duke Mathematical Journal, vol. 6 (1940), pp. 80–88.
2. *On 0-regular surface transformations*, Transactions of the American Mathematical Society, vol. 47 (1940), pp. 95–113.

Radó, T.

1. (With Reichelderfer, P.) *Cyclic transitivity*, Duke Mathematical Journal, vol. 6 (1940), pp. 474–485.

Rechovsky, H.

1. *Über rationale Kurven*, Fundamenta Mathematicae, vol. 15 (1930), pp. 18–37.

Reichelderfer, P.

1. See Radó [1].

Roberts, J. H.

1. *Concerning non-dense plane continua*, Transactions of the American Mathematical Society, vol. 32 (1930), pp. 6–30.
2. (With Steenrod, N. E.) *Monotone transformations of two-dimensional manifolds*, Annals of Mathematics, (2), vol. 39 (1938), pp. 851–862.
3. *Concerning metric collections of continua*, American Journal of Mathematics, vol. 53 (1931), pp. 422–426.
4. *A non-dense plane continuum*, Bulletin of the American Mathematics Society, vol. 37 (1931), pp. 720–722.

Rozanska, J.

1. *Über stetige Abbildungen eines Elements*, Fundamenta Mathematicae, vol. 28 (1937), pp. 219–232.

Rutt, N. E.

1. *Concerning the cut points of a continuous curve when the arc-curve AB contains exactly N independent arcs*, American Journal of Mathematics, vol. 51 (1929), pp. 217–246.

Schoenflies, A.

1. *Die Entwickelung der Lehre von den Punktmannigfaltigkeiten*, Leipzig, 1908.

Schweigert, G. E.

1. *The analysis of certain curves by means of derived local separating points*, American Journal of Mathematics, vol. 58 (1936), pp. 329–335.
2. *A note on the limit of orbits*, Bulletin of the American Mathematical Society, vol. 46 (1940), pp. 963–969.
3. See Hall [3, 4].

Sierpinski, W.

1. *Un théorème sur les ensembles fermés*, Bulletin de l'Académie des Sciences, Cracovie, 1918, pp. 49–51.
2. *Sur une condition pour qu'un continu soit une courbe jordanienne*, Fundamenta Mathematicae, vol. 1 (1920), pp. 44–60.

Simond, R. G.

1. *Relations between certain continuous transformations*, Duke Mathematical Journal, vol. 4 (1938), pp. 575–589.

Smith, P. A.

1. *A theorem on fixed points for periodic transformations*, Annals of Mathematics, (2), vol. 35 (1934), pp. 572–578.
2. *Transformations of finite period*, ibid., vol. 39 (1938), pp. 127–164 and vol. 40 (1939), pp. 690–711.

Steenrod, N. E.

1. See Roberts [2].

Stoïlow, S.

1. *Sur les transformations continues et la topologie des fonctions analytiques*, Annales de l'École Normale Supérieure, III, vol. 45 (1928), pp. 347–382.
2. *Leçons sur les Principes Topologiques de la Théorie des Fonctions Analytiques*, Gauthier-Villars, Paris, 1938.

Tietze, H.

1. *Über stetige Kurven, Jordansche Kurvenbogen und geschlossene Jordansche Kurven*, Mathematische Zeitschrift, vol. 5 (1919), pp. 284–291.

Torhorst, M.

1. *Über den Rand der einfach zusammenhängenden ebenen Gebiete.* Mathematische Zeitschrift, vol. 9 (1921), pp. 44–65.

Tychonoff, A.

1. *Über einen Metrisationsatz von P. Urysohn*, Mathematische Annalen, vol. 95 (1926), pp. 139–142.

Urysohn, P.

1. *Über die Metrisation der kompakten topologischen Räume*, Mathematische Annalen, vol. 92 (1924), pp. 275–293.
2. *Mémoire sur les multiplicités cantoriennes*, Verhandelingen Akademie van Wetenschappen, Amsterdam, vol. 13 (1928), no. 4.
3. *Mémoire sur les multiplicites cantoriennes*, Fundamenta Mathematicae, vol. 7 (1925), pp. 29–137 and vol. 8 (1926), pp. 225–351.
4. See Alexandroff [7].

Veblen, O.

1. *A system of axioms for geometry*, Transactions of the American Mathematical Society, vol. 5 (1904), pp. 343–384.
2. *Theory on plane curves in non-metrical analysis situs*, ibid., vol. 6 (1905), pp. 83–98.
3. *Analysis Situs*, American Mathematical Society Colloquium Publications, vol. 5, Part 2, 1922.

Vietoris, L.

1. *Über stetige Abbildungen einer Kugelfläche*, Proceedings, Akademie van Wetenschappen, Amsterdam, vol. 29 (1926), pp. 443–453.
2. *Über den hoheren Zusammenhang kompakten Räume und eine Klasse von zusammenhangstreuen Abbildungen*, Mathematische Annalen, vol. 97 (1927), pp. 454–472.

Wallace, A. D.

1. *Quasi-monotone transformations*, Duke Mathematical Journal, vol. 7 (1940), pp. 136–145.
2. *On 0-regular transformations*, American Journal of Mathematics, vol. 62 (1940), pp. 277–284.
3. *Relatively non-alternating transformations*, Bulletin of the American Mathematical Society, abstract 46-1-131.

Wardwell, J. F.

1. *Continuous transformations preserving all topological properties*, American Journal of Mathematics, vol. 58 (1936), pp. 709–726.

Ważewski, T.

1. *Sur les courbes de Jordan ne renfermant aucune courbe fermée de Jordan*, Annales de la Société Polonaise de Mathématique, vol. 2 (1923), pp. 49–170.

Wheeler, C. H. III.

1. *A type of homogeneity for continuous curves*, American Journal of Mathematics, vol. 59 (1937), pp. 315–326.

White, P. A.

1. *On certain relatively non-alternating transformations*, Bulletin of the American Mathematical Society, abstract 46-5-347.

Whitney, H.

1. *Planar graphs*, Fundamenta Mathematicae, vol. 21 (1933), pp. 73–84.
2. *Regular families of curves*, Proceedings of the National Academy of Sciences, vol. 18 (1932), pp. 275–278 and pp. 340–342.

Whyburn, G. T.

1. *Semi-locally-connected sets*, American Journal of Mathematics, vol. 61 (1939), pp. 733–749.

2. *A note on spaces having the S property*, ibid., vol. 54 (1932), pp. 536–538.
3. *A junction property of locally connected sets*, ibid., vol. 53 (1931), pp. 753–756.
4. *Non-separated cuttings of connected point sets*, Transactions of the American Mathematical Society, vol. 33 (1931), pp. 444–454.
5. *Concerning the cut points of continua*, ibid., vol. 30 (1928), pp. 597–609.
6. *On the divisibility of locally connected spaces*, Bulletin of the American Mathematical Society, vol. 37 (1931), pp. 734–736.
7. *Some properties of continuous curves*, ibid., vol. 33 (1927), pp. 305–308.
8. *Concerning connected and regular point sets*, ibid., vol. 33 (1927), pp. 685–689.
9. *Cut points of connected sets and of continua*, Transactions of the American Mathematical Society, vol. 32 (1930), pp. 147–154.
10. *Local separating points of continua*, Monatshefte für Mathematik und Physik, vol. 36 (1929), pp. 305–314.
11. *Concerning the structure of a continuous curve*, American Journal of Mathematics, vol. 50 (1928), pp. 167–194.
12. *Cyclicly connected continuous curves*, Proceedings of the National Academy of Science, vol. 13 (1927), pp. 31–38.
13. *Cyclic elements of higher order*, American Journal of Mathematics, vol. 56 (1934) pp. 133–146.
14. *On the structure of continua*, Bulletin of the American Mathematical Society, vol. 42 (1936), pp. 49–73.
15. *On the cyclic connectivity theorem*, ibid., vol. 37 (1931), pp. 429–433.
16. *On multicoherence*, ibid., abstract 46-7-371.
17. *Concerning hereditarily locally connected continua*, American Journal of Mathematics, vol. 53 (1931), pp. 374–384.
18. *Concerning accessibility in the plane and regular accessibility in n-dimensions*, Bulletin of the American Mathematical Society, vol. 34 (1928), pp. 504–510.
19. *Concerning the open subsets of a plane continuous curve*, Proceedings of the National Academy of Sciences, vol. 13 (1927), pp. 650–657.
20. *Concerning the complementary domains of continua*, Annals of Mathematics, (2), vol. 29 (1928), pp. 399–411.
21. *Concerning continua in the plane*, Transactions of the American Mathematical Society, vol. 29 (1927), pp. 369–400.
22. *Non-alternating transformations*, American Journal of Mathematics, vol. 56 (1934), pp. 294–302.
23. *A relation between interior and non-alternating transformations*, Bulletin of the American Mathematical Society, vol. 46 (1940), pp. 320–321.
24. *Interior transformations on certain curves*, Duke Mathematical Journal, vol. 4 (1938), pp. 607–612.
25. *Semi-closed sets and collections*, ibid., vol. 2 (1936), pp. 685–690.
26. *Interior transformations on compact sets*, ibid., vol. 3 (1937), pp. 370–381.
27. *Interior transformations on surfaces*, American Journal of Mathematics, vol. 60 (1938), pp. 477–490.
28. *A certain transformation on metric spaces*, ibid., vol. 54 (1932), pp. 367–376.
29. *On irreducibility of transformations*, ibid., vol. 61 (1939), pp. 820–822.
30. *Arc-preserving transformations*, ibid., vol. 59 (1936), pp. 305–312.
31. *The mapping of Betti groups under interior transformations*, Duke Mathematical Journal, vol. 4 (1938), pp. 1–8.
32. *On sequences and limiting sets*, Fundamenta Mathematicae, vol. 25 (1935), pp. 408–426.
33. *Regular convergence and monotone transformations*, American Journal of Mathematics, vol. 57 (1935), pp. 902–906.
34. *Completely alternating transformations*, Fundamenta Mathematicae, vol. 27 (1936), pp. 140–146.

35. *A theorem on interior transformations*, Bulletin of the American Mathematical Society, vol. 44 (1938), pp. 414–416.

36. *Interior surface transformations*, Duke Mathematical Journal, vol. 4 (1938), pp. 626–634.

37. *The existence of certain transformations*, ibid., vol. 5 (1938), pp. 647–655.

38. *Non-alternating interior retracting transformations*, Annals of Mathematics, (2) vol. 40 (1939), pp. 914–921.

39. (i) *Almost periodicity*, (ii) *Orbit decompositions*, and (iii) *Regular almost periodicity*, Bulletin of the American Mathematical Society, abstracts 47-5-290, 47-5-291 and 47-5-292.

40. See Kuratowski [10].

Wilder, R. L.

1. *Concerning continuous curves*, Fundamenta Mathematicae, vol. 7 (1925), pp. 340–377.
2. *On connected and regular point sets*, Bulletin of the American Mathematical Society, vol. 34 (1928), pp. 649–655.
3. *Characterizations of continuous curves that are perfectly continuous*, Proceedings of the National Academy of Sciences, vol. 15 (1929), pp. 614–621; and *Concerning perfect continuous curves*, ibid., vol. 16 (1930), pp. 233–240.
4. *Sets which satisfy certain avoidability conditions*, Časopis pro Pěstování Matematiky a Fysiky, vol. 67 (1938), pp. 185–198.

Wilson, W. A.

1. *On upper semi-continuous decompositions of compact continua*, American Journal of Mathematics, vol. 54 (1932), pp. 377–386.
2. *On unicoherency about a simple closed curve*, ibid., vol. 55 (1933), pp. 135–145.

Young, W. H.

1. See, for example, Hausdorff [1], p. 136.

Youngs, J. W. T.

1. *K-cyclic elements*, American Journal of Mathematics, vol. 62 (1940), pp. 449–456.
2. *Arc-spaces*, Duke Mathematical Journal, vol. 7 (1940), pp. 68–84.

Zarankiewicz, C.

1. *Sur les points de division dans les ensembles connexes*, Fundamenta Mathematicae, vol. 9 (1927), pp. 124–171.
2. *Über die Zerschneidungspunkte der zusammenhangende Mengen*, ibid., vol. 12 (1928), pp. 121–125.

Zippin, L.

1. *On continuous curves and the Jordan curve theorem*, American Journal of Mathematics, vol. 52 (1930), pp. 331–350.
2. *A characterization of the closed 2-cell*, ibid., vol. 55 (1933), pp. 207–217.
3. *Independent arcs of a continuous curve*, Annals of Mathematics, (2), vol. 34 (1933), pp. 95–113.

Zoretti, L.

1. *Sur les fonctions analytiques uniformes*, Journal de Mathématiques Pures et Appliquées, (6), vol. 1 (1905), pp. 9–11.

INDEX

NOTES

1. The conclusions in (7.3) on p. 147 hold without any restriction on A provided R is assumed conditionally compact. The thus broadened (7.3) then yields the further corollary.

(7.32) *Local compactness is invariant under interior transformations.*

Local compactness of A also should be assumed in (7.31).

2. The theorem in (3.1) on p. 189 may be strengthened by allowing A to be any *generalized continuum* instead of restricting it to *continua*. This improved form of (3.1) applies more directly in the proof of (3.42) on page 191 and facilitates the reading of that proof.

3. In the proof of (ii) on pgs. 194–195, there is no advantage in having the set M connected; and some simplification results if it is taken merely to be compact and locally connected.

4. In the proof of (iv) on p. 195, the points r and s may and should be taken on the arc uqv in the order u, r, q, s, v.

5. The proof of (v) on p. 195 is sketchy and difficult to follow. It should be replaced by the following detailed argument:

By (iv), every point of K is of order $\leqq 2$. Suppose, first, that some $x \in K$ is of order 1 in K. Let V be an $\epsilon/2$-neighborhood of x in B such that $F(V) \cdot K = b \in K$. By (ii) and (iv) there exists an $\epsilon/2$-neighborhood W of b in B whose boundary is a simple arc rs with $rs \cdot K = r + s$ and so that $\overline{W}$ does not contain x. Since x is of order 1 in K, it is clear that for ϵ sufficiently small one end point of rs, say r, is in V and the other in $B - \overline{V}$, as otherwise we could obtain from V and W an ϵ-neighborhood of x with boundary disjoint from K. Let t be an interior point of rs such that $rt \subset V$. Then if T denotes the sum of a finite number of locally connected subcontinua of B so chosen that $T \supset F(V) - W \cdot F(V)$, $T \cdot (K + x) = 0$, $\delta(T + V) < 2\epsilon/3$ and $M = T + ts$, then M is a compact locally connected set ϵ-separating x in B and such that $M \cdot K = s$. But then the reasoning given under (ii) and (iv) shows that M contains a simple arc X *both* of whose end points belong to K and this is impossible.

Now in case some $x \in K$ were of order 0, we need only choose V as above so that $F(V) \cdot K = 0$ and let M be a compact locally connected set in B containing $F(V)$, disjoint from $K + x$ and so that $\delta(V + M) < \epsilon$. Then by the reasoning under (ii) and (iv) we would obtain in M a simple closed curve X which, by (iii), would show that x is a regular point, contrary to $x \in K$.

6. The set $f^{-1}(K)$ used under (viii) on p. 196 is necessarily a finite graph as there stated because it contains only a finite number of simple closed curves and has no end points.

7. In the proof of (4.6) on p. 249, it is tacitly assumed that the mapping f^n is pointwise almost periodic at p. This can be verified fairly easily under the assumptions made in (4.6). Also, it is a consequence of more general propositions along this line established later by W. H. Gottschalk. See, for example, Bulletin of the American Mathematical Society, vol. 50 (1944), p. 223.

8. In line 3 of the proof of (5.2), p. 71, to see that X is separated between a and b, let $X = X_1 + X_2$ be any separation where $b \in X_2$. If $a \in X_2$, then since X_1 is nonempty so that not all points of $E(a, b)$ can be in X_2, there is a point c of $E(a, b)$ in X_1. Then if Q is the component of $M - c$ containing b, the set $X_b = Q \cdot X_2$ is both open and closed in X and does not contain a.

9. The author is indebted to Dr. R. J. Bean for pointing out that the result (7.2), p. 136, in earlier editions of this book is incorrectly worded. The words *for each* $b \in B$ should be inserted between "that" and "there" in the second line of the statement of this result in the earlier editions.

10. In the proof of (3.4), p. 216, to see that M is not unicoherent about J, we reason as follows. For each point p of the intersection of A (resp. B) with the open arc $\widehat{ayb}$ let R_p be a region in M about p of diameter $< 1/3$ the distance from p to $axb + B$ (resp. A) and let R_p^n be a finite union of the R_p covering $ayb \cdot (A + B) - V_{1/n}(a + b)$. Then the sets $H' = H - \sum_{n=1}^{\infty} R_p^n$, $K' = K + \sum_{n=1}^{\infty} \overline{R_p^n}$, are closed and we have

$$H' \cdot J = axb, \; K' \cdot J = ayb, \; M = H' + K', \; H' \cdot K' \subset V_{d/3}(H \cdot K)$$

where $d = \rho(A, B)$. Thus a and b lie in different components of $H' \cdot K'$.

ERRATA

p. 19. Substitute the following for the final paragraph of the proof of (13.2).

Consider first the case of compact M. Now let C_1 and D_1 be components of $M_1 = M - V_{\epsilon/2}(x)$ intersecting both $F[V_{\epsilon/2}(x)]$ and $M - V_\epsilon(x)$ and let

$$M_1 = C_1^0 + D_1^0$$

be a separation of M_1 between C_1 and D_1. Then for at least one of the sets $C_1^0 + \overline{V_{\epsilon/2}(x)}$ and $D_1^0 + \overline{V_{\epsilon/2}(x)}$, say X_1, it must be true that for every $i > 2$, infinitely many components of $X_1 - V_{\epsilon/i}(x)$ intersect $M - V_\epsilon(X)$. We suppose this holds for $X_1 = C_1^0 + V_{\epsilon/2}(x)$ and let K_1 be a component of $D_1 - D_1 \cdot V_{3\epsilon/4}(x)$ intersecting $M - V_\epsilon(x)$.

Then let C_2 and D_2 be components of $M_2 = X_1 - V_{\epsilon/3}(x)$ intersecting both $F[V_{\epsilon/3}(x)]$ and $M - V_\epsilon(x)$ and let

$$M_2 = C_2^0 + D_2^0$$

be a separation of M_2 between C_2 and D_2. Proceeding as before we find $X_2 = C_2^0 + \overline{V_{\epsilon/3}(x)}$ [or $= D_2^0 + \overline{V_{\epsilon/3}(x)}$] so that infinitely many components of $X_2 - V_{\epsilon/i}(x)$, $i > 3$, intersect $M - V_\epsilon(x)$ and let K_2 be a component of $D_2 - D_2 \cdot V_{2\epsilon/3}(x)$ intersecting $M - V_\epsilon(x)$. Continuing this process indefinitely we obtain an infinite sequence $K_1, K_2, \cdots$ of disjoint continua each intersecting $M - V_\epsilon(x)$, with $K_n \cdot F[V_{\frac{2\epsilon}{n+1}}(x)] \neq 0$ and

$$K_n \subset D_n, \quad \sum_{n+1}^{\infty} K_i \subset C_n^0.$$

Thus if $[K_{n_i}]$ is a convergent subsequence of $[K_n]$ with limit K, we have $K \subset \prod C_n^0$ so that $K \cdot K_n = 0$ for all n; and thus K is a continuum of convergence of M containing x.

Next suppose, contrary to our theorem, that x is on no continuum of convergence of M. Then M is locally connected at x and, accordingly, for any $\epsilon > 0$, x is interior to the component C_ϵ of $\overline{V_\epsilon(x)}$ containing x. Suppose ϵ chosen so that C_ϵ is compact. By the compact case treated above, C_ϵ must be semi-locally connected at x. Thus we can find an arbitrarily small $\sigma < \epsilon$ such that $V_\sigma(x) \subset C_\epsilon$ and so that $C_\epsilon - V_\sigma(x) = N$ has only a finite number k of components. But then $M - V_\sigma(x) = N + M - C_\epsilon$ can have at most k components since $M - C_\epsilon$ and $V_\sigma(x)$ are separated and $M - N = (M - C_\epsilon) + V_\sigma(x)$. Whence, M is s.l.c. at x contrary to hypothesis.